Telomeres

SCIENCES

Biology, Field Director – Marie-Christine Maurel

Molecular, Structural and Biochemical Biology,
Subject Head – Carole Saintomé

Telomeres

Chromosome Sentinels

Coordinated by
Carole Saintomé

iSTE

WILEY

First published 2023 in Great Britain and the United States by ISTE Ltd and John Wiley & Sons, Inc.

ISTE Ltd
27-37 St George's Road
London SW19 4EU
UK

www.iste.co.uk

John Wiley & Sons, Inc.
111 River Street
Hoboken, NJ 07030
USA

www.wiley.com

Library of Congress Control Number: 2022950709

British Library Cataloguing-in-Publication Data
A CIP record for this book is available from the British Library
ISBN 978-1-78945-097-2

ERC code:
LS1 Molecular Biology, Biochemistry, Structural Biology and Molecular Biophysics
LS1_2 Biochemistry
LS1_3 DNA synthesis, modification, repair, recombination, degradation

Contents

Foreword

Julia Cooper
Department of Biochemistry and Molecular Genetics,
University of Colorado Anschutz Medical Campus, Aurora, CO, USA

Both portentous consequence and great mystery surround our chromosome termini, as suggested by the titles of seminars and reviews about telomeres, "life at the edge of the chromosome", "the end of the world as we know it", "shining light on the dark side of the genome"... Indeed, the history of telomeres, crucially intertwined with that of DNA, the "information manual" foundational to all of life as we know it, continues to be replete with unexpected twists and turns. Beginning with the foundational studies of Barbara McClintock, our understanding of telomere biology has been driven by some of the most legendary figures in science, with a notably high proportion of women. The field continues to attract a diverse and inspirational scientific community, many of whom are authors of this book.

The typical human chromosome comprises 50–250 million base pairs of DNA that not only encode genes, the instructions for synthesizing every protein in our cells, but also spells out instructions regarding which genes should be expressed (transcribed into protein) in which cells, and when each gene should be expressed. For instance, liver cells express an array of genes (and their encoded proteins) that differs from the array of genes expressed in brain cells; likewise, some genes should be expressed exclusively during early development, for example, while others are expressed only upon hormone stimulation. Therefore, the intactness of this genetic information is crucial for human health. Accordingly, many processes exist to repair damage in chromosomes or inaccuracies in their duplication, to ensure that our genetic information is neither lost nor corrupted. However, despite the importance of maintaining flawless genome stability, evolution has played a fascinating and risky game that threatens the intactness of our genomes: in contrast to bacteria which have circular genomes, evolution has saddled all eukaryotic organisms with *linear* chromosomes, endowing each chromosome with *ends*. These DNA *ends* can be degraded by enzymes in our cells that digest DNA, or fused together by enzymes that join DNA ends, creating chaos-inducing "dicentric chromosomes". If the two centromeres in such chromosomes are pulled towards opposite ends of a dividing cell, the chromosomes can be stretched and broken – in catastrophic ways that are still being defined. Averting this disaster requires telomeres, repeated arrays of a double-strand G-rich DNA sequence (10–15 kilobases of $TTAGGG_n$ in humans; approximately 0.004% of our larger chromosomes) with a short single-strand overhang at the extreme chromosome end, along with proteins that bind the double- and single-stranded regions. The outsized effect of telomeres on global genome stability confers part of their fascination – without telomeres, the stability of our genomes is destroyed, with potentially disastrous consequences for human development and survival.

Evolution has not only planted these potentially catastrophic *ends* on each of our chromosomes, but also has created an additional layer of telomeric complexity with the so-called "end replication problem". The end replication problem refers to the integral inability of DNA polymerases, the enzymes that duplicate our chromosomes, to fully replicate the ends of linear chromosomes – this inability leads to the loss of a small amount of DNA from the end of each chromosome with every round of replication. Eventually, telomere length dwindles to a size that cannot support the full protective function of telomeres, so chromosome ends become visible to the cell as damage-induced DNA ends and therefore trigger cessation of the cell cycle. This cessation is known as senescence. Telomerase, a cellular "reverse transcriptase", an enzyme that copies RNA into DNA, offers the perfect solution to the end replication problem. Telomerase adds telomere repeats onto the ends of chromosomes, replenishing those sequences lost due to the end replication problem. Yet, in an ironic evolutionary twist, human telomerase is expressed only in the germline and stem cells – telomerase expression is kept silent in most of our somatic cells, leading to telomere attrition with age. This telomere attrition limits the lifespans of our somatic cells and provides a "tumor suppressor" function by halting the division of cells that may have accumulated numerous mutations over time. However, this programmed senescence of cells once their telomeres erode to a nonfunctional length is a risky gambit. Some cells with excessively short telomeres are mutated such that they cannot halt cell division, so their foreshortened telomeres are degraded or fused despite continued cell growth, producing corrupted genomes that drive cancer formation. Moreover, while telomere attrition-driven senescence may have evolved to keep us cancer-free through our reproductive years, this same senescence is associated with age-related pathologies. Hence, understanding and eventually intervening in telomere biology has clear implications for cancer prevention, degenerative diseases, and extending the human healthspan.

The mechanisms by which telomeres prevent chromosome degradation and fusion also seem counterintuitive at first glance. One might imagine that the telomere complex would act as a fortress, barring entry to DNA-digesting and end-fusion activities. But again, evolution has come up with a more nuanced, flexible and potentially dangerous solution: the very enzymes that can deleteriously degrade and fuse DNA ends are actually recruited to telomeres but then tightly constrained. For instance, enzymes that digest DNA are utilized, under this tight control, to ensure that telomeres continually regenerate the terminal single strand overhang, which in turn binds essential protective proteins. One might suspect that such utilization of potentially dangerous enzymes is risky, and indeed this is manifested, for example, by occasional sporadic loss of a stretch of telomere. In addition, the repetitive nature of telomeric DNA makes it inherently difficult to replicate, necessitating a complex interplay between the "normal" replication machinery and machineries that snap into action when replication is impeded. Much of what we know about these machineries and their specialized actions at telomeres comes from the authors of this book and is covered herein.

The details of how telomeres coordinate these risky events, whether and how telomeric catastrophes can be remedied, and what opportunities are provided to the cell by these unique chromosome structures, have presented numerous surprises and are still being elucidated. My own journey in the telomere field has likewise been full of surprises. For instance, we have found roles for telomeres in choreographing the movements of chromosomes and therefore controlling which regions of the nucleus, and even which other parts of the chromosome, come into contact with telomeres. In turn, contact with telomeres is important; contact between telomeres and the nuclear envelope can influence nuclear envelope breakdown, and contact between telomeres and centromeres can influence centromere disassembly and reassembly. Hence, the telomere field remains ripe with vital, fascinating questions. This book describes many of the surprises and questions currently driving telomere biology, and invites you to join the diverse community of telomere biologists in discovering answers and posing new questions.

February 2023

1

Telomeres: All's Well That Ends Well

Patrizia ALBERTI

Structure and Instability of Genomes, National Museum of Natural History, CNRS, INSERM, Paris, France

1.1. Introduction

Eukaryotes have linear chromosomes, which means that each chromosome has two extremities. Two crucial questions underlie the studies that have led, over the years, to the understanding of the peculiar DNA–protein structures that constitute the extremities of eukaryotic linear chromosomes. The first question concerns the so-called "end-protection problem", raised by the observations of Müller and McClintock in the 1930s–1940s (Müller 1938; McClintock 1941). When a chromosome undergoes a double-strand break, specialized cellular components try to repair it. How do the repair machineries distinguish between the extremities of a double-strand break and the extremities of a linear chromosome? What makes the extremities of a chromosome different from the extremities of a double-strand break? The second question concerns the so-called "end-replication problem", raised by Olovnikov and Watson in the 1970s (Watson 1972; Olovnikov 1973). DNA synthesis by polymerases proceeds from 5′ to 3′ and requires an RNA primer for replication initiation. As a consequence of this directional mechanism, the very 3′ ends of a chromosome cannot be fully replicated, since, once the RNA primer near the 3′ end is removed, it leaves a 3′ single-stranded tail that cannot be replicated. Hence, in the absence of counterbalancing mechanisms, chromosome ends shorten

Telomeres,
coordinated by Carole SAINTOMÉ. © ISTE Ltd 2023.

at each round of DNA duplication. How do eukaryotes resolve these problems? In the late 1970s, researchers began to discover the features that make extremities of chromosomes in eukaryotes so special and understand how these features ensure the proper replication and protection of chromosome ends, preventing inappropriate repair (that would lead to chromosome fusions) and loss of genetic information. Basically, the ends of eukaryotic chromosomes, named telomeres, feature a peculiar DNA composition, a peculiar protein composition and a mechanism that is able to elongate them. In a large variety of eukaryotes, telomeres exhibit a further characteristic: owing to their particular DNA composition, one of the two telomeric strands can fold into four-stranded structures, named "quadruplexes of guanines" or "G-quadruplexes" or simply "G4s".

In this chapter, I will provide an overview of the history and structures of telomeres, from ciliates (where the main features of telomeres were discovered) to other eukaryotes. Section 1.2 retraces the pathways of two independent discoveries: the discovery of G-quadruplexes and the discovery of the peculiar properties of the ends of chromosomes. We will see when and where G-quadruplexes and telomeres finally met. Section 1.3 provides an overview of DNA sequences, proteins and structures of telomeres in model eukaryotes. My aim is to show how telomeric architectures in eukaryotes present common aspects modulated by species-specific aspects and constitute a fascinating example of the unity and diversity of life. Section 1.4 focuses on G-quadruplexes at telomeres: their structure and what they represent for proteins that deal with them. G-quadruplexes will be the starting and end points of this overview.

1.2. The beginning of the end

In this section, I will retrace the first steps of two independent research topics: one that led to the discovery of G-quadruplexes and the other to the discovery of the peculiar properties of the ends of chromosomes. It will be a journey in the past, from chemistry and biophysics to biology, an example of how unrelated research pathways (G-quadruplexes and telomeres) can at some point cross each other, paving the way for a deeper understanding.

1.2.1. *Chemists and biophysicists at work: G-quartets and G-quadruplexes disclose themselves*

"DNA comes in many forms" (Rich 1993). Although DNA is a linear polymer basically composed of only four types of nucleotides, it exhibits extraordinary properties of self-assembly and can fold into many conformations. The structure of the DNA double-helix was published in 1953 and immediately opened the way to the understanding of how genetic material is duplicated (Watson and Crick 1953). Just a few years after the elucidation of the structure of DNA in its double-helix form, a new class of DNA structures, G-quadruplexes, began to emerge in the studies of chemists and biophysicists. Nevertheless, as we will see, it took many years before biologists become interested in these structures.

The fact that concentrated solutions of guanylic acid (i.e. monomers of guanosine monophosphate (GMP)) form a gel has been known since the early 1900s (it was reported by the biochemist Ivar Bang in 1910). The nature of this gel was unveiled only in the early 1960s, when X-ray diffraction patterns of fibers from dried GMP gels clearly showed "a helical structure with four units per turn of the helix" (Gellert et al. 1962). It was proposed that GMP self-assembled into a cyclic arrangement of four hydrogen-bonded guanines and that these tetrads (later named "G-quartets") stacked on top of each other, forming a sort of cylindrical structure with a hole (a channel) in the middle (Figure 1.1) (Gellert et al. 1962).

The idea of a structure based on tetrads of guanines had already been proposed a few years earlier by Alexandre Rich to interpret X-ray diffraction pattern of fibers of polyinosinic acid (poly(rI), an RNA polymer made of deaminated guanines) (Rich 1958), proposed and rejected in a single article! Rich proposed and wrongly rejected the hypothesis of a tetrahelical structure on the basis of a right remark: he thought that the large hole in the center of a four-stranded structure, if unfilled, would introduce considerable instability. Subsequent studies on polyguanylic acid (poly(rG), an RNA polymer made of guanines) supported the formation of a four-stranded helical structure based on tetrads of guanines (Zimmermann et al. 1975), and the hole in the middle of the structure turned out to be one of the key factors in the stability of this class of structures. Let us see why.

One of the first studies on a guanosine gel showed that its stability strongly depended on the nature of the cation present in the solution. Intriguingly, when plotting the melting temperature of the gel as a function of the ionic radius of several monocations (radius $Li^+ < Na^+ < K^+ < Rb^+ < NH_4^+ < Cs^+$), the result was a bell-shaped curve with a maximum corresponding to the potassium cation (Chantot and Guschlbauer 1969): the potassium cation (K^+) was the one that stabilized the gel most strongly, while the smaller cations lithium (Li^+) and sodium (Na^+) and the larger cations rubidium (Rb^+), ammonium (NH_4^+) and cesium (Cs^+) stabilized the gel to a lesser extent. This puzzling behavior was referred to as "cation anomaly" because it contrasted with what was observed for polyphosphate long-chains and calf thymus DNA, for which the binding constants of alkali metal ions decreased slightly with increasing metal ion radius (Strauss and Ross 1959; Ross and Scruggs 1964). The "cation anomaly" displayed by guanine tetrahelical structures was understood only 10 years later, when researchers grasped that, depending on their dimensions, cations can reside in the electronegative central cavity of the tetrahelical structure and coordinate (more or less tightly, depending on their dimensions) with the carbonyl oxygens of the guanines, thus stabilizing the structures (Figure 1.1) (Miles and Frazier 1978; Pinnavaia et al. 1978). This explained why the stabilizing effect of alkali metal ions on guanine tetrahelical structures (gel and polymers) followed this ranking: $K^+ > Na^+, Rb^+ >> Li^+, Cs^+$: in order to be tightly coordinated to tetrads of guanines, a cation must be small enough to fit in the electron-rich cavity and large enough to bridge the guanine carbonyl oxygens. Na^+ can fit in a tetrad, the larger K^+ can lay between two tetrads, Li^+ is too small to coordinate tightly with the carbonyl oxygens, Cs^+ is too large to fit in the cavity and to allow the optimal distance between two tetrads. It did not escape the attention of researchers that, curiously, the cation that stabilized G-quadruplexes the most was the physiologically relevant cation potassium (Miles and Frazier 1978).

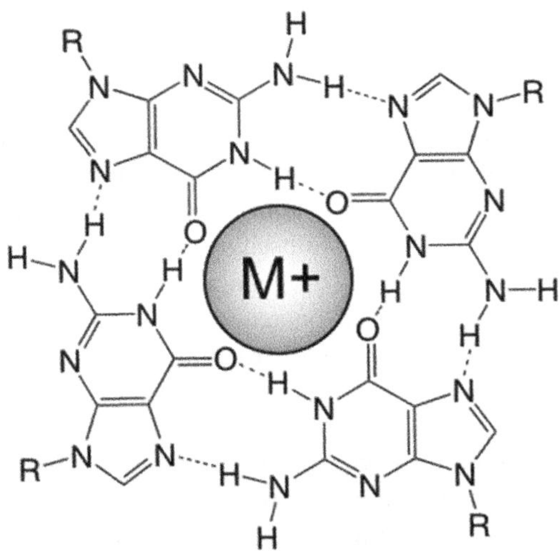

Figure 1.1. *G-quartet: The four guanines auto-assembly through the formation of eight hydrogen bonds between N(1) and N(2) as donors and O(6) and N(7) as acceptors. G-quartets stack on top of each other forming a helical structure. The auto-assembly is driven by the presence of a cation (M^+) that fits into the central cavity and coordinates with the carbonyl oxygens of guanines*

Twenty years have passed from the first publication of an X-ray diffraction image of a G-quadruplex structure in 1958 (Rich 1958) to the understanding of the "cation anomaly" exhibited by this class of structures. We leave here the pathway of the discovery of G-quadruplexes to follow the pathway of the discovery of the peculiar characteristics of telomeres that formally started with an article published in 1978. As we will see, these two pathways, G-quadruplexes and telomeres, will merge in the late 1980s, when G-quadruplexes met telomeres and definitively entered the world of biology.

1.2.2. *Biologists at work: telomeres disclose themselves*

The history of telomeres began in the 1930s–1940s, with the studies of Hermann Müller and Barbara McClintock. Herman Müller (awarded the Nobel Prize in 1946 for the discovery of the production of mutations by means of X-ray irradiation) was working on the fruit fly *Drosophila* and observed that, when treated with X-rays, the ends of chromosomes were protected from rearrangements occurring elsewhere in the chromosomes (Müller 1938). Barbara McClintock was studying the behavior of broken ends of chromosomes in the maize plant. In her complex research (that led to the discovery of mobile genetic elements, for which she was awarded the Nobel Prize in 1983), she observed that broken ends of chromosomes fused with one another under certain conditions, but did not fuse under other conditions where their ends were somehow "healed" and behaved in every respect as normal ends (McClintock 1941). These observations highlighted that the ends of chromosomes must have special features that distinguish them from double-strand breaks, so they deserve a special name. The term "telomeres" was coined by Müller in a lecture given in 1938 at the Woods Hole Marine Biological Laboratory: "The terminal gene must have a special function, that of sealing the end of the chromosome [...]. This gene may accordingly be distinguished by a special term, the telomere" (from the ancient Greek τέλος and μέρος, meaning the terminal part of the chromosome). Although the term "gene" was not appropriate, this sentence marked the birth of the concept of telomeres. Thereafter, very little research was done on telomeres until the 1970s, when studies on DNA molecules in ciliates (unicellular eukaryotes of the phylum Ciliophora) shed light on the ends of chromosomes. But first let us see why telomeres disclosed their nature precisely in ciliates.

Ciliates have two different sorts of nuclei: a micronucleus containing the chromosomes and a macronucleus containing many copies of gene-sized linear DNA molecules (also referred to as minichromosomes). Macronuclear DNA molecules are derived from the fragmentation of micronuclear chromosomes, followed by fragment amplification. The micronucleus (or germline nucleus) is transcriptionally inactive or weakly active, while the macronucleus (or somatic nucleus) is transcriptionally active and supports vegetative cell growth. The macronucleus of the holotrichous ciliates *Tetrahymena* contains thousands of minichromosomes, while the macronucleus of hypotrichous ciliates (such as *Oxytricha, Stylonichia* and *Euplotes*) contains billions of them (reviewed in Prescott (1994)). Owing to the huge number of gene-sized linear DNA molecules they contain, ciliates turned out to be a suitable model for the initial characterization of telomeres. Three major steps marked the discovery of the special properties of telomeres: the sequencing of telomeric DNA, the discovery of telomerase and the identification of telomere-binding proteins.

***Telomeric DNA*:** In 1975, it was observed, by electron microscopy, that when the macronuclear DNA molecules of ciliates were denatured and allowed to renature, they formed single-stranded circles; this supported the presence of inverted repeats at or near their extremities (Wesley 1975). The sequence of these inverted repeats was unveiled by Elisabeth Blackburn, when she joined the laboratory of Joseph Gall after she had learned to master DNA sequencing in the laboratory of Frederick Sanger. Gall was interested in the mechanism of amplification of the DNA molecule encoding ribosomal RNA (rRNA) in the macronucleus of the holotrichous ciliate *Tetrahymena thermophila* (DNA molecules encoding rRNA in ciliate macronuclei are heavily amplified). By sequencing this molecule, Blackburn found that its ends were composed of 20–70 tandem repeats of the hexanucleotide sequence CCCCAA and determined that the CCCCAA strand ran toward the 5′ end, while the complementary GGGGTT strand ran toward the 3′ end (Blackburn and Gall 1978). A slightly different repeated sequence (GGGGTTTT) was found at the ends of DNA molecules in the macronucleus of a hypotrichous ciliate (Oka et al. 1980). After the article published in 1978 by Blackburn and Gall, followed an article from Prescott's group that confirmed the peculiar DNA sequences of telomeres in hypotrichous ciliates (GGGGTTTT repeats on the strand running toward the 3′ end) and highlighted a second feature of telomeric DNA: the G-rich strand exceeded the complementary strand, resulting in a 3′ single-stranded overhang (Figure 1.2) (Klobutcher et al. 1981).

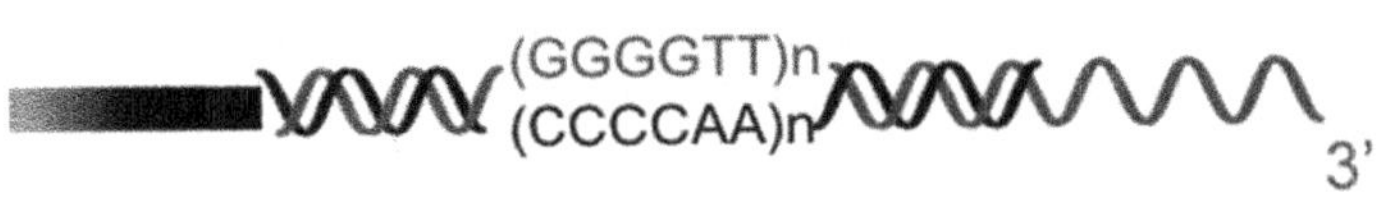

Figure 1.2. *Telomeric DNA in ciliates is composed of GGGGTT repeats (in Tetrahymena species) or GGGGTTTT repeats (in Oxytricha, Stylonichia and Euplotes species) and terminates with a 3′ single-stranded overhang. For a color version of this figure, see www.iste.co.uk/saintome/telomeres.zip*

***Telomerase*:** The understanding of the DNA composition of telomeres paved the way for the discovery of telomerase. Beautiful studies in yeasts showed that telomeric repeats stabilized linear DNA molecules and were added during their replication (Szostak and Blackburn 1982; Pluta et al. 1984; Shampay et al. 1984). These findings prompted the search for the mechanism underlying the addition of telomeric repeats. This led first to the identification, in *Tetrahymena* cell extracts, of an activity adding telomeric repeats to the 3′ end of a synthetic $(TTGGGG)_4$ oligonucleotide (Greider and Blackburn 1985), then to the understanding that this activity was due to a protein-RNA complex, that was named "telomerase" (Greider and Blackburn 1987), and finally to the proof that the RNA component of *Tetrahymena* telomerase carried the 5′CAACCCAA template sequence to add telomeric repeats (Figure 1.3) (Greider and Blackburn 1989). For their works on telomeres and for the discovery of telomerase, Elisabeth Blackburn, Carol Greider and Jack Szostak were awarded the Nobel Prize in 2009.

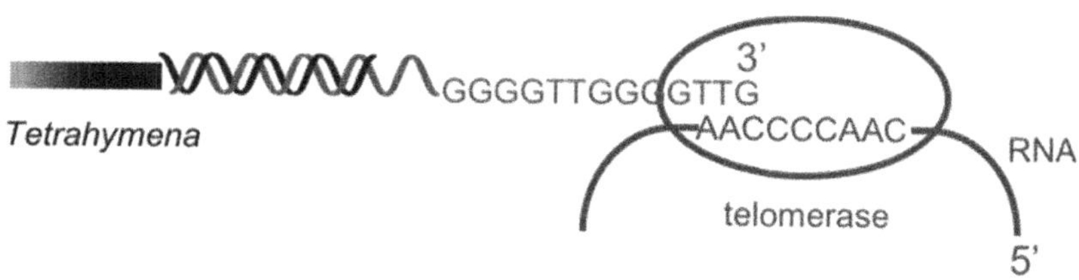

Figure 1.3. *Telomerase adds telomeric repeats to the 3′ overhang. It is composed of a catalytic subunit and an RNA component that carries the template sequence for the synthesis of telomeric repeats. For a color version of this figure, see www.iste.co.uk/saintome/telomeres.zip*

***Telomere-binding proteins*:** Telomeric DNA does not stand alone: it associates with specific proteins. Once again, this major feature of telomeres was revealed by studies in ciliates. Once it was discovered that the termini of the DNA molecules in *Tetrahymena* macronucleus were composed of GGGGTT tandem repeats (Blackburn and Gall 1978), Blackburn examined whether they were packaged as nucleosomes. The study revealed that they were packaged by proteins other than histones (Blackburn and Chiou 1981). This led, step by step, to the identification and characterization of two proteins, named TEBPα and TEBPβ (for telomere end binding protein), that specifically associate and protect the 3′ single-stranded overhang ($T_4G_4T_4G_4$-3′) of the gene-sized linear DNA molecules in *Oxytricha* macronuclei (Figure 1.4) (Gottschling and Zakian 1986; Price and Cech 1987).

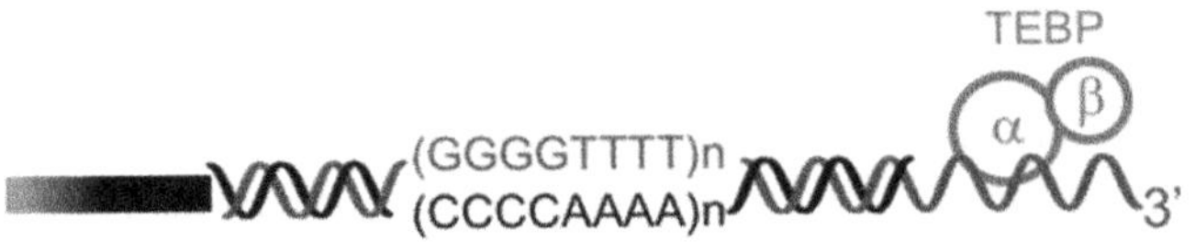

Figure 1.4. *Telomere-binding proteins: TEBPα and TEBPβ protect the 3′ single-stranded overhang of telomeres in hypotrichous ciliates. For a color version of this figure, see www.iste.co.uk/saintome/telomeres.zip*

***The exceptions that confirm the rule*:** As outlined above, ciliates, owing to the huge number of gene-sized linear DNA molecules in their macronuclei, made it possible to understand the peculiar features of telomeres. These peculiar features are basically the following: (i) a DNA composed of short tandem repeats, (ii) a skew in base composition resulting in a G-rich strand (named "G-strand") and a C-rich strand (named "C-strand"), (iii) a defined strand polarity where the G-strand runs toward the 3′ end, (iv) the presence of a 3′ single-stranded overhang (named "G-overhang") due to the fact the G-strand exceeds the C-strand, (v) a mechanism (telomerase) able to elongate the G-strand, thus compensating the loss of nucleotides due to DNA replication, (vi) the association of telomeric DNA with specific proteins (Figures 1.2–1.4). However, once the rule was discovered, researchers began to uncover the exceptions. Telomeric sequences composed of short repeats synthesized by telomerase are widespread throughout eukaryotes, yet they are not universal. During evolution, other nucleoprotein structures and elongation mechanisms have emerged to ensure the two crucial functions of telomeres: elongation and protection of chromosome ends. In insects belonging to the order Diptera (insects with a single pair of wings), there is no evidence for a telomerase reverse transcriptase gene (Mason et al. 2016) and the composition of telomeric DNA is more complex. For example, in the fruit fly *Drosophila melanogaster* (where the ends of chromosomes began to disclose their particular properties), telomeres are maintained by telomere-specific non-long terminal repeat (non-LTR) retrotransposons that are reverse transcribed onto the ends of chromosomes to form an array of tandem repeats (reviewed in Mason et al. (2008)). Furthermore, even in eukaryotes with telomerase, alternative mechanisms of lengthening of telomeres (based on homologous recombination) can be activated under particular circumstances, as first detected in yeasts and humans (Lundblad and Blackburn 1993; Bryan et al. 1995). So, telomeres are declined in a diversity of structures (sharing common aspects) that all ensure a unique function, that is, stabilizing the ends of the linear chromosomes of eukaryotes.

Our story of the discovery of the peculiar features of telomeres stops here and gives way again to G-quadruplexes. As we will see in the last section of this chapter, a further feature shared by telomeric DNA in a variety of eukaryotes is the propensity of the G-strand to fold into G-quadruplexes. Let us see how this feature was revealed for the first time in ciliates, once again.

1.2.3. *When and where G-quadruplexes and telomeres met*

The first steps of the discovery of G-quadruplexes were moved at the turn of the late 1950s and early 1960s. For about 30 years, studies on G-quadruplexes were limited to guanosine gels and polyguanosine polymers. To my knowledge, G-quadruplexes formally entered the world of biology (with the name "G4-DNA") in an article published in 1988, when it was observed (*in vitro*) that G-rich sequences in the immunoglobulin switch region self-associated at physiological salt concentrations to make four-stranded structures (Sen and Gilbert 1988). The authors readily suggested that these structures were held together by tetrads of guanines, as with the structures described by chemists and biophysicists for polyguanosine polymers. They speculated that the self-association of G-rich regions into four-stranded structures may have a role *in vivo*.

At that time there had already been several reports about telomeres in ciliates, where biologists groped around G-quadruplexes, without grasping them. Indeed, in the early 1980s, it was observed (*in vitro)* that the gene-sized linear DNA molecules of the macronuclei of hypotrichous ciliates aggregated when incubated in the presence of NaCl and it was proposed that their terminal repeated sequences associated, someway, to form multi-stranded complexes (Lipps 1980; Lipps et al. 1982; Oka and Thomas 1987). It was also shown that the $(TTGGGG)_4$ oligonucleotide (mimicking the telomeric G-strand of *Tetrahymena*) formed an intramolecular structure involving "G•G base pairs"; the authors proposed it was "an unusual hairpin structure" (Henderson et al. 1987). All these pioneering studies groped around G4 without really grasping them, not yet (none of the proposed models were correct).

Two articles finally proposed a consistent model for the structures of ciliate telomeric repeats, based on the formation of tetrads of guanines (Sundquist and Klug 1989; Williamson et al. 1989). In particular, data supported that DNA molecules mimicking the telomeres of *Tetrahymena* (a double-stranded region made of GGGGTT repeats with a single-stranded TTGGGGTTGGGG-3′ overhang) associated through their 3′ overhangs forming a bimolecular quadruplex (Figure 1.5) (Sundquist and Klug 1989). With these two articles, G-quadruplexes definitively met telomeres, at least *in vitro*. Evidence of the existence of G-quadruplexes in the macronuclei of a ciliate was provided 12 years later by immunofluorescence studies with an antibody specifically recognizing G-quadruplexes formed by telomeric repeats (Schaffitzel et al. 2001). It was shown that this antibody localized in the macronucleus of the ciliate, with the exception of the replication band where G4s were supposed to be unfolded to allow replication and telomere elongation. This study is considered the first study providing evidence of G-quadruplex formation *in vivo*. Thus, it was in ciliates that G-quadruplexes definitively went from being an *in vitro* curiosity to an *in vivo* structure.

Over the years, the study of G4s in genomes (mapping and detection, regulation and function) has considerably progressed (reviewed in Varshney et al. (2020)). G4s have become "stars" among nucleic acid non-canonical structures because of their potential involvement in genome regulation and genome stability/instability.

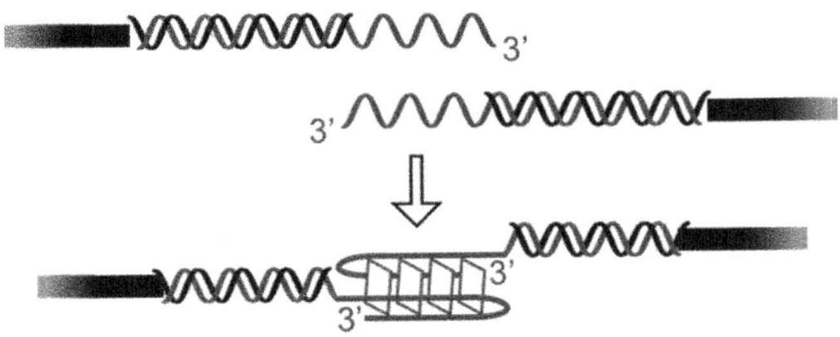

Figure 1.5. *G4s at telomeres of ciliates. The telomeres of ciliates can associate through the formation of G-quadruplexes. In vivo, this association is mediated by the proteins TEBPα/TEBPβ. For a color version of this figure, see www.iste.co.uk/saintome/telomeres.zip*

1.3. Telomeric architectures: from ciliates to other eukaryotes

Telomeres in eukaryotes are like variations played on a single theme in music. The function of telomeres is unique (that of stabilizing the ends of the linear chromosomes), but the structures that ensure this function are multiple. In this section, I will provide an overview of telomere architectures in model eukaryotes: their distinctive DNA sequences, proteins and structures.

1.3.1. *Telomeric DNA and telomeric proteins*

Telomeric DNA: The telomeric sequences of a great variety of eukaryotes are composed of tandem repeats of a short sequence, called "telomeric motifs", made up of five to eight nucleotides and carrying two, three or four consecutives guanines. In the following, I will refer to these short motifs with two, three and four consecutive guanines as G_2, G_3 and G_4 motifs, respectively. Three major telomeric G_2 motifs have been identified: the motif GGCTTA in nematodes and the motifs GGTTA and GGTCA in insects (except dipterans). Most G_3 motifs are of the type $G_3T_{1\text{-}4}A$. GGGTA has been reported to be the telomeric motif of *Giardia intestinalis*. GGGTTA was first identified in human telomeres (Moyzis et al. 1988); it is conserved in vertebrates (Meyne et al. 1989) and is found in many phylogenetically distant eukaryotes. GGGTTTA, first identified in the flowering plant *Arabidopsis thaliana*, is the prevalent motif in plants. The GGGTTTTA motif is found in the unicellular green alga *Chlamydomonas reinhardtii* and in other green algae. More recently, two longer telomeric motifs have been identified in two flowering plants: the GGGTTTTTTA motif in *Cestrum elegans* and the unusual CTCGGTTATGGG motif in *Allium*. The G_4 motifs GGGGTT and GGGGTTTT are the telomeric motifs of ciliates. A G_6 sequence (GGGGGGTTA) has also been indicated as the putative telomeric motif of the unicellular red alga *Cyanidioschyzon merolae*, originally isolated in sulfurous hot springs. The fission yeast *Schizosaccharomyces pombe* has heterogeneous telomeres composed of $G_{2\text{-}8}TTACA_{0\text{-}1}$ repeats (Trujillo et al. 2005).

Dipteran and budding yeast telomeres are a notable exception to the short G-rich repeats found in other eukaryotes. In dipterans, there is no evidence for a telomerase gene (Mason et al. 2016); they have alternative mechanisms of elongation and more complex telomeric sequences. With regard to budding yeasts, the telomeric motifs of about 30 species belonging to the class Saccharomycetes have been identified. *Saccharomyces cerevisiae* and other *Saccharomyces* species have telomeres consisting of $TG_{2\text{-}3}(TG)_{1\text{-}6}$ irregular and G-rich repeats. Species of other genera (such as *Candida*, *Kluyveromyces*) have telomeres consisting of regular repeats of longer motifs (up to 26nt); their telomeric motifs exhibit a remarkable sequence diversity, are not particularly rich in guanines, and have nevertheless conserved the presence of runs of two or three consecutive guanines (e.g. ACGGATTTGATTAGGTATGTGGTGT in *Kluyveromyces lactis*) (all the telomeric motifs identified so far and the references of the original articles are reported in the Telomerase Database http://telomerase.asu.edu; Podlevsky et al. 2008). In *S. cerevisiae,* GGGTTA repeats are found at a high density in the subtelomeric region adjacent to telomeres; this suggests that GGGTTA may have been the ancestral motif prior to the divergence of budding yeast telomeric sequences (Brun et al. 1997).

A second characteristic of telomeric DNA is a single-stranded 3′ overhang (G-overhang). A G-overhang has been detected in many species and appears to be a conserved feature of eukaryotes with telomerase although, as we will see, there are exceptions. The length of telomeres and G-overhangs depends on the species. As examples, the telomeres of the minichromosomes of the ciliate *Tetrahymena thermophila* (GGGGTT motif) consist of about 50 repeats (300 bp) and a highly regulated G-overhang of 14-15 or 20-21 nt (Blackburn and Gall 1978; Henderson and Blackburn 1989; Jacob et al. 2001). The telomeres in the macronucleus of hypotrichous ciliates (GGGGTTTT motif) consist of a few repeats and a G-overhang of 16 nt (Klobutcher et al. 1981). The telomeres of the budding yeast *Saccaromyces cerevisiae* comprise about 300 bp and have a G-overhang of 10–15 nt outside of the S phase (Larrivée et al. 2004); longer overhangs appear in the late S phase (Wellinger et al. 1993). In humans, the telomeric double-stranded region is a few thousand base-pairs long (de Lange et al. 1990), while the telomeric G-overhang is estimated to vary from a few tens to a few hundreds of nucleotides (Makarov et al. 1997; McElligott and Wellinger 1997; Wright el al. 1997). Plants exhibit a wide range of telomere lengths, they can be shorter than 1 kbp (in the moss *Physcomitrella patents*) and longer than 100 kbp (in tobacco plants) (Procházková Schrumpfová et al. 2016), and the termini of their telomeres reserve surprises, as we will see.

***Telomeric proteins*:** Telomeric DNA is capped by proteins, specifically present at telomeres, that are essential for the stability of telomeric DNA. In mammals, the capping complex is named "Shelterin" (de Lange 2005); it took about 15 years to identify all the components of the complex. A detailed description of Shelterin in humans is provided in Chapter 3. Here, I will provide a very basic overview of telomere capping complexes in a few model eukaryotes to highlight how telomeres share common features modulated by species-specific features (Figure 1.6).

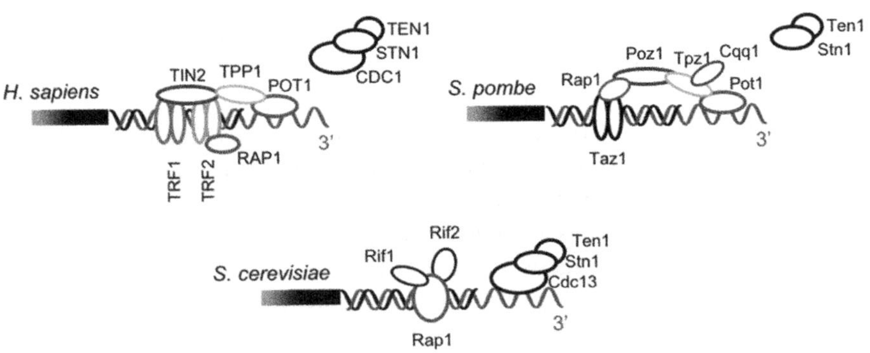

Figure 1.6. *Telomere capping proteins in three model eukaryotes. For a color version of this figure, see www.iste.co.uk/saintome/telomeres.zip*

We have already mentioned that the G-overhang in hypotrichous ciliates is bound by two proteins, TEBPα and TEBPβ (Figure 1.5). TEBPα specifically interacts with single-stranded telomeric DNA; TEBPβ alone binds weakly to DNA, but heterodimerizes with TEBPα in the presence of DNA, forming a ternary complex (Fang and Cech 1993a). Two complexes, POT1-TPP1 and Pot1-Tpz1, homologs of TEBPα–TEBPβ, specifically bind the G-overhang in humans and in the fission yeast *Schizzosaccharomyces pombe*, respectively (Baumann and Cech 2001). TPP1 alone does not bind to DNA, but the complex POT1–TPP1 has a greater affinity for DNA than POT1 alone (Wang et al. 2007; Xin et al. 2007). The interaction between POT1 and TPP1 is required for localization of POT1 to telomeres (Liu et al. 2004; Ye et al. 2004). *S. pombe* Tpz1 binds to the telomeric protein Ccq1 (a telomerase recruiter and checkpoint controller (Tomita and Cooper 2008)). In the budding yeast *Saccharomyces cerevisiae*, the G-overhang is protected by a different protein, Cdc13, that, with the proteins Stn1 and Ten1, forms the trimeric complex CST (Lin and Zakian 1990; Grandin et al. 1997, 2001). All these single-stranded DNA binding proteins contain OB-fold domains that mediate protein/DNA and protein/protein interactions (Chen et al. 2017; Rice et al. 2017).

The double-stranded region of telomeres is specifically bound by TRF1 and TRF2 (telomeric repeat binding factor 1 and 2) in humans, by Taz1 in *S. pombe* (Cooper at al. 1997) and by Rap1 (repressor-activator protein 1) in *S. cerevisiae* (Conrad et al. 1990). All these proteins interact with the double-stranded DNA through Myb domains: *S. cerevisiae* Rap1 contains two Myb domains, while TRF1, TRF2 and Taz1 contain a single Myb domain and bind to DNA in the form of homodimers (Bianchi et al. 1997; Spink et al. 2000). In *S. cerevisiae*, Rap1 recruits to telomeres Rif1 and Rif2 (Rap1-interaction factors) (Levy and Blackburn 2004). Homologs of Rap1 are also present in the Shelterin complexes of humans and *S. pombe*, but they do not directly interact with telomeric DNA and are recruited to telomeres through their interaction with TRF2 and Taz1 (Li et al. 2000; Kanoh and Ishikawa 2001). A recent study supports that human Rap1 is specifically involved in protecting critically short telomeres (Lototska et al. 2020).

In humans, TIN2 bridges the single-stranded DNA binding complex POT1-TPP1 to the double-stranded DNA binding proteins TRF1 and TRF2; in *S. pombe*, Poz1 bridges Tpz1 to Rap1 (Miyoshi et al. 2008).

A CST complex (CTC1, STN1, TEN1) is also present at human telomeres. Human CST is involved in several processes: in the fill-in of the C-strand (after resection by nucleases to generate the G-overhang), in telomere replication, in the rescue of stalled replication forks at both telomeric and non-telomeric sites; CST can also inhibit telomerase activity through primer sequestration (Chen et al. 2012;

Huang et al. 2012; Stewart et al. 2012; Wang et al. 2012; Zaug et al. 2021). In *S. pombe,* only Stn1 and Ten1 have been identified.

The telomeric architecture of budding yeasts diverges remarkably from the one of other eukaryotes and is a fascinating example of the coevolution of telomeric capping proteins and telomeric DNA (Steinberg-Neifach and Lue 2015).

In addition to the Shelterin complex, many factors involved in DNA damage signaling and double-strand break repair pathways are paradoxically present at telomeres, where they participate in telomere processing under the tight control of Shelterin subunits (reviewed in de Lange (2018)). In particular, TRF2 prevents inappropriate activation of the DNA damage response mediated by the ATM kinase and the Non-Homologous End Joining (NHEJ) pathway. POT1 represses activation of the DNA damage response mediated by the ATR kinase, likely by excluding the binding of the single-stranded protein RPA from the G-overhang (Gong and de Lange 2010).

Proteins are not the only factors present at telomeres. The telomeric C-strand is transcribed, starting from subtelomeric regions, into non-coding RNA molecules comprising GGGUUA repeats and named TERRA (for "telomeric repeat-containing RNA"). They are a component of telomeric heterochromatin and participate to telomere maintenance (reviewed in Azzalin and Lingner (2015) and Chapter 10).

Telomeres come in many forms: At this point, a reader may think that the general statement "telomeric DNA is composed of tandem repeats of a short motif and ends with a 3′ G-overhang" holds true in all eukaryotes in which telomerase is present. It is not so, life is richer, as some examples here below illustrate. In the nematode *Caenorhabditis elegans*, both 3′ G-overhangs and 5′ C-overhangs have been detected at the ends of chromosomes, and two single-stranded DNA binding proteins involved in telomere homeostasis have been identified (Raices et al. 2008). Both proteins contain a single OB-fold domain: one protein (named CeOB1) binds the G-rich strand, and the other protein (named CeOB2) binds the C-rich strand. Recent studies provide further insights into the compositions of the telomeric capping complex in the nematode *C. elegans* (Dietz et al. 2021; Yamamoto et al. 2021). Telomeric 5′ C-overhangs have also been detected in human tumor cells using ALT (an alternative mechanism of telomere lengthening based on homologous recombination) and in murine cells (Oganesian and Karlseder 2011). It has been proposed that the presence of telomeric 5′ C-overhangs is a mark of recombination-based telomere elongation.

In the plant *Arabidopsis thaliana*, half of the telomeres end with blunt ends or very short G-overhangs (1-3 nt); blunt-ended telomeres are protected, counterintuitively, by the complex Ku70/80, a key component of the classical NHEJ pathway (Kazda et al. 2012; Valuchova et al. 2017). The telomeric proteins in the model plant *Arabidopsis thaliana* (not presented here) are a further example of variations on a common theme (reviewed in Procházková Schrumpfová et al. (2016)).

The insect *Bombyx mori* (whose larva is the silkworm) has a telomerase gene, but it has hybrid telomeres consisting of GGTTA repeats and telomere-specific retrotransposons inserted into GGTTA repeats. Differently from other insects, no telomerase activity has been detected in this lepidopterous, despite the presence of a telomerase gene. For these unusual features, it has been speculated that *Bombyx mori* may be in transition between a telomerase and a retrotransposon-based architecture of telomere protection (reviewed in Mason et al. (2016)).

1.3.2. *Telomeric structures*

Telomeres are not just linear structures made of DNA repeated sequences that associate with proteins. They fold into dynamic structures that participate in telomere protection and processing.

***G4s in ciliates*:** As previously mentioned, in hypotrichous ciliates, the ends of the gene-sized DNA molecules of the macronuclei aggregate via the formation of bimolecular G4s that are resolved during replication (Figure 1.5). *In vivo*, aggregation is regulated in a cell-cycle dependent manner by TEBPα,β proteins (Paeschke et al. 2005). In the proposed model, dimers of TEBPα promote telomere pairing by interacting with the single-stranded G-overhangs, while TEBPβ (recruited by TEBPα) act as chaperons for G4 formation and ensure the stable association of telomeres (Paeschke et al. 2005), as suggested by early *in vitro* studies (Fang and Cech 1993b). G4 unfolding occurs when telomerase is recruited by TEBPβ during S phase (Paeschke et al. 2008).

A back-folded structure in budding yeast? Several studies in *Saccharomyces cerevisiae* have suggested that protein–protein interactions make its telomeres fold back onto themselves and onto subtelomeric regions (Strahl-Bolsinger et al. 1997); nevertheless, this hypothetical structure has been questioned in a recent study, in which the authors conclude that, if it forms *in vivo*, it must be rare and transient (Pasquier and Wellinger 2020).

***T-loops*:** In a variety of eukaryotes, telomeres fold into a structure named a "t-loop" for telomeric loop (Griffith et al. 1999). A t-loop forms by the invasion of the G-overhang into the telomeric duplex region, where it base-pairs with the C-strand displacing 75–200 nt of the G-strand (Figure 1.7). T-loop formation relies on the peculiar features of telomeric DNA: tandem repeats and a single-stranded overhang (Stansel et al. 2001).

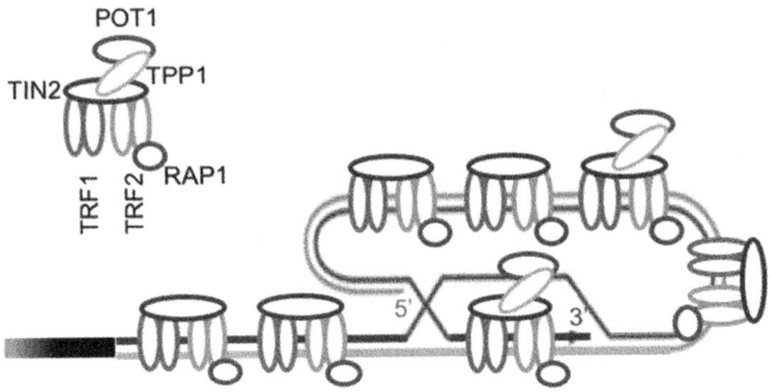

Figure 1.7. *Schematic illustration of a t-loop. Shelterin complexes are also represented. Bridging of the displaced G-strand to the double-stranded base of the t-loop by a Shelterin complexes is putative. For a color version of this figure, see www.iste.co.uk/saintome/telomeres.zip*

The t-loop structure has been proposed to be the primordial structure that allowed the stabilization of the ends of linear chromosomes when they emerged, providing a means to protect and elongate the ends of chromosomes before the emergence of telomere-binding proteins and telomerase (de Lange 2004, 2015). T-loops formation in murine and human cells was first reported in an article published in 1999, which was a collaborative study from Griffith and Titia de Lange laboratories (Griffith et al. 1999). The authors already had this kind of structure in mind and employed a psoralen/UV cross-linking strategy to test their hypothesis. Cell nuclei were treated with psoralen (a compound that intercalates into double-stranded DNA and, on irradiation with UV light, covalently links pyrimidines on the two strands, preferentially at TpA sites). After UV irradiation, telomeric DNA was separated from bulk genomic DNA because of a unique feature of GGGTTA repeats: their lack of recognition sites for restriction endonucleases. This makes it possible to cut bulk genomic DNA into small fragments using several restriction enzymes, while preserving telomeres intact. Electron microscopy analysis of telomere-enriched fractions from psoralen treated nuclei revealed the formation of large duplex loops, the sizes of which correlated with telomere lengths. It is interesting to note that t-loop-like structures had already been observed 20 years before. They were observed by electron microscopy at the ends of the linear mitochondrial DNA of a ciliate, in a fascinating study where some peculiar features of telomeres were indirectly deduced (Goldbach et al. 1979). So, it was in a ciliate, once again, that telomeric loops disclosed themselves, ahead of time. After the detection of t-loops in murine and human cells, t-loops were detected in many other eukaryotes (in *Trypanosoma brucei*, in plants, in vertebrates, in the nematode *Caenorhabditis elegans*, etc.; reviewed in Tomaska et al. (2020)). Notably, t-loops were also detected in the micronucleus of a hypotrichous ciliate (Murti and Prescott 1999). So, in hypotrichous ciliates, while the short telomeres of the gene-sized DNA molecules in the macronucleus associate through the formation of bimolecular G4s (Figure 1.5), the long telomeres of the chromosomes in the micronucleus are likely to fold into a t-loop structure. As discussed above, the structure of telomeres in the budding yeast *Saccharomyces cerevisiae* is still elusive. Its short telomeres (300–500 bp) and the heterogeneous repeats may make t-loops difficult to form.

***Lighting up t-loops*:** In more recent years, super resolution light microscopy has made it possible to detect t-loop by fluorescence in situ hybridization with no need for DNA fragmentation and purification (Doksani et al. 2013). The improvements in t-loop detection and quantification, along with the development of suitable genetic models, has allowed us to gain a deeper understanding of the factors that regulate t-loop assembly/disassembly and stability, in particular of the role of the Shelterin subunit TRF2. Early studies had shown that TRF2 promotes t-loop formation *in vitro* (Griffith et al. 1999; Stansel et al. 2001). More recent studies in mammalian cells have shown that TRF2 is the only Shelterin subunit necessary and sufficient for t-loop formation (Doksani et al. 2013; Timashev and de Lange 2020). Human TRF2 is composed of a N-terminal basic domain, a dimerization domain (TRFH), a RAP1 binding domain, a TIN2 binding domain and a C-terminal DNA binding Myb domain. The TRFH dimerization domain is involved in t-loop formation (Benarroch-Popivker et al. 2016; Van Ly et al. 2016). Dimers of TRF2 wrap about 90 bp of telomeric DNA around their dimerized TRFH domains; this activity appears to be important for t-loop formation, since, in human cells, the ectopic expression of TRF2 mutants with impaired capacity to wrap DNA results in a decrease in the frequency of t-loops compared to the ectopic expression of wild-type TRF2 (Benarroch-Popivker et al. 2016). Wrapping of DNA around TRF2 is thought to induce unwinding of the neighboring unbound DNA, thereby facilitating the invasion of the 3′ overhang in the double-stranded telomeric DNA (Amiard et al. 2007). The N-terminal basic domain of TRF2 prevents t-loop cleavage (Schmutz et al. 2017): it binds to the three-way junction at the base of the t-loop; this binding has been proposed to mask the 5′ recessed end of the telomere from PARP1 and to prevent branch migration. TRF2 also mediates the recruitment of the helicase RTEL1 to telomeres, a helicase involved in the disassembly of t-loops (Vannier et al. 2012; Sarek et al. 2015). A recent study shows that the recruitment and release of RTEL1 at telomeres is regulated by a phosphorylation site in TRF2: following dephosphorylation of TRF2 in S phase, RTEL1 is recruited to telomeres and t-loops are unfolded; following phosphorylation of TRF2 after replication, RTEL1 is released from telomeres and t-loops refold (Sarek et al. 2019).

T-loops provide a structural solution to protect the ends of chromosomes. What do t-loops protect telomeres from? A recent study provides a partial answer to this question. As previously mentioned, factors involved in DNA damage signaling and DNA repair are paradoxically present at telomeres and necessary for their correct processing. TRF2 has long been known to prevent inappropriate ATM and NHEJ activation. It has long been supposed that TRF2 repressed both ATM and NHEJ by remodeling telomeres into t-loops (Doksani et al. 2013). A recent study in murine and human cells provides evidence that ATM activation at telomeres actually coincides with a structural change of chromosome ends from a t-loop to a linear structure. This suggests that the t-loop acts as a conformational switch that, under the control of TRF2, regulates ATM activation during the cell cycle (Van Ly et al. 2018). Unlike ATM, the mechanism through which TRF2 prevents inappropriate NHEJ activity at telomeres has been shown to be independent of t-loop formation (Benarroch-Popivker et al. 2016; Van Ly et al. 2018).

Since the detection of t-loops in mammalian cells, progress has been made in the understanding of the dynamic structure of telomeres. The 20-year history of the t-loop is retraced in a compelling review (Tomaska et al. 2020). Much remains to be done to unravel the high-order architecture of an entire telomere and its dynamics during the cellular cycle.

1.4. G-quadruplexes at telomeres

In this section, I will focus on another level of telomere structuring: the DNA secondary structures that can form on the telomeric G-strand. I will first shortly present the DNA secondary structures formed by telomeric motifs identified so far. Then I will discuss what telomeric G4s represent for proteins that deal with them (unfriendly or friendly structures?).

1.4.1. *Secondary structures formed by repeats of telomeric motifs*

As previously presented, with the exception of dipterans (that have lost telomerase and have complex telomeric sequences), the telomeric strand running toward the 3′ end is composed of repeats of a motif, species-specific, comprising consecutive guanines. The presence of runs of consecutive guanines and the presence of K^+ in cells make this strand (the G-strand) potentially prone to fold into G4s.

We have already seen that the short telomeric G-overhangs of ciliates minichromosomes (composed of about two telomeric repeats) associate and form bimolecular G4s (Figure 1.5). This higher order structure, which links telomeres to each other, has been reported only for these particular DNA molecules. In this section, I will focus on intramolecular G4s that can form at longer G-overhangs and, more generally, all along a telomeric G-strand.

***Structures formed by four repeats of telomeric motifs*:** To make an intramolecular G4, four runs of consecutive guanines are needed. In potassium, four repeats of G_2, G_3 and G_4 telomeric motifs fold into stable G4s (De Cian et al. 2008; Tran et al. 2011). The melting temperature of the G4s they form in KCl 100 mM increases with the number of consecutive guanines (from 45°C for the insect motif GGTTA to more than 80°C for the ciliate GGGGTT and GGGGTTTT motifs). For G_3 motifs, the melting temperatures decrease with the number of consecutive thymines (from 68°C for the *Giardia* motif GGGTA to 58°C for the green algae motif GGGTTTTA).

The G4 structures formed in solution by four repeats of several telomeric motifs have been determined by nuclear magnetic resonance. In Figure 1.8, the G4 structures formed by four repeats of the GGGTTA motif are represented (GGGTTA is the most studied motif, because it is characteristic of human telomeres). They illustrate the polymorphism of this class of structures and how even a single sequence can fold into a variety of G4 conformations well. The G4 structures formed in solution by repeats of the G_2 motif of insects (GGTTA), the G_3 motif of the parasite *Giardia* (GGGGTA) and the G_4 motifs of ciliates (GGGGTT and GGGGTTTT) have also been determined (Amrane et al. 2009; Hu et al. 2009; Wang and Patel 1994, 1995).

The telomeric motifs of yeasts (that are highly diverse in sequence) do not all share the potential to fold into G4s. The G-rich repeats of telomeres in *Saccharomyces* species can fold into stable G4s (Tran et al. 2011). A study has suggested that G4s can protect the G-overhang of *Saccharomyces cerevisiae* telomeres when the binding of the Cdc13 protein is impaired (Smith et al. 2011). Nevertheless, when considering the long and diverse telomeric motifs identified so far in other genera of Saccharomycetes, the potential to fold into G4s seems to be preserved in some genera, but not in others (Tran et al. 2011; P.A. unpublished results).

***Structures formed by longer telomeric sequences: G4s like beads on a string*:** Four repeats of G_2, G_3 and G_4 telomeric motifs fold into G4s. But telomeric sequences are much longer than four repeats. What structures are formed by longer telomeric sequences? Do they fold into regular or irregular G4 structures? How stable are these structures? Several studies have addressed this question, focusing on GGGTTA repeats. Early studies on an increasing number of GGGTTA repeats suggested the possibility that a number of repeats multiple of 4 folded into contiguous G4s (Vorlickova et al. 2005; Yu et al. 2006). Subsequent studies on 8 and 12 GGGTTA repeats clearly showed the formation of two and three contiguous G4s, respectively; nevertheless, it was proposed that the G4 units were not identical nor independent and that unfavorable interactions could limit complete folding of the telomeric overhang into contiguous G4s (Bauer et al. 2011; Petraccone et al. 2011). A successive study demonstrated that sequences composed of a number of GGGTTA multiple of 4 form contiguous G4 units that do not interact with each other and have similar stabilities, independently of the number of G4 units (Figure 1.9) (Bugaut and Alberti 2015). A study by optical tweezers, published at the same time, supported the formation of non-interacting G4 units (Abraham Punnoose et al. 2014) and a recent study on electron microscopy highlighted the compaction of 20 kb of GGGTTA repeats into bead-on-a-string filaments consistent with the formation of G4 units (Kar et al. 2018).

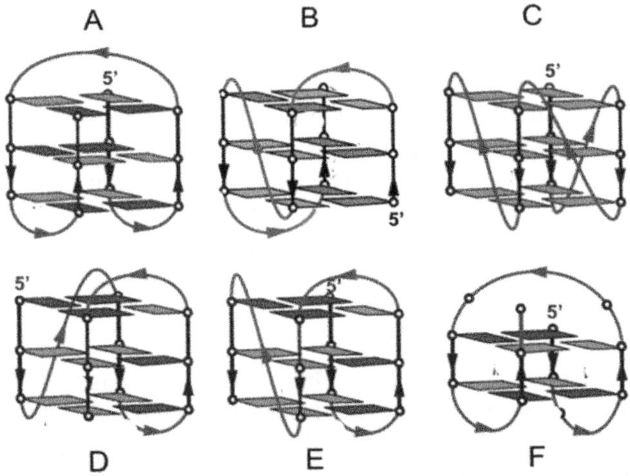

Figure 1.8. *Single G4s formed by GGGTTA repeats. Schematic illustration of the G4 structures formed in solution by the sequence (GGGTTA)3GGG (reviewed in Phan (2010)). They differ in the number of G-tetrads and in two related features: the relative orientation of the four strands and the arrangement of the loops. Structures A and B form in Na$^+$, both rely on three G-tetrads (all the 16 guanines are involved in G-tetrads). Structures D, E and F form in K$^+$: structures D and E rely on three G-tetrads, while structure F relies on two G-tetrads (12 of the 16 guanines are involved in G-tetrads, 4 guanines are in the loops). Despite a lower number of G-tetrads, structure F is as stable as structures D and E (Lim et al. 2009). Structure C forms in the presence of K$^+$ and of polyethylene glycol*

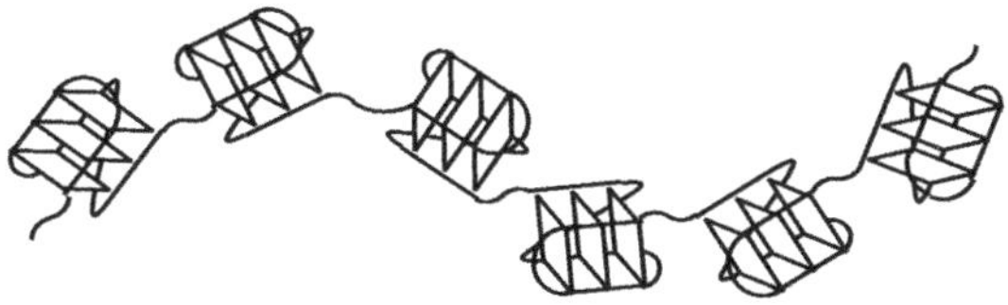

Figure 1.9. *Contiguous G4s formed by GGGTTA repeats. TTA(GGGTTA)$_{4N}$ sequences fold into N contiguous G4 units non-interacting with each other and of similar stability (Bugaut and Alberti 2015). Each G4 unit is formed by the sequence (GGGTTA)$_3$GGG and G4 units are connected by TTA trinucleotides*

The ability to fold into contiguous G4 units is not exclusive of GGGTTA repeats; it is conserved in G_2, G_3 and G_4 telomeric motifs identified so far (P.A. unpublished results), with a notable exception: while four repeats of the telomeric motif GGCTTA of nematodes can fold into a G4 (Tran et al. 2011), longer sequences fold into hairpins (Skolakova et al. 2015; P.A. unpublished results).

1.4.2. *G4s at telomeres: friendly or unfriendly structures?*

Contiguous G4s can form on the single-stranded G-overhang (when telomeres are in a linear form and not folded into a t-loop), but also all along the G-strand in the double-stranded region of telomeres during replication or during the transcription of the telomeric C-strand, when the double-helix is open and portions of the G-strand are transiently exposed as a single-strand (Figure 1.10).

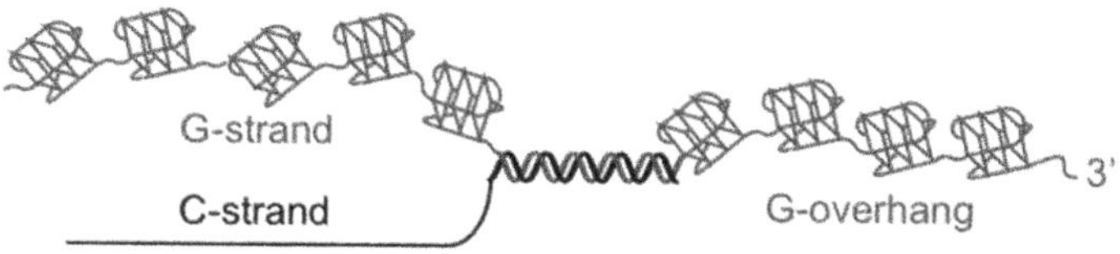

Figure 1.10. *G4s at telomeres. For a color version of this figure, see www.iste.co.uk/saintome/telomeres.zip*

During replication, G4s form at unwound DNA prior to nascent DNA synthesis (Lee et al. 2021). Replication of telomeres starts, most frequently, from replication origins situated in the subtelomeric regions (Sfeir et al. 2009; Drosopoulos et al. 2012), so the G-strand is most frequently replicated discontinuously (it is the template for the synthesis of the lagging strand) all along its length. Discontinuous replication makes the template G-strand more likely to fold into G4s. In the following, I will mainly focus on G4s at human telomeres and discuss them from two points of view: why they are considered problematic structures, in particular for the replication machinery, and what features make them easily manageable structures for proteins that deal with them.

Telomeric G4s: unfriendly structures? Many studies have provided evidence of the formation of G4s at human telomeres (for a review, see Bryan Molecules (2020)). For example, G4s in human chromosomes and in particular at telomeres have been indirectly visualized using a radiolabeled G4-ligand (Granotier et al. 2005) or an antibody recognizing G4s (Biffi et al. 2013); they have also been pulled down with a tagged G4-ligand (Müller et al. 2010). These studies suggest that G4s can transiently form at telomeres and be trapped by G4 ligands or antibodies. Telomeric G4s are commonly considered problematic structures, in particular for the replication machinery (Sfeir et al. 2009). This point of view mainly stems from the observation of aberrant telomeric phenotypes in cells treated with G4 stabilizing ligands and in cells where proteins involved in telomere maintenance and able to unfold G4s are depleted or not functional. Treating human cells with G4 ligands results in troubles at telomeres, such as the displacement of TRF2 and POT1 from telomeres, the degradation of the G-overhang, the induction of DNA damage signals at telomeres and impaired telomere replication (Gomez et al. 2004; Tahara et al. 2006; Gomez et al. 2006a, 2006b; Rodriguez et al. 2008; Rizzo et al. 2009). In mammalian cells, at least three helicases are important for telomere stability: WRN (mutated in Werner's syndrome), BLM (mutated in Bloom syndrome) and RTEL1 (regulator of telomere length 1, mutated in a severe form of dyskeratosis congenita). Impaired activity or depletion of one of these helicases results in telomeric anomalies that indicate troubles in the replication of telomeres, in particular of the G-strand (Crabbe et al. 2004; Sfeir et al. 2009; Vannier et al. 2012, 2013; Zimmermann et al. 2014; Drosopoulos et al., 2015). These anomalies are exacerbated in the presence of G4 stabilizing ligands (Rizzo et al. 2009; Vannier et al. 2012, Vannier et al. 2013; Drosopoulos et al., 2015). Cells lacking BLM or WRN also display an increase in G4s (detected by antibodies) in the nucleus and in particular at telomeres (Drosopoulos et al. 2015). On the basis of these studies, it has been proposed that WRN and BLM (able to unfold G4s *in vitro* (Sun et al. 1998; Mohaghegh et al. 2001), as well as RTEL1, contribute (among other roles) to telomere stability by unwinding G4s on the telomeric G-strand.

Besides helicases, at least two single-stranded DNA binding complexes are faced with G4s on the G-strand during replication: the CST complex (mentioned in the paragraph "Telomeric proteins") and the heterotrimeric replication protein A (RPA). RPA acts genome wide; it protects single-stranded DNA transiently exposed during DNA metabolism and participates in the orchestration of adapted metabolic pathways (replication, repair, recombination) (Chen and Wold 2014). RPA is present at human telomeres during replication (Verdun and Karlseder 2006); in *Schizosaccharomyces pombe*, RPA transiently accumulates on the template strand (the G-strand) during the synthesis of the lagging strand (Moser et al. 2009). Depletion of CST in human cells or impaired binding of RPA in *S. pombe* result in defective replication of telomeres, in particular of the G-strand (Huang 2012; Audry et al. 2015). Troubles in replication are exacerbated by G4 ligands (Zhang et al. 2019). *In vitro*, both RPA and CST bind telomeric repeats structured in G4s, unfolding them (Salas et al. 2006; Bhattacharjee et al. 2017). On the basis of these studies, it has been proposed that RPA and CST keep the template G-strand unfolded during replication (Audry et al. 2015; Zhang et al. 2019). Overall, these kinds of approaches support the fact that stabilization or inefficient unfolding of G4s result in telomere instability. So, it is in perturbed systems that telomeric G4s reveal their challenging character.

Telomeric G4s: friendly structures? I now propose to change the point of view. In non-perturbed conditions (in the absence of G4 stabilizing ligands and when all proteins involved in telomere maintenance are functional), telomeres are stable. This indicates that telomeric repeats and their potential to fold into G4s are well managed by cellular machineries. Contiguous G4s formed by GGGTTA repeats exhibit at least two characteristics that make them easily manageable structures compared to other G4s: (i) they are stable, but not too stable; (ii) they are *identical* and *independent* structural units. These two features make telomeric G4s easily manageable structures for proteins that have to unfold them, as discussed hereafter for the protein RPA.

(i) In physiological conditions, contiguous G4s formed by GGGTTA repeats are stable (their melting temperature is around 50°C, in KCl 100 mM) (Bugaut and Alberti 2015), but not too stable to prevent binding of RPA (Lancrey et al. 2018). Consistently, *in vitro*, RPA is sufficient to allow progression of DNA synthesis by a polymerase through a G4 formed by GGGTTA repeats, while a helicase is required for progression through a more stable G4 (Sparks et al. 2019). Likely, three interdependent factors make telomeric G4s more easily unfolded by RPA compared to more stable G4s found in other genomic loci: lower thermodynamic stability, higher loop accessibility (Qureshi et al. 2012) and faster unfolding dynamics (Tran et al. 2021). As for RPA, the structuration of GGGTTA repeats into G4s does not impair binding of POT1-TPP1 (Taylor et al. 2011).

(ii) Contiguous G4s formed by GGGTTA repeats have identical stability and do not interact each other (Figure 1.9) (Bugaut and Alberti 2015). So, proteins that bind the G-strand (such as RPA, CST, POT1) and proteins that progress along telomeres (such as helicases) are faced with *independent* and *identical* (in term of stability) structural units. This implies that they bind the G-strand, or progress along it, with an efficiency that is independent of the number of G4 units (namely of the length of the portion of the G-strand structured into G4s), as demonstrated for the interaction of human RPA with GGGTTA repeats of increasing length (Lancrey et al. 2018). As a counterpoint, the binding efficiency of RPA to GGGTTA repeats strongly decreases in a particular non-physiological condition where the telomeric G4 units interact with each other (Saintomé et al. 2016; Lancrey et al. 2018).

Overall, the above *in vitro* studies illustrate the easy manageable features of telomeric G4s for proteins that have to deal with them.

In cells, helicase activity facilitates the loading of RPA on single-stranded DNA folded into stable G4s (Lee et al. 2021) and several helicases are involved in the resolution of G4s (reviewed in Lerner and Sale (2019)). *In vitro*, binding of RPA to a telomeric sequence folded into a G4 does not require any helicase activity, contrary to other more stable G4s (Lee et al. 2021). What is the scenario in cells? Does RPA bind telomeric sequences once they are unfolded by helicases or does it directly unfold G4s? RPA interacts with many helicases, including WRN and BLM (reviewed in Awate and Brosh (2017)). Likely, the resolution of G4s at telomeres is mediated by an interplay between RPA and helicases. I think that, in this interplay, the ability of RPA to unfold secondary structures at telomeres is important, otherwise hairpin structures (that are readily unfolded by helicases) should not pose a problem at telomeres (and elsewhere), contrary to what is suggested by the studies presented here below.

Telomeric G4s: friendly or unfriendly, in any case better than hairpins: Let us see what happens when other structures form at human telomeres. In the proximity of the subtelomeric regions of human telomeres, canonical GGGTTA repeats are interspersed with variant repeats (GGGTGA, GGGTCA, GGGGTT, etc.). The variant motif GGGCTA (estimated to be present in 7% of human telomeres) exhibits a length-dependent instability: sequences of more than 10 GGGCTA repeats gain or lose repeats, likely through replication (Mendez-Bermudez et al. 2011). A recent *in vitro* study shows that GGGCTA repeats can form both G4s and hairpins; the hairpin fraction increases when the number of GGGCTA increases, along with their thermal stability. While G4s formed by GGGTTA and GGGCTA repeats are efficiently unfolded by RPA, GGGCTA hairpins resist to unfolding. Overall, this study points to hairpin structures and impaired binding of RPA as possible origins of the length-dependent instability of GGGCTA repeats in human telomeres (Figure 1.11) (Chatain et al. 2021a). From the perspective of evolution, this study also suggests that the instability of repeated sequences folding into hairpins may have led to a counterselection of hairpin-prone telomeric motifs.

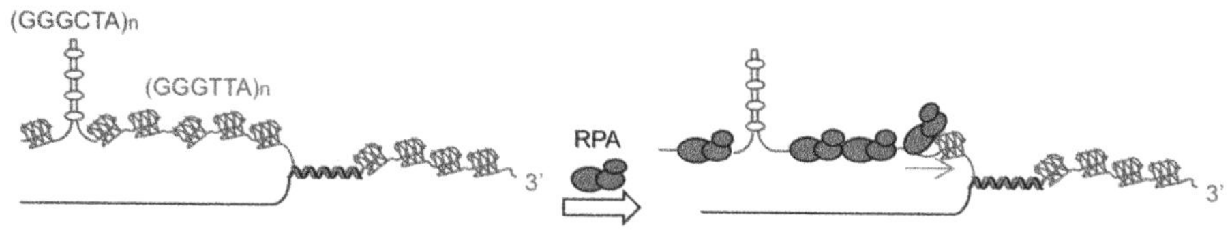

Figure 1.11. *Hairpins at telomeres. The unstable telomeric variant motif GGGCTA can form hairpins. GGGCTA hairpins are poorly unfolded by RPA. For a color version of this figure, see www.iste.co.uk/saintome/telomeres.zip*

Terminus: When considering telomeres in eukaryotes other than dipteran and budding yeasts, natural selection has led to a strong asymmetry in the two telomeric strands. Why this strong asymmetry? Why a prevalence of motifs that make one strand fold into G4s? Is their folding potential just a side-effect of the co-evolution of telomerase-generated repeated sequences and telomeric proteins or does it confer some kind of advantage? It is difficult to assess. *In vitro* studies suggest that G4s may play a role at the G-overhang. The telomeric G-overhang is the only region of a chromosome where G4s do not have to compete with the double-helix to exist (at least when the telomere is in a linear form and not structured into a t-loop). *In vitro* experiments suggest that G4s at the G-overhang may favor the binding of POT1-TPP1 over the binding of RPA (Ray et al. 2014), affect the mode of binding of multiple POT1-TPP1 to the G-overhang (Chatain et al. 2021b) and modulate the activity of telomerase (Jansson et al. 2019). For the moment, we have at least highlighted some friendly features of telomeric G4s that make them easily manageable structures for proteins that deal with them. It is as if *Nature* said: "If secondary structures at telomeres can't be avoided, then let's at least make these secondary structures the easiest to manage". The GGGCTA motif (unstable and folding into hairpins poorly unfolded by RPA) is an enlightening counterpoint to the canonical GGGTTA motif (stable and folding into identical and independent G4 units efficiently unfolded by RPA). Perhaps the exceptions will help to understand the reasons for the common theme (the G4 folding potential of telomerase-generated sequences), and one of the exceptions is the telomeric sequence of nematodes (the ones that fold into hairpins). How do nematodes manage possible hairpin structures at their telomeres? Doing science is like eating cherries (once you eat one, you cannot stop): each answer raises new questions.

1.5. References

Abraham Punnoose, J., Cui, Y., Koirala, D., Yangyuoru, P.M., Ghimire, C., Shrestha, P., Mao, H. (2014). Interaction of G-quadruplexes in the full-length 3′ human telomeric overhang. *J. Am. Chem. Soc.*, 136, 18062.

Amiard, S., Doudeau, M., Pinte, S., Poulet, A., Lenain, C., Faivre-Moskalenko, C., Angelov, D., Hug, N., Vindigni, A., Bouvet, P. et al. (2007). A topological mechanism for TRF2-enhanced strand invasion. *Nat. Struct. Mol. Biol.*, 14, 147.

Amrane, S., Ang, R.W., Tan, Z.M., Li, C., Lim, J.K., Lim, J.M., Lim, K.W., Phan, A.T. (2009). A novel chair-type G-quadruplex formed by a *Bombyx mori* telomeric sequence. *Nucleic Acids Res.*, 37, 931.

Audry, J., Maestroni, L., Delagoutte, E., Gauthier, T., Nakamura, T.M., Gachet, Y., Saintomé, C., Géli, V., Coulon, S. (2015). RPA prevents G-rich structure formation at lagging-strand telomeres to allow maintenance of chromosome ends. *EMBO J.*, 34, 1942.

Awate, S. and Brosh Jr., R.M. (2017). Interactive roles of DNA helicases and translocases with the single-stranded DNA binding protein RPA in nucleic acid metabolism. *Int. J. Mol. Sci.*, 18(6), 1233.

Azzalin, C.M. and Lingner, J. (2015). Telomere functions grounding on TERRA firma. *Trends Cell Biol.*, 25, 29.

Bauer, L., Tlučková, K., Tóhová, P., Viglaský, V. (2011). G-quadruplex motifs arranged in tandem occurring in telomeric repeats and the insulin-linked polymorphic region. *Biochemistry*, 50, 7484.

Baumann, P. and Cech, T.R. (2001). Pot1, the putative telomere end-binding protein in fission yeast and humans. *Science*, 292, 1171.

Benarroch-Popivker, D., Pisano, S., Mendez-Bermudez, A., Lototska, L., Kaur, P., Bauwens, S., Djerbi, N., Latrick, C.M., Fraisier, V., Pei, B. et al. (2016). TRF2-mediated control of telomere DNA topology as a mechanism for chromosome-end protection. *Mol. Cell*, 61, 274.

Bhattacharjee, A., Wang, Y., Diao, J., Price, C.M. (2017). Dynamic DNA binding, junction recognition and G4 melting activity underlie the telomeric and genome-wide roles of human CST. *Nucleic Acids Res.*, 45, 12311.

Bianchi, A., Smith, S., Chong, L., Elias, P., de Lange, T. (1997). TRF1 is a dimer and bends telomeric DNA. *EMBO J.*, 16, 1785.

Biffi, G., Tannahill, D., McCafferty, J., Balasubramanian, S. (2013). Quantitative visualization of DNA G-quadruplex structures in human cells. *Nat. Chem.*, 5, 182.

Blackburn, E.H. and Chiou, S.S. (1981). Non-nucleosomal packaging of a tandemly repeated DNA sequence at termini of extrachromosomal DNA coding for rRNA in Tetrahymena. *Proc. Natl. Acad. Sci. U.S.A.,* 78, 2263.

Blackburn, E.H. and Gall, J.G. (1978). A tandemly repeated sequence at the termini of the extrachromosomal ribosomal RNA genes in Tetrahymena. *J. Mol. Biol.,* 120, 33.

Brun, C., Marcand, S., Gilson, E. (1997). Proteins that bind to double-stranded regions of telomeric DNA. *Trends Cell Biol.,* 7, 317.

Bryan T.M. (2020). G-Quadruplexes at telomeres: Friend or foe? *Molecules,* 25, 3686.

Bryan, T.M., Englezou, A., Gupta, J., Bacchetti, S., Reddel, R.R. (1995). Telomere elongation in immortal human cells without detectable telomerase activity. *EMBO J.*, 14, 4240.

Bugaut, A. and Alberti, P. (2015). Understanding the stability of DNA G-quadruplex units in long human telomeric strands. *Biochimie,* 113, 125.

Chantot, J. and Guschlbauer, W. (1969). Physicochemical properties of nucleosides 3. Gel formation by 8-bromoguanosine. *FEBS Lett.,* 4, 173.

Chatain, J., Blond, A., Phan, A.T., Saintomé, C., Alberti, P. (2021a). GGGCTA repeats can fold into hairpins poorly unfolded by replication protein A: A possible origin of the length-dependent instability of GGGCTA variant repeats in human telomeres. *Nucleic Acids Res.,* 49, 7588.

Chatain, J., Hatem, G., Delagoutte, E., Riou, J.F., Alberti, P., Saintomé, C. (2021b). Multiple hPOT1-TPP1 cooperatively unfold contiguous telomeric G-quadruplexes proceeding from 3′ toward 5′, a feature due to a 3′-end binding preference and to structuring of telomeric DNA. *Nucleic Acids Res.,* 49, 10735.

Chen, R. and Wold, M.S. (2014). Replication protein A: Single-stranded DNA's first responder: Dynamic DNA-interactions allow replication protein A to direct single-strand DNA intermediates into different pathways for synthesis or repair. *BioEssays,* 36, 1156.

Chen, L.Y., Redon, S., Lingner, J. (2012). The human CST complex is a terminator of telomerase activity. *Nature*, 488, 540.

Chen, C., Gu, P., Wu, J., Chen, X., Niu, S., Sun, H., Wu, L., Li, N., Peng, J., Shi, S. et al. (2017). Structural insights into POT1-TPP1 interaction and POT1 C-terminal mutations in human cancer. *Nature Commun.,* 8, 14929.

Conrad, M.N., Wright, J.H., Wolf, A.J., Zakian, V.A. (1990). RAP1 protein interacts with yeast telomeres *in vivo*: Overproduction alters telomere structure and decreases chromosome stability. *Cell,* 63, 739.

Cooper, J.P., Nimmo, E.R., Allshire, R.C., Cech, T.R. (1997). Regulation of telomere length and function by a Myb-domain protein in fission yeast. *Nature,* 385, 744.

Crabbe, L., Verdun, R.E., Haggblom, C.I., Karlseder, J. (2004). Defective telomere lagging strand synthesis in cells lacking WRN helicase activity. *Science,* 306, 1951.

De Cian, A., Grellier, P., Mouray, E., Depoix, D., Bertrand, H., Monchaud, D., Teulade-Fichou, M.P., Mergny, J.L., Alberti, P. (2008). Plasmodium telomeric sequences: Structure, stability and quadruplex targeting by small compounds. *Chembiochem,* 9, 2730.

Dietz, S., Almeida, M.V., Nischwitz, E., Schreier, J., Viceconte, N., Fradera-Sola, A., Renz, C., Ceron-Noriega, A., Ulrich, H.D., Kappei, D. et al. (2021). The double-stranded DNA-binding proteins TEBP-1 and TEBP-2 form a telomeric complex with POT-1. *Nat. Commun.,* 12, 2668.

Doksani, Y., Wu, J.Y., de Lange, T., Zhuang, X. (2013). Super-resolution fluorescence imaging of telomeres reveals TRF2-dependent T-loop formation. *Cell,* 155, 345.

Drosopoulos, W.C., Kosiyatrakul, S.T., Yan, Z., Calderano, S.G., Schildkraut, C.L. (2012). Human telomeres replicate using chromosome-specific, rather than universal, replication programs. *J. Cell Biol.,* 197, 253.

Drosopoulos, W.C., Kosiyatrakul, S.T., Schildkraut, C.L. (2015). BLM helicase facilitates telomere replication during leading strand synthesis of telomeres. *J. Cell Biol.,* 210, 191.

Fang, G. and Cech, T.R. (1993a). Oxytricha telomere-binding protein: DNA-dependent dimerization of the alpha and beta subunits. *Proc. Natl. Acad. Sci. U.S.A.*, 90, 6056.

Fang, G. and Cech, T.R. (1993b). The beta subunit of Oxytricha telomere-binding protein promotes G-quartet formation by telomeric DNA. *Cell,* 74, 875.

Gellert, M., Lipsett, M.N., Davies D.R. (1962). Helix formation by guanylic acid. *Proc. Natl. Acad. Sci. U.S.A.,* 48, 2013.

Goldbach, R.W., Bollen-de Boer, J.E., van Bruggen, E.F., Borst, P. (1979). Replication of the linear mitochondrial DNA of *Tetrahymena pyriformis. Biochim Biophys Acta,* 562, 400.

Gomez, D., Paterski, R., Lemarteleur, T., Shin-Ya, K., Mergny, J.L., Riou, J.F. (2004). Interaction of telomestatin with the telomeric single-strand overhang. *J. Biol. Chem.*, 279, 41487.

Gomez, D., O'Donohue, M.F., Wenner, T., Douarre, C., Macadré, J., Koebel, P., Giraud-Panis, M.J., Kaplan, H., Kolkes, A., Shin-ya, K., Riou, J.F. (2006a). The G-quadruplex ligand telomestatin inhibits POT1 binding to telomeric sequences *in vitro* and induces GFP-POT1 dissociation from telomeres in human cells. *Cancer Res.,* 66, 6908.

Gomez, D., Wenner, T., Brassart, B., Douarre, C., O'Donohue, M.F., El Khoury, V., Shin-Ya, K., Morjani, H., Trentesaux, C., Riou, J.F. (2006b). Telomestatin-induced telomere uncapping is modulated by POT1 through G-overhang extension in HT1080 human tumor cells. *Journal Biol. Chem.*, 281, 38721.

Gong, Y. and de Lange, T. (2010). A Shld1-controlled POT1a provides support for repression of ATR signaling at telomeres through RPA exclusion. *Mol. Cell,* 40, 377.

Gottschling, D.E. and Zakian, V.A. (1986). Telomere proteins: Specific recognition and protection of the natural termini of Oxytricha macronuclear DNA. *Cell*, 47, 195.

Grandin, N., Reed, S.I., Charbonneau, M. (1997). Stn1, a new *Saccharomyces cerevisiae* protein, is implicated in telomere size regulation in association with Cdc13. *Genes Dev.,* 11, 512.

Grandin, N., Damon, C., Charbonneau, M. (2001). Ten1 functions in telomere end protection and length regulation in association with Stn1 and Cdc13. *EMBO J.*, 20, 1173.

Granotier, C., Pennarun, G., Riou, L., Hoffschir, F., Gauthier, L.R., De Cian, A., Gomez, D., Mandine, E., Riou, J.F., Mergny, J.L. et al. (2005). Preferential binding of a G-quadruplex ligand to human chromosome ends. *Nucleic Acids Res.,* 33, 4182.

Greider, C.W. and Blackburn, E.H. (1985). Identification of a specific telomere terminal transferase activity in Tetrahymena extracts. *Cell*, 43, 405.

Greider, C.W. and Blackburn, E.H. (1987). The telomere terminal transferase of Tetrahymena is a ribonucleoprotein enzyme with two kinds of primer specificity. *Cell*, 51, 887.

Greider, C.W. and Blackburn, E.H. (1989). A telomeric sequence in the RNA of Tetrahymena telomerase required for telomere repeat synthesis. *Nature*, 337, 331.

Griffith, J.D., Comeau, L., Rosenfield, S., Stansel, R.M., Bianchi, A., Moss, H., de Lange, T. (1999). Mammalian telomeres end in a large duplex loop. *Cell,* 97, 503.

Hänsel, R., Löhr, F., Trantirek, L., Dötsch, V. (2013). High-resolution insight into G-overhang architecture. *J. Am. Chem. Soc.,* 135, 2816.

Henderson, E. and Blackburn, E.H. (1989). An overhanging 3′ terminus is a conserved feature of telomeres. *Mol. Cell Biol.,* 9, 345.

Henderson, E., Hardin, C.C., Walk, S.K., Tinoco, I., Blackburn, E.H. (1987). Telomeric DNA oligonucleotides form novel intramolecular structures containing guanine-guanine base pairs. *Cell*, 51, 899.

Hu, L., Lim, K.W., Bouaziz, S., Phan, A.T. (2009). Giardia telomeric sequence d(TAGGG)4 forms two intramolecular G-quadruplexes in K+ solution: Effect of loop length and sequence on the folding topology. *J. Am. Chem. Soc.,* 131, 16824.

Huang, C., Dai, X., Chai, W. (2012). Human Stn1 protects telomere integrity by promoting efficient lagging-strand synthesis at telomeres and mediating C-strand fill-in. *Cell Res.,* 22, 1681.

Jacob, N.K., Skopp, R., Price, C.M. (2001). G-overhang dynamics at Tetrahymena telomeres. *EMBO J.*, 20, 4299.

Jansson, L.I., Hentschel, J., Parks, J.W., Chang, T.R., Lu, C., Baral, R., Bagshaw, C.R., Stone, M.D. (2019). Telomere DNA G-quadruplex folding within actively extending human telomerase. *Proc. Natl. Acad. Sci. U.S.A.*, 116, 9350.

Kanoh, J. and Ishikawa, F. (2001). spRap1 and spRif1, recruited to telomeres by Taz1, are essential for telomere function in fission yeast. *Curr. Biol.*, 11, 1624.

Kar, A., Jones, N., Arat, N.Ö., Fishel, R., Griffith, J.D. (2018). Long repeating (TTAGGG) *n* single-stranded DNA self-condenses into compact beaded filaments stabilized by G-quadruplex formation. *J. Biol. Chem.*, 293(24), 9473.

Kazda, A., Zellinger, B., Rössler, M., Derboven, E., Kusenda, B., Riha, K. (2012). Chromosome end protection by blunt-ended telomeres. *Genes Dev.*, 26, 1703.

Kim, S.H., Kaminker, P., Campisi, J. (1999). TIN2, a new regulator of telomere length in human cells. *Nat. Genet.*, 23, 405.

Klobutcher, L.A., Swanton, M.T., Donini, P., Prescott, D.M. (1981). All gene-sized DNA molecules in four species of hypotrichs have the same terminal sequence and an unusual 3′ terminus. *Proc. Natl. Acad. Sci. U.S.A.*, 78, 3015.

Lancrey, A., Safa, L., Chatain, J., Delagoutte, E., Riou, J.F., Alberti, P., Saintomé, C. (2018). The binding efficiency of RPA to telomeric G-strands folded into contiguous G-quadruplexes is independent of the number of G4 units. *Biochimie*, 146, 68.

de Lange, T. (2004). T-loops and the origin of telomeres. *Nat. Rev. Mol. Cell Biol.*, 5, 323.

de Lange, T. (2005). Shelterin: The protein complex that shapes and safeguards human telomeres. *Genes Dev.*, 19, 2100.

de Lange, T. (2015). A loopy view of telomere evolution. *Front. Genet.*, 6, 321.

de Lange, T. (2018). Shelterin-mediated telomere protection. *Annu. Rev. Genet.*, 52, 223.

de Lange, T., Shiue, L., Myers, R.M., Cox, D.R., Naylor, S.L., Killery, A.M., Varmus, H.E. (1990). Structure and variability of human chromosome ends. *Mol. Cell. Biol.*, 10, 518.

Larrivée, M., LeBel, C., Wellinger, R.J. (2004). The generation of proper constitutive G-tails on yeast telomeres is dependent on the MRX complex. *Genes Dev.*, 18, 1391.

Lee, W., Yin, Y., Morten, M.J., Tonzi, P., Gwo, P.P., Odermatt, D.C., Modesti, M., Cantor, S.B., Gari, K., Huang, T.T., Rothenberg, E. (2021). Single-molecule imaging reveals replication fork coupled formation of G-quadruplex structures hinders local replication stress signaling. *Nat. Commun.*, 12, 2525.

Lerner, L.K. and Sale, J.E. (2019). Replication of G quadruplex DNA. *Genes*, 10, 95.

Levy, D.L. and Blackburn, E.H. (2004). Counting of Rif1p and Rif2p on *Saccharomyces cerevisiae* telomeres regulates telomere length. *Mol. Cell. Biol.*, 24, 10857.

Li, B., Oestreich, S., de Lange, T. (2000). Identification of human Rap1: Implications for telomere evolution. *Cell,* 101, 471.

Lim, K.W., Amrane, S., Bouaziz, S., Xu, W., Mu, Y., Patel, D.J., Luu, K.N., Phan, A.T. (2009). Structure of the human telomere in K+ solution: A stable basket-type G-quadruplex with only two G-tetrad layers. *J. Am. Chem. Soc.*, 131, 4301.

Lin, J.J. and Zakian, V.A. (1996). The Saccharomyces CDC13 protein is a single-strand TG1-3 telomeric DNA-binding protein *in vitro* that affects telomere behavior *in vivo. Proc. Natl. Acad. Sci. U.S.A.,* 93, 13760.

Lipps H.J. (1980). *In vitro* aggregation of the gene-sized DNA molecules of the ciliate *Stylonychia mytilus*. *Proc. Natl. Acad. Sci. U.S.A.,* 77, 4104.

Lipps, H.J., Gruissem, W., Prescott, D.M. (1982). Higher order DNA structure in macronuclear chromatin of the hypotrichous ciliate *Oxytricha nova*. *Proc. Natl. Acad. Sci. U.S.A.,* 79, 2495.

Liu, D., Safari, A., O'Connor, M.S., Chan, D.W., Laegeler, A., Qin, J., Songyang, Z. (2004). PTOP interacts with POT1 and regulates its localization to telomeres. *Nat. Cell Biol.,* 6, 673.

Lototska, L., Yue, J.X., Li, J., Giraud-Panis, M.J., Songyang, Z., Royle, N.J., Liti, G., Ye, J., Gilson, E., Mendez-Bermudez, A. (2020). Human RAP1 specifically protects telomeres of senescent cells from DNA damage. *EMBO Rep.,* 21, e49076.

Lundblad, V. and Blackburn, E.H. (1993). An alternative pathway for yeast telomere maintenance rescues est1- senescence. *Cell,* 73, 347.

Makarov, V.L., Hirose, Y., Langmore, J.P. (1997). Long G tails at both ends of human chromosomes suggest a C strand degradation mechanism for telomere shortening. *Cell,* 88, 657.

Mason, J.M., Frydrychova, R.C., Biessmann, H. (2008). Drosophila telomeres: An exception providing new insights. *BioEssays,* 30, 25.

Mason, J.M., Randall, T.A., Capkova Frydrychova, R. (2016). Telomerase lost? *Chromosoma*, 125, 65.

McClintock, B. (1941). The stability of broken ends of chromosomes in Zea mays. *Genetics,* 26, 234.

McElligott, R. and Wellinger, R.J. (1997). The terminal DNA structure of mammalian chromosomes. *EMBO J,* 16, 3705.

Mendez-Bermudez, A., Hills, M., Pickett, H.A., Phan, A.T., Mergny, J.L., Riou, J.F., Royle, N.J. (2009). Human telomeres that contain (CTAGGG)n repeats show replication dependent instability in somatic cells and the male germline. *Nucleic Acids Res.,* 37, 6225.

Meyne, J., Ratliff, R.L., Moyzis, R.K. (1989). Conservation of the human telomere sequence (TTAGGG)n among vertebrates. *Proc. Natl. Acad. Sci. U.S.A.*, 86, 7049.

Miles, H.T. and Frazier, J. (1978). Poly(I) helix formation. Dependence on size-specific complexing to alkali metal ions. *J. Am. Chem. Soc.,* 100, 8037.

Miyoshi, T., Kanoh, J., Saito, M., Ishikawa, F. (2008). Fission yeast Pot1-Tpp1 protects telomeres and regulates telomere length. *Science,* 320, 1341.

Mohaghegh, P., Karow, J.K., Brosh Jr., R.M., Bohr, V.A., Hickson, I.D. (2001). The Bloom's and Werner's syndrome proteins are DNA structure-specific helicases. *Nucleic Acids Res.,* 29, 2843.

Moser, B.A., Subramanian, L., Chang, Y.T., Noguchi, C., Noguchi, E., Nakamura, T.M. (2009). Differential arrival of leading and lagging strand DNA polymerases at fission yeast telomeres. *EMBO J.,* 28, 810.

Moyzis, R.K., Buckingham, J.M., Cram, L.S., Dani, M., Deaven, L.L., Jones, M.D., Meyne, J., Ratliff, R.L., Wu, J.R. (1988). A highly conserved repetitive DNA sequence, (TTAGGG)n, present at the telomeres of human chromosomes. *Proc. Natl. Acad. Sci. U.S.A.,* 85, 6622.

Müller, H.J. (1938). The remaking of chromosomes. *Collecting Net.,* 13,181.

Müller, S., Kumari, S., Rodriguez, R., Balasubramanian, S. (2010). Small-molecule-mediated G-quadruplex isolation from human cells. *Nat. Chem.,* 2, 1095.

Murti, K.G. and Prescott, D.M. (1999). Telomeres of polytene chromosomes in a ciliated protozoan terminate in duplex DNA loops. *Proc. Natl. Acad. Sci. U.S.A.*, 96, 14436.

Oganesian, L. and Karlseder, J. (2011). Mammalian 5′ C-rich telomeric overhangs are a mark of recombination-dependent telomere maintenance. *Mol. Cell,* 42, 224.

Oka, Y. and Thomas, C.A. (1987). The cohering telomeres of Oxytricha. *Nucleic Acids Res.,* 15, 8877.

Oka, Y., Shiota, S., Nakai, S., Nishida, Y., Okubo, S. (1980). Inverted terminal repeat sequence in the macronuclear DNA of *Stylonychia pustulata. Gene*, 10, 301.

Olovnikov, A.M. (1973). A theory of marginotomy. The incomplete copying of template margin in enzymic synthesis of polynucleotides and biological significance of the phenomenon. *J. Theor. Biol.,* 41, 181–190.

Paeschke, K., Simonsson, T., Postberg, J., Rhodes, D., Lipps, H.J. (2005). Telomere end-binding proteins control the formation of G-quadruplex DNA structures *in vivo. Nat. Struct. Mol. Biol.,* 12, 847.

Paeschke, K., Juranek, S., Simonsson, T., Hempel, A., Rhodes, D., Lipps, H.J. (2008). Telomerase recruitment by the telomere end binding protein-beta facilitates G-quadruplex DNA unfolding in ciliates. *Nat. Struct. Mol. Biol.,* 15, 598.

Pasquier, E. and Wellinger, R.J. (2020). *In vivo* chromatin organization on native yeast telomeric regions is independent of a cis-telomere loopback conformation. *Epigenetics Chromatin,* 13, 23.

Petraccone, L., Spink, C., Trent, J.O., Garbett, N.C., Mekmaysy, C.S., Giancola, C., Chaires, J.B. (2011). Structure and stability of higher-order human telomeric quadruplexes. *J. Am. Chem. Soc.,* 133, 20951.

Phan A.T. (2010). Human telomeric G-quadruplex: Structures of DNA and RNA sequences. *FEBS J.,* 277, 1107.

Pinnavaia, T.J., Marshall, C.L., Mettler, C.M., Fisk, C.L., Miles, H.T., Becker, E.D. (1978). Alkali metal ion specificity in the solution ordering of a nucleotide, 5′-guanosine monophosphate. *J. Am. Chem. Soc.,* 100, 3625.

Pluta, A.F., Dani, G.M., Spear, B.B., Zakian, V.A. (1984). Elaboration of telomeres in yeast: Recognition and modification of termini from Oxytricha macronuclear DNA. *Proc. Natl. Acad. Sci. U.S.A.,* 81, 1475.

Podlevsky, J.D., Bley, C.J., Omana, R.V., Qi, X., Chen, J.J. (2008). The telomerase database. *Nucleic Acids Res.,* 36, D339.

Prescott, D.M. (1994). The DNA of ciliated protozoa. *Microbiol. Rev.,* 58, 233.

Price, C.M. and Cech, T.R. (1987). Telomeric DNA-protein interactions of Oxytricha macronuclear DNA. *Genes Dev.*, 1, 783.

Procházková Schrumpfová, P., Schořová, Š., Fajkus, J. (2016). Telomere- and telomerase-associated proteins and their functions in the plant cell. *Front Plant Sci.,* 7, 851.

Procházková Schrumpfová, P., Fojtová, M., Fajkus, J. (2019). Telomeres in plants and humans: Not so different, not so similar. *Cells*, 8, 58.

Qureshi, M.H., Ray, S., Sewell, A.L., Basu, S., Balci, H. (2012). Replication protein A unfolds G-quadruplex structures with varying degrees of efficiency. *J. Phys. Chem. B.,* 116, 5588.

Raices, M., Verdun, R.E., Compton, S.A., Haggblom, C.I., Griffith, J.D., Dillin, A., Karlseder, J. (2008). *C. elegans* telomeres contain G-strand and C-strand overhangs that are bound by distinct proteins. *Cell*, 132, 745.

Ray, S., Bandaria, J.N., Qureshi, M.H., Yildiz, A., Balci, H. (2014). G-quadruplex formation in telomeres enhances POT1/TPP1 protection against RPA binding. *Proc. Natl. Acad. Sci. U.S.A.,* 111, 2990.

Rice, C., Shastrula, P.K., Kossenkov, A.V., Hills, R., Baird, D.M., Showe, L.C., Doukov, T., Janicki, S., Skordalakes, E. (2017). Structural and functional analysis of the human POT1-TPP1 telomeric complex. *Nat. Commun.,* 8, 14928.

Rich, A. (1958). The molecular structure of polyinosinic acid. *Biochim Biophys Acta,* 29, 502–509.

Rich, A. (1993). DNA comes in many forms. *Gene,* 135, 99.

Rizzo, A., Salvati, E., Porru, M., D'Angelo, C., Stevens, M.F., D'Incalci, M., Leonetti, C., Gilson, E., Zupi, G., Biroccio, A. (2009). Stabilization of quadruplex DNA perturbs telomere replication leading to the activation of an ATR-dependent ATM signaling pathway. *Nucleic Acids Res.,* 37, 5353.

Rodriguez, R., Müller, S., Yeoman, J.A., Trentesaux, C., Riou, J.F., Balasubramanian, S. (2008). A novel small molecule that alters shelterin integrity and triggers a DNA-damage response at telomeres. *J. Am. Chem. Soc.,* 130, 15758.

Ross, P.D. and Scruggs, R.L. (1964). Electrophoresis of DNA. II. Specific interactions of univalent and divalent cations with DNA. *Biopolymers,* 2, 79.

Saintomé, C., Amrane, S., Mergny, J.L., Alberti, P. (2016). The exception that confirms the rule: A higher-order telomeric G-quadruplex structure more stable in sodium than in potassium. *Nucleic Acids Res.,* 44, 2926.

Salas, T.R., Petruseva, I., Lavrik, O., Bourdoncle, A., Mergny, J.L., Favre, A., Saintomé, C. (2006). Human replication protein A unfolds telomeric G-quadruplexes. *Nucleic Acids Res.,* 34, 4857.

Sarek, G., Vannier, J.B., Panier, S., Petrini, J., Boulton, S.J. (2015). TRF2 recruits RTEL1 to telomeres in S phase to promote t-loop unwinding. *Mol. Cell,* 57, 622.

Sarek, G., Kotsantis, P., Ruis, P., Van Ly, D., Margalef, P., Borel, V., Zheng, X.F., Flynn, H.R., Snijders, A.P., Chowdhury, D. et al. (2019). CDK phosphorylation of TRF2 controls t-loop dynamics during the cell cycle. *Nature,* 575, 523.

Schaffitzel, C., Berger, I., Postberg, J., Hanes, J., Lipps, H.J., Plückthun, A. (2001). *In vitro* generated antibodies specific for telomeric guanine-quadruplex DNA react with Stylonychia lemnae macronuclei. *Proc. Natl. Acad. Sci. U.S.A.,* 98, 8572.

Schmutz, I., Timashev, L., Xie, W., Patel, D.J., de Lange, T. (2017). TRF2 binds branched DNA to safeguard telomere integrity. *Nat. Struct. Mol. Biol.,* 24, 734.

Sen, D. and Gilbert, W. (1988). Formation of parallel four-stranded complexes by guanine-rich motifs in DNA and its implications for meiosis. *Nature,* 334, 364.

Sfeir, A., Kosiyatrakul, S.T., Hockemeyer, D., MacRae, S.L., Karlseder, J., Schildkraut, C.L., de Lange, T. (2009). Mammalian telomeres resemble fragile sites and require TRF1 for efficient replication. *Cell,* 138, 90.

Shampay, J., Szostak, J.W., Blackburn, E.H. (1984). DNA sequences of telomeres maintained in yeast. *Nature,* 310, 154.

Školáková, P., Foldynová-Trantírková, S., Bednářová, K., Fiala, R., Vorlíčková, M., Trantírek, L. (2015). Unique C. elegans telomeric overhang structures reveal the evolutionarily conserved properties of telomeric DNA. *Nucleic Acids Res.*, 43, 4733.

Smith, J.S., Chen, Q., Yatsunyk, L.A., Nicoludis, J.M., Garcia, M.S., Kranaster, R., Balasubramanian, S., Monchaud, D., Teulade-Fichou, M.P., Abramowitz, L. et al. (2011). Rudimentary G-quadruplex-based telomere capping in *Saccharomyces cerevisiae*. *Nat. Struct. Mol. Biol.,* 18, 478.

Sparks, M.A., Singh, S.P., Burgers, P.M., Galletto, R. (2019). Complementary roles of Pif1 helicase and single stranded DNA binding proteins in stimulating DNA replication through G-quadruplexes. *Nucleic Acids Res.,* 47, 8595.

Spink, K.G., Evans, R.J., Chambers, A. (2000). Sequence-specific binding of Taz1p dimers to fission yeast telomeric DNA. *Nucleic Acids Res.,* 28, 527.

Stansel, R.M., de Lange, T., Griffith, J.D. (2001). T-loop assembly *in vitro* involves binding of TRF2 near the 3′ telomeric overhang. *EMBO J.,* 20, 5532.

Steinberg-Neifach, O. and Lue, N.F. (2015). Telomere DNA recognition in Saccharomycotina yeast: Potential lessons for the co-evolution of ssDNA and dsDNA-binding proteins and their target sites. *Front. Genet.,* 6, 162.

Stewart, J.A., Wang, F., Chaiken, M.F., Kasbek, C., Chastain 2nd, P.D., Wright, W.E., Price, C.M. (2012). Human CST promotes telomere duplex replication and general replication restart after fork stalling. *EMBO J.*, 31, 3537.

Strahl-Bolsinger, S., Hecht, A., Luo, K., Grunstein, M. (1997). SIR2 and SIR4 interactions differ in core and extended telomeric heterochromatin in yeast. *Genes Dev.,* 11, 83.

Strauss, U.P. and Ross, P.D. (1959). Counterion binding by polyelectrolytes III. Stability constants for the binding of univalent cations by $P0_3^-$-groups of polyphosphates from electrophoresis measurements. *J. Am. Chem. Soc.,* 81, 5295.

Sun, H., Karow, J.K., Hickson, I.D., Maizels, N. (1998). The Bloom's syndrome helicase unwinds G4 DNA. *J. Biol. Chem.,* 273, 27587.

Sundquist, W.I. and Klug, A. (1989). Telomeric DNA dimerizes by formation of guanine tetrads between hairpin loops. *Nature*, 342, 825.

Szostak, J.W. and Blackburn, E.H. (1982). Cloning yeast telomeres on linear plasmid vectors. *Cell*, 29, 245.

Tahara, H., Shin-Ya, K., Seimiya, H., Yamada, H., Tsuruo, T., Ide, T. (2006). G-quadruplex stabilization by telomestatin induces TRF2 protein dissociation from telomeres and anaphase bridge formation accompanied by loss of the 3′ telomeric overhang in cancer cells. *Oncogene,* 25, 1955.

Taylor, D.J., Podell, E.R., Taatjes, D.J., Cech, T.R. (2011). Multiple POT1-TPP1 proteins coat and compact long telomeric single-stranded DNA. *J. Mol. Biol.*, 410, 10.

Timashev, L.A. and de Lange, T. (2020). Characterization of t-loop formation by TRF2. *Nucleus,* 11, 164.

Tomáška, L., Cesare, A.J., AlTurki, T.M., Griffith, J.D. (2020). Twenty years of t-loops: A case study for the importance of collaboration in molecular biology. *DNA Repair,* 94, 102901.

Tomita, K. and Cooper, J.P. (2008). Fission yeast Ccq1 is telomerase recruiter and local checkpoint controller. *Genes Dev.,* 22, 3461.

Tran, P.L., Mergny, J.L., Alberti, P. (2011). Stability of telomeric G-quadruplexes. *Nucleic Acids Res.,* 39, 3282.

Tran, P.L., Rieu, M., Hodeib, S., Joubert, A., Ouellet, J., Alberti, P., Bugaut, A., Allemand, J.F., Boulé, J.B., Croquette, V. (2021). Folding and persistence times of intramolecular G-quadruplexes transiently embedded in a DNA duplex. *Nucleic Acids Res.,* 49, 5189.

Trujillo, K.M., Bunch, J.T., Baumann, P. (2005). Extended DNA binding site in Pot1 broadens sequence specificity to allow recognition of heterogeneous fission yeast telomeres. *J. Biol. Chem.*, 280, 9119.

Valuchova, S., Fulnecek, J., Prokop, Z., Stolt-Bergner, P., Janouskova, E., Hofr, C., Riha, K. (2017). Protection of arabidopsis blunt-ended telomeres is mediated by a physical association with the Ku heterodimer. *Plant Cell,* 29, 1533.

Van Ly, D., Low, R., Frölich, S., Bartolec, T.K., Kafer, G.R., Pickett, H.A., Gaus, K., Cesare, A.J. (2018). Telomere loop dynamics in chromosome end protection. *Mol. Cell,* 71, 510.

Vannier, J.B., Pavicic-Kaltenbrunner, V., Petalcorin, M.I., Ding, H., Boulton, S.J. (2012). RTEL1 dismantles T loops and counteracts telomeric G4-DNA to maintain telomere integrity. *Cell,* 149, 795.

Vannier, J.B., Sandhu, S., Petalcorin, M.I., Wu, X., Nabi, Z., Ding, H., Boulton, S.J. (2013). RTEL1 is a replisome-associated helicase that promotes telomere and genome-wide replication. *Science,* 342, 239.

Varshney, D., Spiegel, J., Zyner, K., Tannahill, D., Balasubramanian, S. (2020). The regulation and functions of DNA and RNA G-quadruplexes. *Nat. Rev. Mol. Cell Biol.,* 21, 459.

Verdun, R.E. and Karlseder, J. (2006). The DNA damage machinery and homologous recombination pathway act consecutively to protect human telomeres. *Cell,* 127, 709.

Vorlickova, M., Chladkova, J., Kejnovska, I., Fialova, M., Kypr, J. (2005). Guanine tetraplex topology of human telomere DNA is governed by the number of (TTAGGG) repeats. *Nucleic Acids Res.,* 33, 5851.

Wang, Y. and Patel, D.J. (1994). Solution structure of the Tetrahymena telomeric repeat d(T2G4)4 G-tetraplex. *Structure, 2,* 1141.

Wang, Y. and Patel, D.J. (1995). Solution structure of the Oxytricha telomeric repeat d[G4(T4G4)3] G-tetraplex. *J. Mol. Biol.,* 251, 76.

Wang, F., Podell, E.R., Zaug, A.J., Yang, Y., Baciu, P., Cech, T.R., Lei, M. (2007). The POT1-TPP1 telomere complex is a telomerase processivity factor. *Nature,* 445, 506.

Wang, F., Stewart, J.A., Kasbek, C., Zhao, Y., Wright, W.E., Price, C.M. (2012). Human CST has independent functions during telomere duplex replication and C-strand fill-in. *Cell Rep.*, 2, 1096.

Watson, J.D. (1972). Origin of concatemeric T7 DNA. *Nat New Biol.,* 239, 197.

Watson, J.D. and Crick F.H. (1953). Molecular structure of nucleic acids; a structure for deoxyribose nucleic acid. *Nature,* 171, 737.

Wellinger, R.J., Wolf, A.J., Zakian, V.A. (1993). Saccharomyces telomeres acquire single-strand TG1-3 tails late in S phase. *Cell*, 72, 51.

Wesley R.D. (1975). Inverted repetitious sequences in the macronuclear DNA of hypotrichous ciliates. *Proc. Natl. Acad. Sci. U.S.A.,* 72, 678.

Williamson, J.R., Raghuraman, M.K., Cech, T.R. (1989). Monovalent cation-induced structure of telomeric DNA: The G-quartet model. *Cell*, 59, 871.

Wright, W.E., Tesmer, V.M., Huffman, K.E., Levene, S.D., Shay, J.W. (1997). Normal human chromosomes have long G-rich telomeric overhangs at one end. *Genes Dev.,* 11, 2801.

Xin, H., Liu, D., Wan, M., Safari, A., Kim, H., Sun, W., O'Connor, M.S., Songyang, Z. (2007). TPP1 is a homologue of ciliate TEBP-beta and interacts with POT1 to recruit telomerase. *Nature,* 445, 559.

Xu, Y., Ishizuka, T., Kurabayashi, K., Komiyama, M. (2009). Consecutive formation of G-quadruplexes in human telomeric-overhang DNA: A protective capping structure for telomere ends. *Angew. Chem. Int. Ed. Engl.,* 48, 7833.

Yamamoto, I., Zhang, K., Zhang, J., Vorontsov, E., Shibuya, H. (2021). Telomeric double-strand DNA-binding proteins DTN-1 and DTN-2 ensure germline immortality in *Caenorhabditis elegans*. *eLife,* 10, e64104.

Ye, J.Z., Hockemeyer, D., Krutchinsky, A.N., Loayza, D., Hooper, S.M., Chait, B.T., de Lange, T. (2004). POT1-interacting protein PIP1: A telomere length regulator that recruits POT1 to the TIN2/TRF1 complex. *Genes Dev.,* 18, 1649.

Yu, H.Q., Miyoshi, D., Sugimoto, N. (2006). Characterization of structure and stability of long telomeric DNA G-quadruplexes. *J. Am. Chem. Soc.,* 128, 15461.

Zaug, A.J., Lim, C.J., Olson, C.L., Carilli, M.T., Goodrich, K.J., Wuttke, D.S., Cech, T.R. (2021). CST does not evict elongating telomerase but prevents initiation by ssDNA binding. *Nucleic Acids Res.*, 49, 11653.

Zhang, M., Wang, B., Li, T., Liu, R., Xiao, Y., Geng, X., Li, G., Liu, Q., Price, C.M., Liu, Y., Wang, F. (2019). Mammalian CST averts replication failure by preventing G-quadruplex accumulation. *Nucleic Acids Res.*, 47, 5243.

Zimmerman, S.B., Cohen, G.H., Davies, D.R. (1975). X-ray fiber diffraction and model-building study of polyguanylic acid and polyinosinic acid. *J. Mol. Biol.,* 92,181.

Zimmermann, M., Kibe, T., Kabir, S., de Lange, T. (2014). TRF1 negotiates TTAGGG repeat-associated replication problems by recruiting the BLM helicase and the TPP1/POT1 repressor of ATR signaling. *Genes Dev.,* 28, 2477.

2

Methods to Measure Telomere Lengths

Lauréline ROGER

Structure and Instability of Genomes, National Museum of Natural History, CNRS, INSERM, Paris, France

2.1. Why does telomere length matter?

Telomeres are nucleoprotein structures that cap and protect the natural ends of linear chromosomes. One of their main functions is to prevent chromosome ends from being recognized as DNA double strand breaks (DSBs) by the DNA repair machinery (Müller 1938; McClintock 1939).

In human somatic cells, telomeres shorten at each cell division because of the incomplete replication of linear chromosomes, the so-called "end-replication problem" (Olovnikov 1971; Harley et al. 1990; Baird et al. 2003). Telomere erosion ultimately leads to a partial loss of telomere function, resulting in their recognition as DNA breaks (Fagagna et al. 2003).

This triggers the activation of p53 and Rb cell cycle check points, which leads to a stable cell cycle arrest, called senescence (Fagagna et al. 2003). In the absence of fully functional DNA damage checkpoints, the cells keep dividing and the telomeres shorten up to a length at which they become fully dysfunctional (Counter et al. 1992; Capper et al. 2007; Letsolo et al. 2010). Dysfunctional telomeres are capable of fusing with each other and with broken DNA ends at non-telomeric loci. These

For a color version of all the figures in this chapter, see www.iste.co.uk/saintome/telomeres.zip.

Telomeres,
coordinated by Carole SAINTOMÉ.

fusions lead to the formation of dicentric chromosomes and result in a state of genomic instability, which is known as a telomere induced-crisis (McClintock 1939; Sabatier 2005; Murnane 2006; Capper et al. 2007; Bhargava et al. 2020) (Figure 2.1).

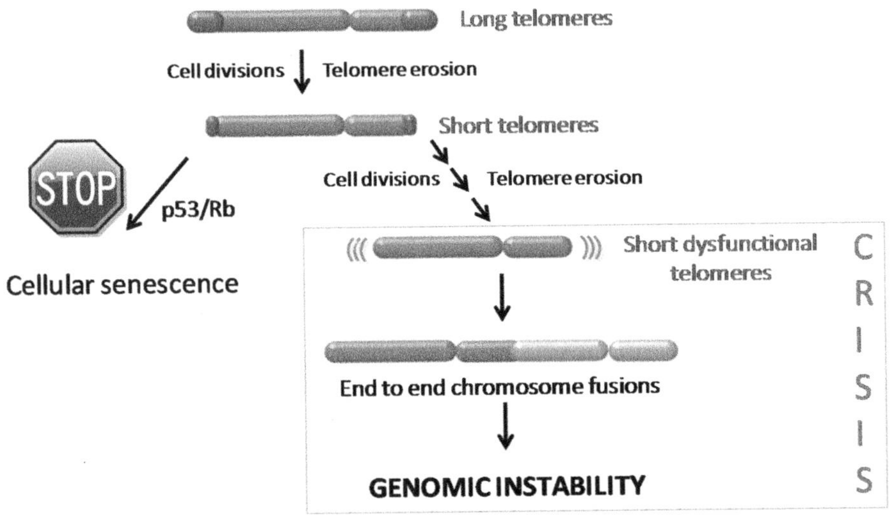

Figure 2.1. *Telomeres shorten at each cell division to a length at which they partially lose their capping function and are recognized as DNA breaks, which triggers a p53/Rb-dependant cell cycle arrest called senescence. In the absence of fully functional DNA damage checkpoints, the cells keep dividing and the telomeres shorten up to a length at which they become fully dysfunctional and capable of fusing with each other. End to end chromosome fusions lead to chromosome rearrangements and fuel genomic instability. This state is called telomere crisis*

Telomere crisis is characterized by slower cell growth and massive cell death, thus creating a selective pressure for the clonal outgrowth of rare cells that manage to maintain their telomere length (Figure 2.2).

Mounting evidence suggests that telomere dysfunction could be a major contributor to the chromosomal rearrangements that occur during tumor progression.

In vivo, this view is supported by late-generation, telomerase-deficient, p53-mutant mice models, in which telomere dysfunction is associated with the emergence of carcinomas harboring chromosomal abnormalities (Artandi et al. 2000). In humans, telomere dysfunction has been associated with the progression of solid and blood cancers (Lin et al. 2010; Roger et al. 2013; Letsolo et al. 2017).

Furthermore, several studies have demonstrated that telomere length can be used as a prognostic and/or predictive biomarker in some types of human cancers such as, among others, chronic lymphocytic leukemia (CLL) and breast cancer (Lin et al. 2014; Simpson et al. 2015; Strefford et al. 2015; Hyatt et al. 2017; Norris et al. 2019).

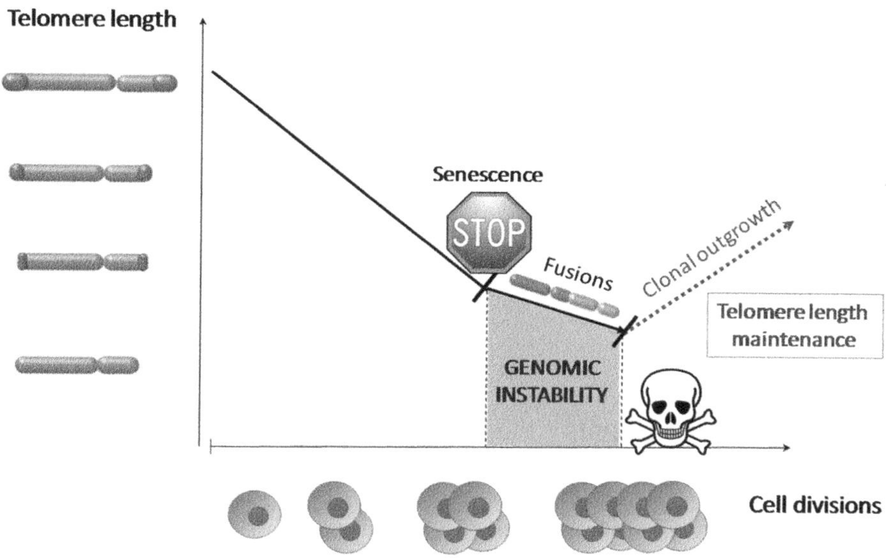

Figure 2.2. *Telomere crisis is characterized by massive cell death and creates a selective pressure for the clonal outgrowth of rare cells that acquired a way to maintain their telomere length*

2.2. Methods for measuring telomere length

Due to the role of telomere erosion/lengthening in cancer progression and other pathologies (covered in Chapters 11 and 12), considerable efforts have been made to develop methods to measure telomere length. Several techniques now exist, each of them with their own advantages and disadvantages.

2.2.1. *Terminal restriction fragment analysis (TRF)*

Terminal restriction fragment analysis (TRF) was the first method to be used to measure telomere length and is still considered by many to be the gold standard of techniques (Moyzis et al. 1988; Allshire et al. 1989; Harley et al. 1990). TRF is a southern blot-based approach in which telomeric DNA is detected with a telomere-specific probe (Figure 2.3). TRF analysis relies on the absence of restriction enzyme cut sites within the TTAGGG repeat array.

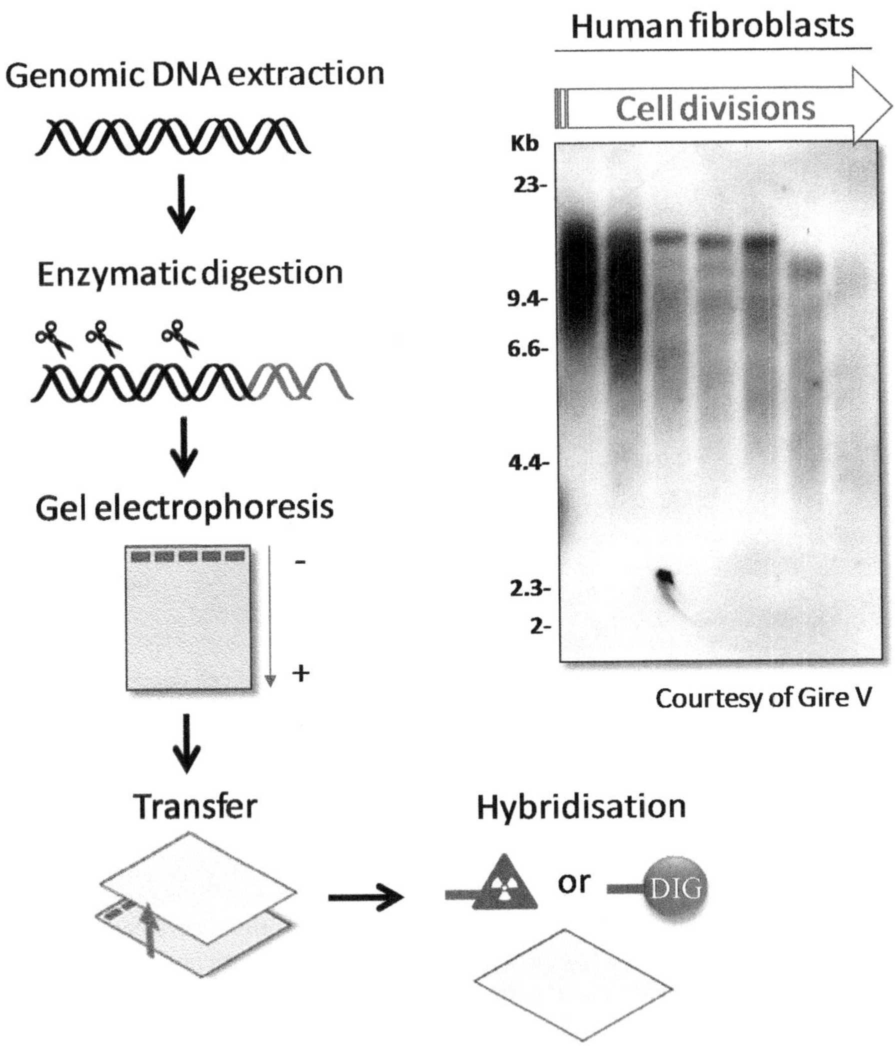

Figure 2.3. *TRF methodology and example of TRF in human fibroblasts illustrating telomere erosion as a function of cell divisions*

In this process, high molecular weight and good quality DNA is first extracted. The DNA is subsequently digested using a mix of restriction enzymes, which cut frequently in the genome, so that only intact telomeric DNA remains. *Hinf*I and *Rsa*I are typically used. *Hph*I and *Mnl*I have also been used, as these offer the advantage of cutting at the telomere variant sequences TGAGGG and TCAGGG, respectively (Britt-Compton et al. 2006). The digested DNA samples are resolved by agarose gel electrophoresis and then denatured. Finally, after transfer to a membrane, the DNA fragments are detected by hybridization with a probe containing telomere repeats. Alternatively, direct in-gel hybridization is possible. Radiolabeled or digoxigenine-labeled probes can be used. The signal obtained is a heterogeneous smear of telomeric DNA. The measurement of the smear intensity and its comparison with a DNA ladder gives the average telomere length of the sample analyzed.

Advantages of TRF are as follows:

– TRF analysis can be applied to different kinds of samples – cells, blood and tissues – which can be either fresh or frozen.

– It can be used on different organisms (e.g. yeast, mammals and plants among others).

– It provides the average telomere length of the entire genome of a cell population.

Disadvantages of TRF are as follows:

– TRF analysis requires a large amount of DNA (usually between 1 and 5 μg) as well as high-quality DNA, which can limit the type of samples that can be analyzed. It is impossible, or extremely challenging, to analyze small tissue samples of less than 10^5 cells. This may be an issue with clinical samples, or when the size of the model studied limits the amount of blood that can be drawn.

– This technique is time consuming (typically 5 days) and labor-intensive and thus not adapted for large cohorts or epidemiological studies, for instance.

– TRF analysis yields a heterogeneous smear, as telomere length varies between chromosome ends and from one cell to another.

– The shortest telomeres cannot be detected by TRF. Indeed, the intensity of the signal depends on the quantity of probes that hybridize the telomere, which is in turn dependant on the number of telomere repeats. The shorter the telomere, the lower the signal.

– TRFs can also contain a variable amount of telomere variants and/or subtelomeric DNA. Indeed, the beginning of the telomere alternates between arrays of pure telomeric sequence and telomere variants (Baird et al. 1995; Letsolo et al. 2010). The size of this region varies between chromosome ends and can vary between individuals. Adjacent to the variants repeat region is the subtelomere, which contains variable repetitive sequences (Mefford and Trask 2002; Riethman et al. 2005; Ambrosini et al. 2007). As such, the distance between the first restriction site and the beginning of the telomere can vary by up to 2 kb (Figure 2.4). Consequently, the amount of non-telomeric DNA in the TRF will also depend on the restriction enzymes used.

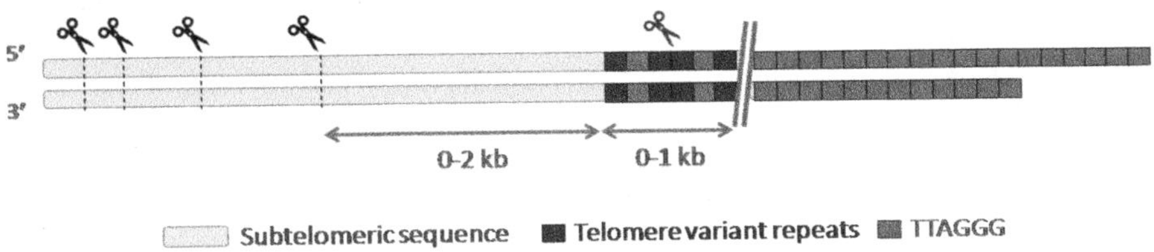

Figure 2.4. *TRFs can contain a variable amount of telomere variants (up to 1 kb) and/or subtelomeric DNA (up to 2 kb) and the distance between the first restriction site and the beginning of the telomere can vary by up to 2kb. Restriction enzymes are represented as scissors. The red scissors represent restriction enzymes that cut at the variant repeats*

2.2.2. *Single telomere length analysis*

Single telomere length analysis (STELA) is a single molecule PCR-based approach that measures the full spectrum of telomere lengths at a single chromosome terminus (Figure 2.5) (Baird et al. 2003). It relies on the fact that all telomeres terminate with a G-rich 3' overhang. In a first step, a linker called "Telorette" specifically anneals with the 3'-overhang of the telomere. The Telorette is composed of seven bases complementary to the TTAGGG repeats followed by a 20-nucleotides non-complementary tail to the 3' overhang of the telomere. In the second step, the PCR is performed using a primer called "Teltail" that is identical to the 20-nucleotides tail, together with a primer located in the subtelomeric region close to the beginning of the telomere, and which is specific to a given chromosome end. The genomic DNA is analyzed with several PCR reactions (at least six per sample) for statistical reasons. Beforehand, the DNA is diluted such that each PCR reaction contains 4–30 amplifiable molecules. The PCR products are then resolved by agarose gel electrophoresis and detected by southern-blot with a telomere probe or a telomere-adjacent specific probe. Each discrete band detected on the hybridized membrane represents a single telomere from the chromosome end analyzed. The size of each band is calculated in comparison with a DNA ladder using an analysis software such as, for example, Total Lab Quant.

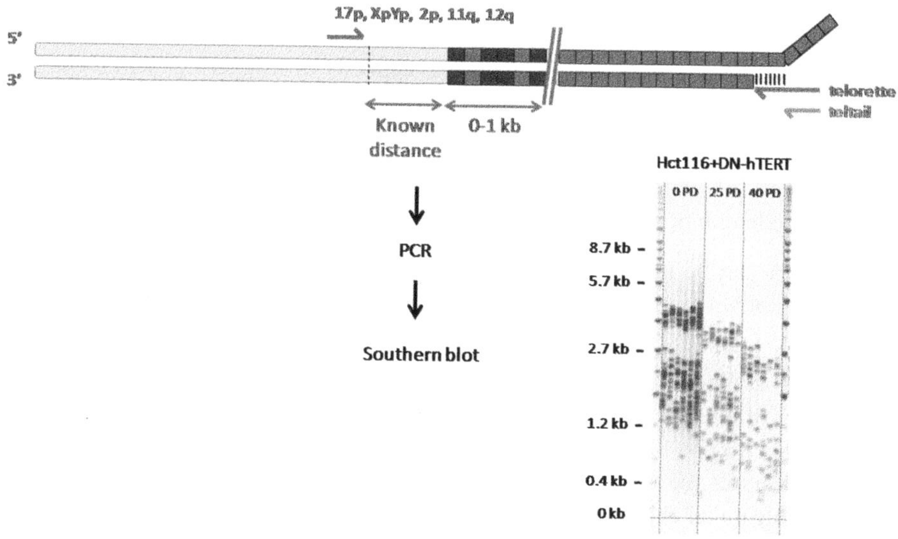

Figure 2.5. *STELA methodology and example of STELA at 17p telomere in HCT116 carcinoma cell line expressing a dominant-negative telomerase (DN-hTERT). HCT116 are cancerous cells and express telomerase thus maintaining the length of their telomeres. Over-expression of DN-hTERT inhibits telomerase and leads to telomere shortening at each cell division. The cells were analyzed at three different time points (after 0 population doublings [PD] (first time point of the cell culture experiment), after 25 PD and after 40 PD). For each sample, six PCRs were undertaken (each lane is a PCR). Each band visualized is a telomere*

Importantly, a single molecule PCR assay has been developed by Baird DM to detect telomere-telomere fusion events between specific chromosome arms (Capper et al. 2007; Letsolo et al. 2010). The fusion events detected can be characterized further by sequencing.

Advantages of STELA:

– STELA is extremely resolutive. It measures precisely the size of individual telomeres up to 20 kb. The size of the subtelomeric region contained is known, as it is fixed by the position of the chromosome-specific upstream primer, and the number of variant repeats can be determined (Capper et al. 2007).

– STELA only requires a small quantity of DNA (in the range of the picogram) but requires high molecular weight and good quality DNA (i.e. intact telomeres with an intact 3' overhang).

– STELA can accurately measure the length of the shortest telomeres in the size range of dysfunctional telomeres that are capable of triggering senescence or chromosomal rearrangements. It can even detect rare short telomeres that are distinct from the bulk population.

– By using the subtelomere sequence polymorphism of some chromosomes such as XpYp, the length of specific allele can be measured (Baird et al. 2003).

– STELA can be used on other organisms provided that the telomere adjacent sequence is known (e.g. *C. Elegans*) (Cheung 2004).

Disadvantages of STELA:

– Only a limited number of chromosome ends can be analyzed by STELA in humans (e.g. XpYp, 17p, 2p, 11q, 12q) (Britt-Compton et al. 2006). Indeed, this technique requires the identification of a unique telomere adjacent sequence, which is not possible on all chromosome ends given the complexity of human subtelomeres (i.e. they contain duplications, repetitive sequences and sequences that are shared among several chromosome ends) (Mefford and Trask 2002; Riethman et al. 2005; Ambrosini et al. 2007).

– The fact that STELA is a PCR-based approach limits the analysis to telomeres not longer than 20 kb. This prevents the use of STELA for the analysis of samples from model organisms with extremely long telomeres such as *Mus Musculus*.

– STELA is labor intensive and time consuming (typically 4 days to result) as well as technically challenging, and as such not suited to the analysis of large number of samples. To circumvent this limitation, a modified version of STELA has been developed. This assay, called high-throughput STELA (HT-STELA), allows the analysis of a huge cohort of samples (larger than 100) and has been specifically designed to be used in clinical studies (Norris et al. 2019).

2.2.3. *Telomere shortest length assay*

Telomere shortest length assay (TeSLA) is based on an adaptation of the STELA methodology. It can measure telomere lengths from less than 1 kb to 18 kb at all chromosome ends (Lai et al. 2017). In a first step, genomic DNA is ligated to an adapter that anneals to the 3' overhang of the telomere (Figure 2.6). As in STELA, this adapter contains a seven-nucleotide sequence, which is complementary to the overhang followed by a unique sequence. In the second step, the genomic ligated

DNA is digested with a mix of restriction enzymes (*Bfa*I/*CviA*II/*Mse*I/*Nde*I) that will digest non-telomeric DNA and some of the variants repeats. In the third step, the digested extremities are ligated with double stranded adapters (both the proximal end of the telomeres and the genomic DNA fragments are ligated to the adapters). Subsequently, a PCR is performed to amplify the telomeric DNA. The PCR products are then resolved by agarose gel electrophoresis and detected by southern blot with a telomere probe. As in STELA, each individual band detected is a telomere.

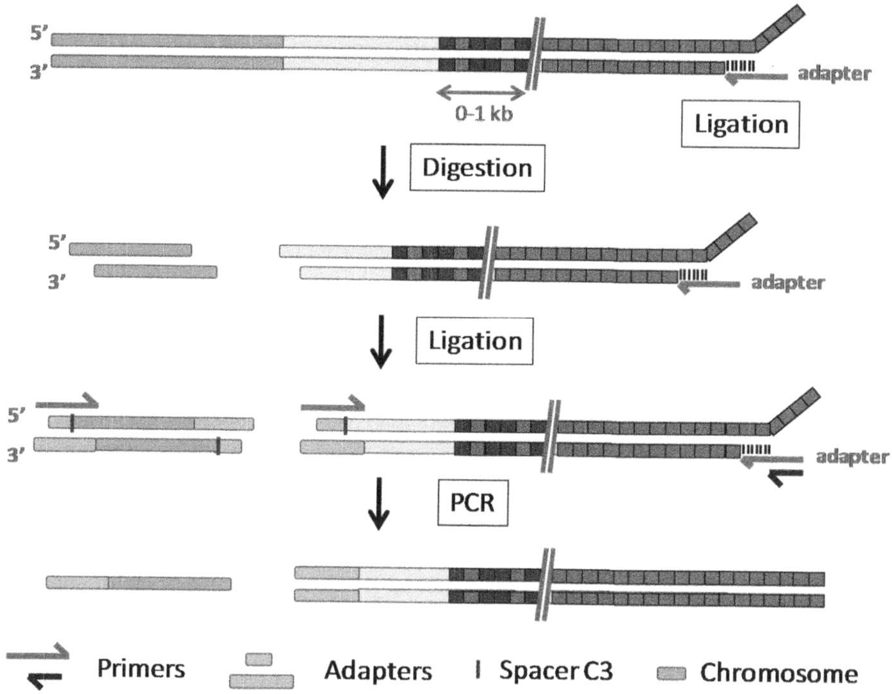

Figure 2.6. *TeSLA methodology: In a first step, genomic DNA is ligated to an adapter. In a second step, the genomic ligated DNA is digested. In a third step, digested extremities are ligated to double stranded adapters. Finally, a PCR is performed to amplify telomeric DNA. The double stranded adapter contains a spacer C3, which blocks the progression of the polymerase. As a result, genomic DNA can only be amplified at one copy*

Advantages of TeSLA are as follows:

– The main advantage of TeSLA is that it can detect the shortest telomeres from all chromosome ends.

– TeSLA requires small amount of DNA.

– TeSLA can be applied to different organisms (Lai et al. 2017).

Disadvantages of TeSLA are as follows:

– TeSLA measures telomeres from all chromosome ends but cannot identify to which chromosome end the telomeres that are detected belong.

– TeSLA is time consuming and technically challenging.

– The telomere length distribution is heterogeneous as the size of telomeres varies between chromosome ends, between homologous chromosomes and from one cell to another.

– As in TRF a variable amount of subtelomeric DNA and/or variant repeats is included in the length measured. Not all the telomere variants are digested by the restriction enzymes used.

2.2.4. *Quantitative-polymerase chain reaction*

Quantitative-polymerase chain reaction (Q-PCR) can detect and precisely quantify the initial amount of the target sequence. Telomere length measurement by Q-PCR has been rendered possible by the design of a pair of primers, which contain mismatches in order to prevent primer dimer formation (Cawthon 2002). Indeed, primer dimers are formed when the two primers anneal to each other and are amplified. This would be the case with telomeric primers as the G-rich primer and C-rich primer are intrinsically complementary. However, the pair of primers designed by Cawthon et al. contain mismatches across their length. As a result, the DNA polymerase can extend the 3' end of the primer only when it is hybridized to the telomere but cannot when the primers are hybridized to each other. The shortest possible products (i.e. the sum of the length of the two primers) are primarily amplified and are directly proportional to the number of telomere repeats (Figure 2.7). The telomere signal is normalized by the signal of a single copy gene, thus generating a telomere/single copy gene ratio (T/S ratio). The single copy gene is either amplified in a different tube using the same quantity of DNA from the same sample or in the same tube by multiplexe monochrome (MM) Q-PCR (Cawthon 2009). MM Q-PCR has the advantage of avoiding errors due to pipetting volume discrepancies between the telomere and single copy gene reaction mixes. The normalization with the single copy gene is crucial in order to account for possible variations between samples such as the quality of the DNA. This ratio is proportional to the average telomere length of the sample of interest. The T/S ratio of the sample of interest is then compared with the T/S ratio of a reference DNA sample, thus giving a relative telomere length (relative TL). If the telomere length of the sample of interest is the same as the telomere length of the reference sample, the relative telomere length will be 1. If the telomere length of the unknown sample is longer than the telomere length of the reference sample, the relative telomere length will higher than 1. Likewise, if the telomere length of the unknown sample is shorter than the telomere length of the reference sample, the relative telomere length will be smaller than 1.

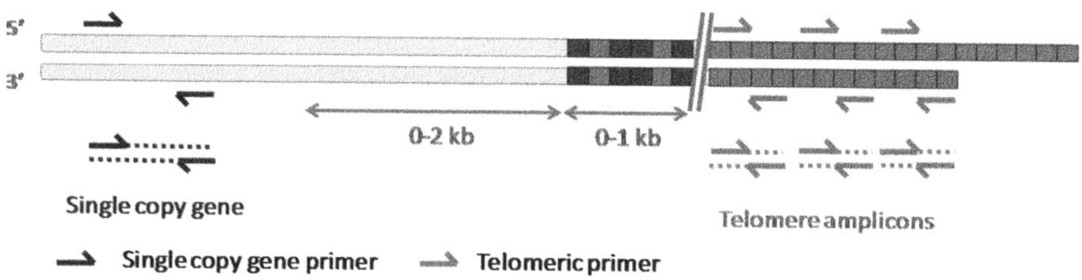

Figure 2.7. *Q-PCR methodology: Single copy genes primers are represented in dark blue and telomeric primers in green. Dark blue dashed lines represent the single copy gene amplicon and green dashed lines represent telomere amplicons. The shortest possible telomere amplicons are primarily amplified*

Advantages of Q-PCR are as follows:

– Q-PCR is a sensitive technique as quantities of DNA as small as a nanogram can be analyzed.

– Q-PCR does not require high molecular weight and good quality DNA. As a result, any kind of sample can be analyzed, even fixed paraffin-embedded material or degraded DNA.

– This method is fast and relatively easy to set up, which makes it suitable for epidemiological studies and analysis of a large number of samples.

– Q-PCR telomere length measurement can be used on several species.

Disadvantages of Q-PCR are as follows:

– Q-PCR provides a relative telomere length and not the actual length in kilobases. Thus, it requires a calibration step in order to give an estimation of the actual telomere length.

– It does not provide any information on the telomere length distribution of the cell population analyzed.

– This method relies on the fact that the control gene is unique in the genome. If the control gene has been duplicated or lost, the T/S ratio will be altered. This restricts the use of this method to samples that are diploid and karyotypically stable.

– Inter- and intra-assay variability has been reported (Aviv 2009; Aubert et al. 2012). The intra-assay variability has been improved by the use of MM Q-PCR. The inter-assay variability between labs may come from the use of different reagents or the choice of the control gene.

2.2.5. *Quantitative-fluorescence in situ hybridization*

Quantitative-fluorescence in situ hybridization (Q-FISH) of telomeres uses fluorescently labeled peptide nucleotide acid (PNA) telomere probes that hybridize to denatured telomeric DNA of metaphase chromosomes in order to measure the length of individual telomeres (Lansdorp 1996) (Figure 2.8). The number of probes hybridized to the telomeric repeats is proportional to the telomere length. The fluorescence signal of individual chromosome ends is quantified and the telomere length is expressed as telomere fluorescence units (TFU, with each unit corresponding to 1 kb of telomere). To convert arbitrary fluorescence units into TFU, the use of a standard of known telomere length is required. Typically, the analysis of 20 metaphases is required for reliable telomere length measurements. Telomere Q-FISH can also be done on cells in interphase. But contrary to metaphase FISH which give access to the length of individual telomeres, interphase Q-FISH will only provide the mean telomere length per nuclei.

Importantly Q-FISH can also be used to detect telomere fusions (Sfeir and de Lange 2012) and phenotypes of dysfunctional telomeres such as fragile telomeres, telomere sister chromatids exchanges and extra-chromosomal telomere signal (Sfeir et al. 2009; Sfeir and de Lange 2012; Doksani and de Lange 2016) (Figure 2.9).

Advantages of Q-FISH are as follows:

– Q-FISH is resolutive. It can measure the telomere length of individual chromosome arms for all chromosomes. The length of specific alleles can also be measured (Londoño-Vallejo et al. 2001).

– Q-FISH provides information on the heterogeneity of telomere length in a cell population.

– Q-FISH can be used on several species.

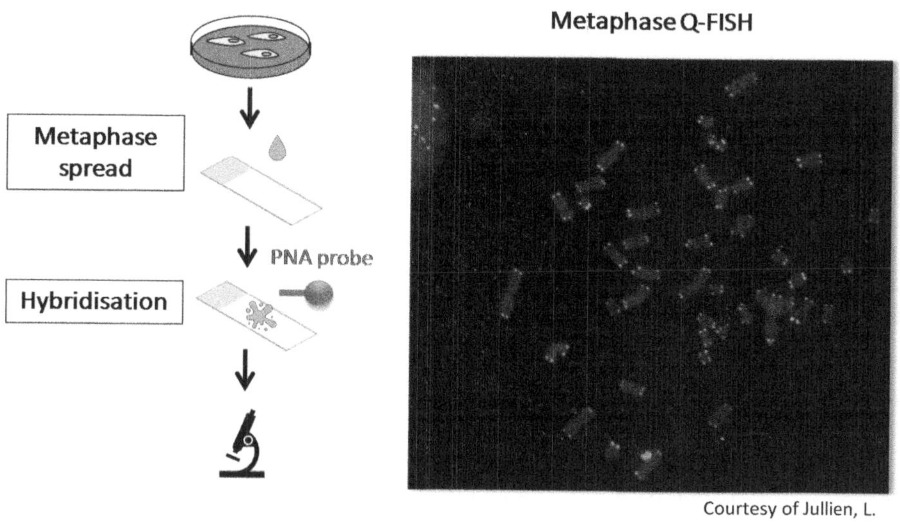

Figure 2.8. *Metaphase Q-FISH methodology: Briefly, metaphase spreads are obtained from cells in culture and hybridized with a fluorescently labeled PNA probe that binds specifically to the telomeres*

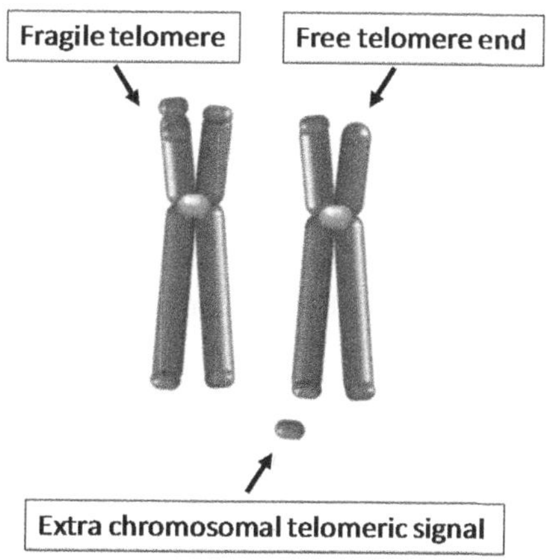

Figure 2.9. *Schematic of dysfunctional telomeres phenotypes. Fragile telomeres appear as a double signal at the chromosome end. No fluorescent signal is detected at free telomere ends. And extra chromosome telomeric signal is when the telomeric fluorescence signal is separated from the chromosome end*

Disadvantages of Q-FISH are as follows:

– Q-FISH is limited in its use to cells that proliferate in culture. It cannot be utilized on clinical samples or non-proliferating cells such as cells in senescence. However, to circumvent this limitation it is possible to undertake interphase Q-FISH although less resolutive (Canela et al. 2007).

– As Q-FISH is hybridization-based, there is a threshold below which telomeres cannot be detected. As a result, it is impossible to know if signal free ends are short telomeres below the detection threshold or chromosome ends where the telomere has been completely lost (Figure 2.9).

2.2.6. *Flow-fluorescence in situ hybridization*

As Q-FISH, flow-fluorescence in situ hybridization (Flow-FISH) uses a fluorescently labeled PNA telomere probe (Rufer et al. 1998; Baerlocher et al. 2006). But the probe is hybridized to the telomeres of interphase cells in suspension and the fluorescence signal is analyzed by flow cytometry. DNA is denatured by heating the cell suspension and is followed by hybridization of the PNA telomere probe to the complementary telomere sequences. The fluorescence is detected by a flow cytometer where a laser excites the fluorophores and the emitted fluorescence is read by a detector. Flow-FISH measures the genome telomere length of single cells.

Advantages of Flow-FISH are as follows:

– Flow-FISH can measure the telomere length of distinct cell populations from the same sample. The cell populations are stained with a specific antibody and further sorted by fluorescence-activated cell sorting (FACS) either before the telomere FISH or after.

– Flow-FISH has been adapted for the analysis of large number of samples (Baerlocher et al. 2006).

– Flow-FISH can be used on several species (Baerlocher et al. 2003).

Disadvantages of Flow-FISH are as follows:

– Flow-FISH is particularly adapted for nucleated blood cells but limited to fresh blood samples. Flow-FISH can be used on adherent cells but only on nuclei preparations rather than whole cells. The reason is that blood cells have higher nucleus/cytoplasm ratio than most adherent cells and it has been shown that the PNA probe can non-specifically bind to cytoplasm components (Wieser et al. 2006).

– Flow-FISH is time consuming and technically challenging.

2.3. References

d'Adda di Fagagna, F., Reaper, P.M., Clay-Farrace, L., Fiegler, H., Carr, P., von Zglinicki, T., Saretzki, G., Carter, N.P., Jackson, S.P. (2003). A DNA damage checkpoint response in telomere-initiated senescence. *Nature*, 426, 194–198.

Allshire, R.C., Dempster, M., Hastie, N.D. (1989). Human telomeres contain at least three types of G-rich repeat distributed non-randomly. *Nucleic Acids Res.*, 17, 4611–4627.

Ambrosini, A., Paul, S., Hu, S., Riethman, H. (2007). Human subtelomeric duplicon structure and organization. *Genome Biol.*, 8, R151.

Artandi, S.E., Chang, S., Lee, S.-L., Alson, S., Gottlieb, G.J., Chin, L., DePinho, R.A. (2000). Telomere dysfunction promotes non-reciprocal translocations and epithelial cancers in mice. *Nature*, 406, 641–645.

Aubert, G., Hills, M., Lansdorp, P.M. (2012). Telomere length measurement – Caveats and a critical assessment of the available technologies and tools. *Mutat. Res. Mol. Mech. Mutagen.*, 730, 59–67.

Aviv, A. (2009). Commentary: Raising the bar on telomere epidemiology. *Int. J. Epidemiol.*, 38, 1735–1736.

Baerlocher, G.M., Mak, J., Röth, A., Rice, K.S., Lansdorp, P.M. (2003). Telomere shortening in leukocyte subpopulations from baboons. *J. Leukoc. Biol.*, 73, 289–296.

Baerlocher, G.M., Vulto, I., de Jong, G., Lansdorp, P.M. (2006). Flow cytometry and FISH to measure the average length of telomeres (flow FISH). *Nat. Protoc.*, 1, 2365–2376.

Baird, D.M., Jeffreys, A.J., Royle, N.J. (1995). Mechanisms underlying telomere repeat turnover, revealed by hypervariable variant repeat distribution patterns in the human Xp/Yp telomere. *EMBO J.*, 14, 5433–5443.

Baird, D.M., Rowson, J., Wynford-Thomas, D., Kipling, D. (2003). Extensive allelic variation and ultrashort telomeres in senescent human cells. *Nat. Genet.*, 33, 203–207.

Bhargava, R., Fischer, M., O'Sullivan, R.J. (2020). Genome rearrangements associated with aberrant telomere maintenance. *Curr. Opin. Genet. Dev.*, 60, 31–40.

Britt-Compton, B., Rowson, J., Locke, M., Mackenzie, I., Kipling, D., Baird, D.M. (2006). Structural stability and chromosome-specific telomere length is governed by cis-acting determinants in humans. *Hum. Mol. Genet.*, 15, 725–733.

Canela, A., Vera, E., Klatt, P., Blasco, M.A. (2007). High-throughput telomere length quantification by FISH and its application to human population studies. *Proc. Natl. Acad. Sci.*, 104, 5300–5305.

Capper, R., Britt-Compton, B., Tankimanova, M., Rowson, J., Letsolo, B., Man, S., Haughton, M., Baird, D.M. (2007). The nature of telomere fusion and a definition of the critical telomere length in human cells. *Genes & Dev.*, 21, 2495–2508.

Cawthon, R.M. (2002). Telomere measurement by quantitative PCR. *Nucleic Acids Res.*, 30, 47e–47.

Cawthon, R.M. (2009). Telomere length measurement by a novel monochrome multiplex quantitative PCR method. *Nucleic Acids Res.*, 37, e21–e21.

Cheung, I. (2004). Strain-specific telomere length revealed by single telomere length analysis in Caenorhabditis elegans. *Nucleic Acids Res.*, 32, 3383–3391.

Counter, C.M., Avilion, A.A., LeFeuvre, C.E., Stewart, N.G., Greider, C.W., Harley, C.B., Bacchetti, S. (1992). Telomere shortening associated with chromosome instability is arrested in immortal cells which express telomerase activity. *EMBO J.*, 11, 1921–1929.

Doksani, Y., and de Lange, T. (2016). Telomere-internal double-strand breaks are repaired by homologous recombination and PARP1/Lig3-dependent end-joining. *Cell Rep.*, 17, 1646–1656.

Harley, C.B., Futcher, A.B., Greider, C.W. (1990). Telomeres shorten during ageing of human fibroblasts. *Nature*, 345(6274), 458-60

Hyatt, S., Jones, R.E., Heppel, N.H., Grimstead, J.W., Fegan, C., Jackson, G.H., Hills, R., Allan, J.M., Pratt, G., Pepper, C. et al. (2017). Telomere length is a critical determinant for survival in multiple myeloma. *Br. J. Haematol.*, 178, 94–98.

Lai, T.-P., Zhang, N., Noh, J., Mender, I., Tedone, E., Huang, E., Wright, W.E., Danuser, G., Shay, J.W. (2017). A method for measuring the distribution of the shortest telomeres in cells and tissues. *Nat. Commun.*, 8, 1356.

Lansdorp, P. (1996). Heterogeneity in telomere length of human chromosomes. *Hum. Mol. Genet.*, 5, 685–691.

Letsolo, B.T., Rowson, J., Baird, D.M. (2010). Fusion of short telomeres in human cells is characterized by extensive deletion and microhomology, and can result in complex rearrangements. *Nucleic Acids Res.*, 38, 1841–1852.

Letsolo, B.T., Jones, R.E., Rowson, J., Grimstead, J.W., Keith, W.N., Jenkins, G.J.S., Baird, D.M. (2017). Extensive telomere erosion is consistent with localised clonal expansions in Barrett's metaplasia. *PLoS One 12*, e0174833.

Lin, T.T., Letsolo, B.T., Jones, R.E., Rowson, J., Pratt, G., Hewamana, S., Fegan, C., Pepper, C., Baird, D.M. (2010). Telomere dysfunction and fusion during the progression of chronic lymphocytic leukemia: evidence for a telomere crisis. *Blood*, 116, 1899–1907.

Lin, T.T., Norris, K., Heppel, N.H., Pratt, G., Allan, J.M., Allsup, D.J., Bailey, J., Cawkwell, L., Hills, R., Grimstead, J.W. et al. (2014). Telomere dysfunction accurately predicts clinical outcome in chronic lymphocytic leukaemia, even in patients with early stage disease. *Br. J. Haematol.*, 167, 214–223.

Londoño-Vallejo, J.A., DerSarkissian, H., Cazes, L., Thomas, G. (2001). Differencces in telomere length between homologous chromosomes in humans. *Nucleic Acids Res.*, 29, 3164–3171.

McClintock, B. (1939). The behavior in successive nuclear divisions of a chromosome broken at meiosis. *Proc. Natl. Acad. Sci.*, 25, 405–416.

Mefford, H.C. and Trask, B.J. (2002). The complex structure and dynamic evolution of human subtelomeres. *Nat. Rev. Genet.*, 3, 91–102.

Moyzis, R.K., Buckingham, J.M., Cram, L.S., Dani, M., Deaven, L.L., Jones, M.D., Meyne, J., Ratliff, R.L., Wu, J.R. (1988). A highly conserved repetitive DNA sequence, (TTAGGG)(n), present at the telomeres of human chromosomes. *Proc. Natl. Acad. Sci. U.S.A.*, 18, 6622–6626.

Müller, J.H. (1938). The remaking of chromosomes. *Collect. Net.*, 13, 181–198.

Murnane, J.P. (2006). Telomeres and chromosome instability. *DNA Repair (Amst)*, 5, 1082–1092.

Norris, K., Hillmen, P., Rawstron, A., Hills, R., Baird, D.M., Fegan, C.D., Pepper, C. (2019). Telomere length predicts for outcome to FCR chemotherapy in CLL. *Leukemia*, 33, 1953–1963.

Olovnikov, A.M. (1971). Principle of marginotomy in template synthesis of polynucleotides. *Dokl. Akad. Nauk SSSR.*, 201, 1496–1499.

Riethman, H., Ambrosini, A., Paul, S. (2005). Human subtelomere structure and variation. *Chromosom. Res.*, 13, 505–515.

Roger, L., Jones, R.E., Heppel, N.H., Williams, G.T., Sampson, J.R., Baird, D.M. (2013). Extensive telomere erosion in the initiation of colorectal adenomas and its association with chromosomal instability. *JNCI J. Natl. Cancer Inst.*, 105, 1202–1211.

Rufer, N., Dragowska, W., Thornbury, G., Roosnek, E., Lansdorp, P.M. (1998). Telomere length dynamics in human lymphocyte subpopulations measured by flow cytometry. *Nat. Biotechnol.*, 16, 743–747.

Sabatier, L. (2005). The loss of a single telomere can result in instability of multiple chromosomes in a human tumor cell line. *Mol. Cancer Res.*, 3, 139–150.

Sfeir, A. and de Lange, T. (2012). Removal of shelterin reveals the telomere end-protection problem. *Science*, 336(80), 593–597.

Sfeir, A., Kosiyatrakul, S.T., Hockemeyer, D., MacRae, S.L., Karlseder, J., Schildkraut, C.L., de Lange, T. (2009). Mammalian telomeres resemble fragile sites and require TRF1 for efficient replication. *Cell*, 138, 90–103. doi: 10.1016/j.cell.2009.06.021.

Simpson, K., Jones, R.E., Grimstead, J.W., Hills, R., Pepper, C., Baird, D.M. (2015). Telomere fusion threshold identifies a poor prognostic subset of breast cancer patients. *Mol. Oncol.*, 9, 1186–1193.

Strefford, J.C., Kadalayil, L., Forster, J., Rose-Zerilli, M.J.J., Parker, A., Lin, T.T., Heppel, N., Norris, K., Gardiner, A., Davies, Z. et al. (2015). Telomere length predicts progression and overall survival in chronic lymphocytic leukemia: data from the UK LRF CLL4 trial. *Leukemia*, 29, 2411–2414.

Wieser, M., Stadler, G., Böhm, E., Borth, N., Katinger, H., Grillari, J., Voglauer, R. (2006). Nuclear flow FISH: Isolation of cell nuclei improves the determination of telomere lengths. *Exp. Gerontol.*, 41, 230–235.

3

The Shelterin Complex

Claire GHILAIN, Eric GILSON and Marie-Josèphe GIRAUD-PANIS
Medicine Faculty, CNRS, INSERM, IRCAN,
Côte d'Azur University, Nice, France

Linearization of the genome in the eukaryotic lineage made hiding chromosomes ends a necessity to avoid their illicit repair and the resulting chromosomal fusions and genomic instability (Mendez-Bermudez et al. 2020). Germ and stem cells are also able to compensate for the inevitable loss of terminal sequence due to incomplete replication of DNA ends. Facing these problems, eukaryotic ancestors devised common answers: a protein complex capping chromosomes termini and processes to elongate DNA ends. In many organisms, the protective cap takes the form of a complex called shelterin, a name coined by Titia de Lange in 2005 (de Lange 2005) and used ever since. On its simplest configuration, the shelterin complex of ciliates such as *Oxytrichia nova* only contains a heterodimeric protein that protects the single stranded terminus of chromosomes (TEBPαβ) (Giraud-Panis et al. 2013). Humans have evolved a more elaborate system containing six proteins (TRF1, TRF2, RAP1, TIN2, TPP1 and POT1; Palm and de Lange 2008; Figure 3.1) but the ultimate goal remains the same: inhibiting activation of the DNA damage response (DDR) and double strand break (DSB) repair as well as recruiting/regulating telomerase, an enzyme that elongates telomeres. Despite

---

For a color version of all the figures in this chapter, see www.iste.co.uk/saintome/telomeres.zip.

Telomeres,
coordinated by Carole SAINTOMÉ. © ISTE Ltd 2023.

overall differences in composition, common structural themes can be found: Myb/SANT (Telobox) domains for specific binding to double stranded (ds) telomeric DNA repeats (Bilaud 1996; Nishikawa 2001), oligonucleotide/ oligosaccharide binding (OB) folds for binding the single stranded telomeric DNA repeats (Lei et al. 2003) and TRFH domains for protein and additional DNA interactions (Fairall et al. 2001; Giraud-Panis et al. 2018). Hence, although this chapter will be mainly focused on the mammalian proteins, most of the points discussed here are also relevant to many other species. However, as demonstrated by Songyang and colleagues (2017a), we have to be careful in drawing a pan-species model for telomeres since differences exist between organisms, even between human and mouse. Furthermore, some exceptions to shelterin do exist, such as in dipterans that do not have G-rich repeats as terminal sequences, having lost telomerase (Mason et al. 2016), but cap their chromosome ends using retrotransposons (Levis et al. 1993). In the dipteran Drosophila, the telomeric complex terminin (Raffa et al. 2011) serves the same purpose as shelterin, although its different molecular mechanism of capping based on heterochromatinization of telomeres stands it apart (Raffa et al. 2011).

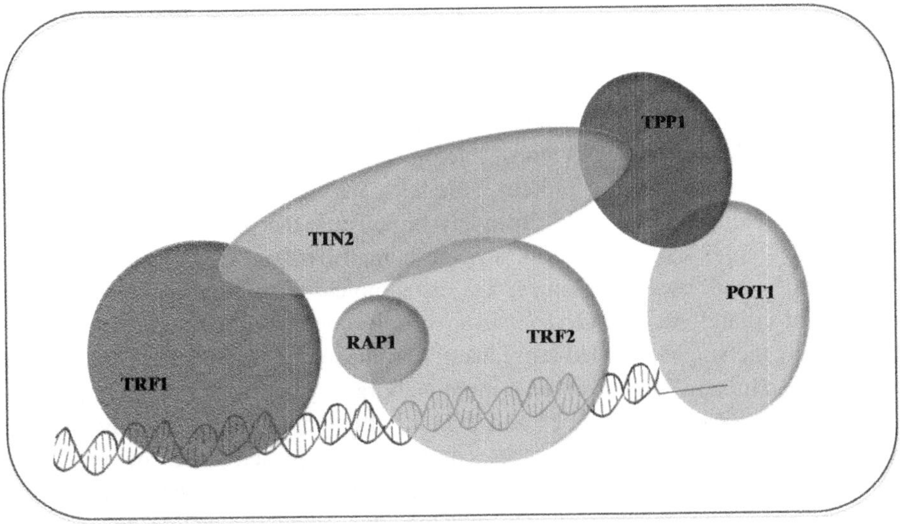

Figure 3.1. *Model of the human Shelterin complex*

As mentioned above, the human shelterin contains six proteins: TRF1, TRF2, RAP1, TIN2, TPP1 and POT1 (Figure 3.1). Of note, two Pot1 proteins (Pot1a and Pot1b) are found in rodents (Hockemeyer et al. 2006) (discussed later in the text). Shelterin proteins can be divided into three groups depending on their structural role on telomeres: the dsDNA binders, TRF1 and the TRF2-RAP1 complex; the TPP1-POT1 complex that deals with the telomeric single strand overhang; TIN2 that bridges the dsDNA binders to TPP1 and thus closes the complex. Each protein has a specific role in telomere biology owing to their different intrinsic biochemical properties and interactors.

3.1. TRF1 and TRF2, structurally alike but...

TRF1, the first shelterin protein discovered in Titia de Lange's laboratory (Zhong et al. 1992; Chong et al. 1995), is a dimeric protein (Bianchi et al. 1997), each subunit being composed of four domains, properties it shares with the other dsDNA binder, TRF2, discovered concomitantly by Titia de Lange's and Eric Gilson's laboratories just a few years later (Bilaud 1996; Bilaud et al. 1997; Broccoli et al. 1997). They are encoded by the *Terf1* (chromosome 8, ENSG00000147601 in humans) and *Terf2* (chromosome 16, ENSG00000132604 in humans) genes. Phylogenetic studies suggest that these genes originated from a common ancestor duplicated during one of the rounds of genome duplication that occurred at the base of the chordates lineage around 500 Myr ago (Poulet et al. 2012). Comparison of the overall shape of these proteins obtained by single molecule electron microscopy (TRF1) (Boskovic et al. 2016) and SAXS (TRF2) (Gaullier et al. 2016) suggest a much more extended conformation for TRF2 which probably contributes to their different behavior with DNA (see DNA wrapping below).

3.1.1. *Domains organization*

Both proteins bind ds telomeric DNA through a C-terminal Myb/SANT domain named Telobox (Bilaud 1996; Bilaud et al. 1997) (Figure 3.2). TRF1 and TRF2 Teloboxes share more than 50% identity (nearly 77% if counting similarities) but a few residues signature allows us to distinguish one from the other (Poulet et al. 2012). Although this domain is structurally very close to non-telomeric Myb/SANT domains (Nishikawa et al. 1998), they are sufficiently different which allows for the separating of the telomeric domains from the others and determining a monophyletic group of Telobox containing proteins (Bilaud 1996; Poulet et al. 2012).

Dimerization of these proteins is mediated by the central TRFH (TTAGGG Repeat Factor Homology), an alpha helix-rich domain which primary sequence largely differs between TRF1 and TRF2 but is remarkably similar in its architecture (Fairall et al. 2001). Besides dimerization, this domain has two crucial functions: it mediates interaction between TRFs and several of their partners and provides a site for a secondary binding of DNA for TRF2. The former is insured by the inner part of the horseshoe structure. Subtle differences between TRF1 and TRF2 in this region allow these proteins to interact with different partners, such as TIN2 for TRF1 and SLX4, Apollo or RAP1 for TRF2 (Chen et al. 2008; Kim et al. 2009; Gaullier et al. 2016). The latter property corresponds to non-specific interactions between the outer surface of the TRFH of TRF2 and DNA to form a complex where ~90 bp of DNA is wrapped around this domain in a right-handed orientation (Amiard et al. 2007; Benarroch-Popivker et al. 2016). This allows TRF2 to recognize positively supercoiled DNA (Ye et al. 2010) to participate in R-loop formation (though TERRA RNA invasion of ds telomeric DNA) (Lee et al. 2018) and to form a lasso-like protective structure called t-loop (Griffith et al. 1999; Benarroch-Popivker et al. 2016; Van Ly et al. 2018) (more details about this structure is given elsewhere in this book). Formerly the TRFH domain of TRF1 has the same capacity to wrap DNA but this function is inhibited in the full length protein by the acidic nature of the N-terminus of the protein (Poulet et al. 2012).

The N-termini of TRF1 and TRF2 are indeed very different. TRF1 bears a very acidic sequence of 76 residues (pHi 3.5, hence its name of A domain). TRF2, on the other hand, bears a 45 residues sequence rich in arginines and glycines and thus very basic (pHi 11.9, hence its name of B domain). Of note, in recent years, TRF2 sequence in databases shows a longer N-terminus (87 residues) due to the use of an upstream ATG codon. It has been proposed that the two isoforms were expressed (de Lange 2018) but the effect of this extension on TRF2 functions is still unknown. As it stands, these 42 additional residues do not overly modify the basic nature of the domain (pHi 11.8 for the 87 residues N-terminus) and the shorter version of the protein is capable of complementing the phenotypes caused by TRF2 deficiency, at least those investigated so far. Considering that the 5′ region of the *Terf1* and *Terf2* genes have similar sequences (notably G-rich), the origin of the N-terminal divergence might be a frameshift in the coding sequence generating D/E- or R/G- rich primary sequences for TRF1 or TRF2, respectively (Poulet et al. 2012). In TRF2, this domain has the capacity to bind branched DNA structures in vitro (Fouche et al. 2006; Poulet et al. 2009), to protect Holliday junctions from resolution by resolvase enzymes in vitro (Poulet et al. 2009) and in cells (Saint-Leger et al. 2014; Schmutz et al. 2017), to bind G-quadruplex DNA structures (G4, DNA conformations formed by Hoogsteen base-paring of runs of four guanines; Neidle 2016), to bind to the telomeric RNA TERRA (Deng et al. 2009) and to promote RNA invasion into ds telomeric DNA to form telomeric R-loops (Lee et al. 2018). The B domain of TRF2 also harbors binding sites for several proteins: DNA polymerase β, FEN1 and ORC1 have been put forward (Atanasiu et al. 2006; Muftuoglu et al. 2006; Deng et al. 2009). TRF1 acidic domain suppresses the capacity of TRF1 TRFH to wrap DNA and also inhibits TRF2-mediated R-loop formation (Lee et al. 2018). In vitro, it modulates the behavior of TRF1 toward nucleosomal arrays probably because of interactions between this very acidic domain and the very basic histone tails (Galati et al. 2015). It is also the site of interactions with Tankyrase 1 (TNKS1) (Li et al. 2016) in human TRF1 (this site is absent in the mouse protein), a telomere specific poly(ADP-ribose) polymerase that parylates TRF1. This parylation causes its detachment from telomeres, thus allowing access to telomerase and to oxidative damage repair proteins (Yang et al. 2017) and permits sister chromatids separation at telomeres during mitosis (Canudas et al. 2007). It is interesting to note that the N-terminus domain of TRFs seems to be an invention of chordates, since other species from Deuterostomia and its sister clade Protostomia (therefore encompassing most of the animal kingdom) have been shown to contain only one TRF protein with a much smaller domain when it exists. The impact of these differences on telomeres dynamics would certainly be interesting to investigate in the future.

Finally, the Hinge (or Linker) domain that lies between the TRFH and the Telobox is also one of the divergent domains of the TRFs, both in terms of size (97 residues for TRF1 versus 190 for TRF2) and function. This size difference likely explains the more extended overall conformation of TRF2 (Boskovic et al. 2016; Gaullier et al. 2016). This domain allows for the binding of the BLM helicase to TRF1, interactions that are important for the replicative role of TRF1 on telomeres (Zimmermann et al. 2014) and is probably responsible for the flexibility in DNA binding exhibited by the protein (Bianchi et al. 1999). In TRF2, this domain bears the binding sites for TIN2 and RAP1 of the shelterin as well a small sequence named iDDR (inhibition of the DDR) that was proposed to be important for the anti-DDR activities of TRF2 (Okamoto et al. 2013).

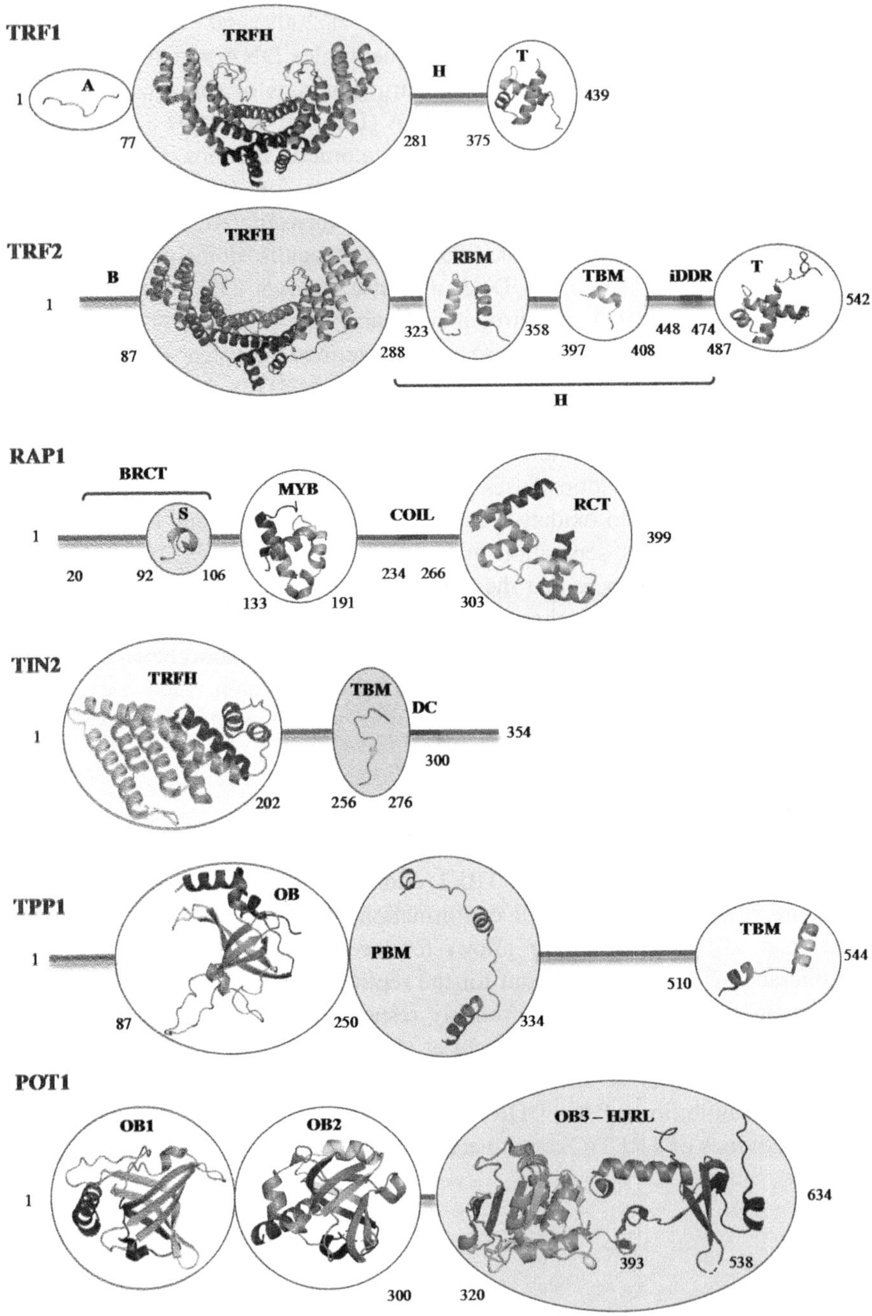

Figure 3.2. *Structural elements and interactions within the Shelterin complex*

COMMENTS ON FIGURE 3.2.– *Known 3D structures are shown in circles which color corresponds to their interacting partner(s) in the Shelterin assembly. Structural domains not involved in Shelterin assembly are shown in transparent circles. All structures were extracted from the PDB database using PyMol and shown as rainbow colored ribbons from blue (N-terminus) to red (C-terminus).*

TRF1

The four domains of TRF1 are labelled A (Acidic domain, structure extracted from its complex with TNKS1, pdb 5hkp), TRFH (TRF Homology domain shown as a dimer, structure extracted from its complex with TIN2, pdb 3bqo), H (Hinge domain) and T (Telobox, pdb 1ity).

TRF2

The four domains of TRF2 are labelled B (Basic domain), TRFH (TRF Homology domain shown as a dimer, structure extracted from its complex with Apollo, pdb 3bua), H (Hinge domain), RBM (RAP1 Binding Motif, structure extracted from its complex with RAP1, pdb 3k6g), TBM (TIN2 Binding Motif, structure extracted from the TRF2-TIN2-TPP1 complex, pdb 5xyf), iDDR (inhibition of the DNA Damage Response) and T (Telobox, pdb 1xg1).

RAP1

The four domains of RAP1 are labelled BRCT (BRCA1 C Terminal domain) which contains a secondary binding site for the TRF2 TRFH domain (S, Structure extracted from its complex with TRF2 TRFH, pdb 4rqi), MYB (Myb domain, pdb 1fex), Coil (Coil domain) and RCT (RAP1 C Terminal domain, structure extracted from its complex with TRF2, pdb 3k6g).

TIN2

The three domains of TIN2 are labelled TRFH (TRF Homology domain, structure extracted from the TRF2-TIN2-TPP1 complex, pdb 5xyf), TBM (TRFH Binding Motif, structure extracted from its complex with TRF1, pdb 3bqo) and DC (Dyskeratosis Congenita hotspot).

TPP1

The three domains of TIN2 are labelled OB (Oligosaccharide Binding Fold, pdb 2i46), PBM (POT1 Binding Motif, pdb 5h65), TBM (TIN2 Binding Motif, structure extracted from the TRF2-TIN2-TPP1 complex, pdb 5xyf).

POT1

The four domains of POT1 are labelled OB1 and OB2 (Oligosaccharide Binding Fold 1 and 2, pdb 1xjv), OB3 and HJRL (Oligosaccharide Binding Fold 3 and Holliday Junction Resolvase Like domain, structures extracted from the complex with TPP1, pdb 5h65).

3.1.2. *Biological roles*

Despite their many similarities, TRF1 and TRF2 play different roles with telomeres. There are multiple reasons for this, such as their different intrinsic biochemical properties that set them apart (like DNA wrapping or branched DNA recognition, see above) or their different sets of cofactors.

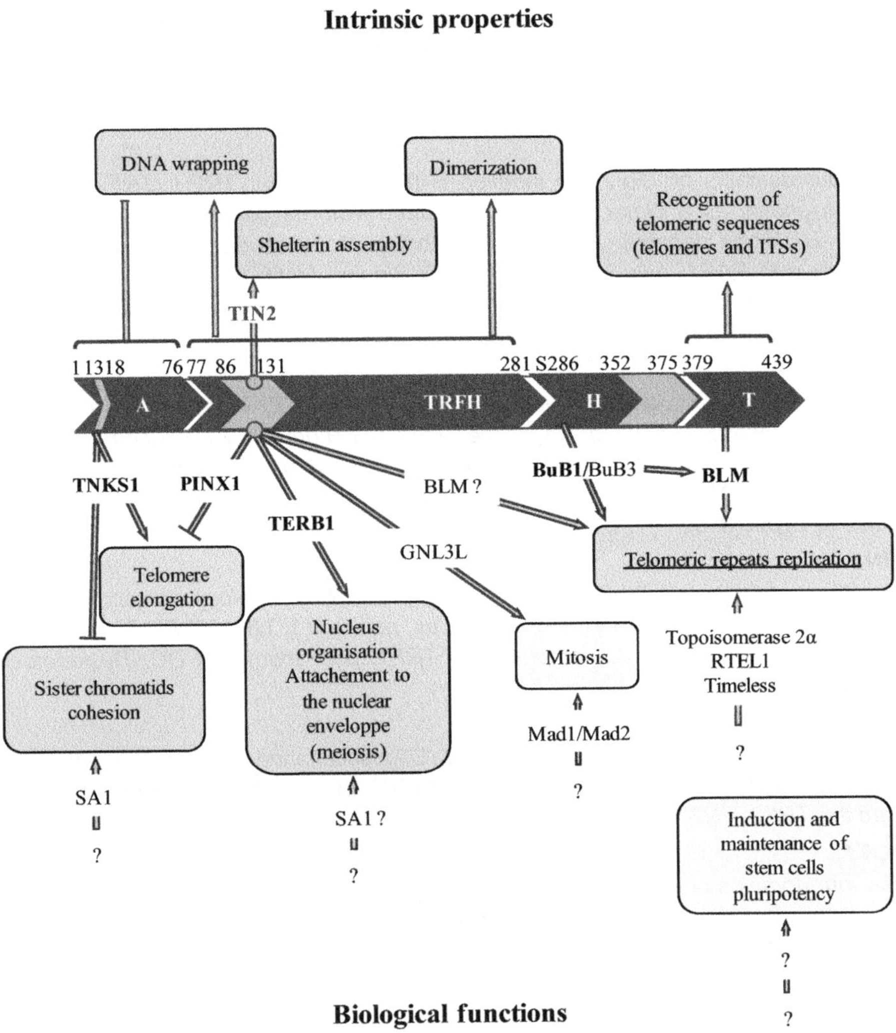

Figure 3.3. *Intrinsic properties and biological functions of TRF1*

COMMENTS ON FIGURE 3.3.– *The four domains of TRF1 are labelled A (Acidic domain), TRFH (TRF Homology domain), H (Hinge domain) and T (Telobox). Orange boxes, green boxes and yellow boxes correspond to intrinsic properties, telomeric functions and extra-telomeric functions respectively. Arrows indicate the domains involved for each property or function and the shape of the arrow stimulatory or inhibiting effect of TRF1. When known, TRF1 partners for the corresponding property or function are noted on the arrow in bold for direct interaction with TRF1 or plain text if indirect. Shelterin partners are labelled in red. Question marks signal an unknown partner or domain involved. The light blue circle signals the protein hub (or docking site) in the TRFH domain.*

3.1.2.1. *TRF1*

Terf1 is an essential gene and its deletion leads to early (blastocyst stage) embryonic lethality in mice (Karlseder et al. 2003) (Figure 3.3). TRF1 does not seem to be directly involved in telomere protection since its inhibition (KD or KO) does not lead to telomere fusion or ATM activation. Even if telomere deprotection can result from the expression of TIN2 mutants unable to bind TRF1 (L247E in mouse TIN2, L260E in human TIN2) (Frescas and de Lange 2014c), the effect of these mutations is TRF2 dependent since it is rescued by overexpressing TRF2 or tethering TIN2 to TRF2 (Frescas and de Lange 2014c). As discussed below, TRF1 emerges as a key factor for other telomere processes other than protection, including replication, telomerase and mitosis.

3.1.2.1.1. TRF1 and telomere length

From the onset, it was known that TRF1 was a negative regulator of telomere length (van Steensel and de Lange 1997). Explaining this, at least in part, it was shown to be inhibiting telomerase (Ancelin et al. 2002) through direct interactions between PINX1 (Zhou and Lu 2001) (an inhibitor of telomerase activity) and, as much later discovered, the TRFH domain of TRF1 (Chen et al. 2008). It also regulates the nuclease XPF, which is responsible for the telomere attrition observed when overexpressing TRF1 in mouse epithelial tissues (Munoz et al. 2009). Although this regulation is likely to be dependent on TRF2, TRF2 interacts with XPF via SLX4 (Wan et al. 2013).

3.1.2.1.2. Replicative role on telomeres

Allowing efficient replication of telomeres is one of the major roles of TRF1. Removal or depletion of TRF1 in mammalian cells causes chromosomal aberrations including, most notably, the appearance of multiple telomeric signals (MTS) on chromosome ends as observed by fluorescence microscopy following fluorescence in situ hybridization (FISH) of telomeres (Okamoto et al. 2008; Martinez et al. 2009;

Sfeir et al. 2009). Because of the resemblance of these defects with those observed on fragile sites when cells are subjected to replication stress, this phenotype was dubbed "fragile telomeres" and has been attributed to replicative problems. As for fragile sites, the exact nature of the lesion is basically unknown, but its replicative origin has found support since (i) activation of the ATR kinase and recruitment of DDR proteins on telomeres after TRF1 depletion was only observed in cells that have gone through S phase in the absence of TRF1 (Sfeir et al. 2009); (ii) functional links and direct interactions were discovered between TRF1 and important players of DNA replication. One of these direct interactors is timeless which is required for efficient replication of hard to replicate regions of the genome such as repeats and heterochromatin (Leman et al. 2012). One can also cite the BLM (Bloom syndrome-associated) and RTEL1 (regulator of telomere length helicase 1) helicases known to be able to resolve G4 structures and therefore to help replication in G-rich regions (Sfeir et al. 2009; Zimmermann et al. 2014; Deng et al. 2013; Drosopoulos et al. 2015). Furthermore, topoisomerase 2α, a topoisomerase essential for relieving topological stress and entanglements during and after S phase, is recruited on telomeres in a TRF1-dependent manner (d'Alcontres et al. 2014). (iii) TRF1 is phosphorylated by BUB1 to promote BLM recruitment and to overcome replication stress at telomeres (Li et al. 2018). Of note, the TRF1 replicative functions are independent from its chromosome terminus-binding since TRF1 is also needed for stabilization of the telomeric sequences found in the long arm of chromosome 2 (2q14) (Bosco et al. 2012).

3.1.2.1.3. TRF1 and mitosis

The early work of Lu and colleagues (1997) in the late 1990s had suggested a role of TRF1 in mitosis under the form of a smaller variant (lacking 20 residues in the linker/hinge domain) named PIN2. Since then several publications corroborated the involvement of TRF1 in mitosis and, more precisely, its link with the mitotic spindle. Overexpression of TRF1 in mouse epithelial tissue increased mitotic spindle aberrations (Munoz et al. 2009) and its depletion in MEFs (mouse embryonic fibroblasts) causes ultrafine anaphase bridges (UFBs) (d'Alcontres et al. 2014). TRF1 interacts with microtubules (Nakamura et al. 2001; Lan et al. 2014), colocalizes with the spindle assembly checkpoint proteins BubR1 (Munoz et al. 2009), interacts with Mad1 and Mad2 (d'Alcontres et al. 2014) and interactions with the mitotic Nek2 kinase were observed by yeast two-hybrid experiments (Prime and Markie 2005). TRF1 also participates in microtubule-kinetochore attachment through regulation of the centromeric function of the Aurora-A and Aurora-B mitotic kinases (Ohishi et al. 2014). Furthermore, GNL3L (guanine nucleotide-binding protein-like 3) binds TRF1 and stabilizes it during mitosis promoting metaphase to anaphase transition (Zhu et al. 2009). The mitotic functions of TRF1 are independent from its binding to telomeric DNA (no involvement of the C-terminal telomeric DNA binding domain) and are not shared with TRF2.

3.1.2.1.4. TRF1 and stemness

Maria A. Blasco et al. have investigated the role of TRF1 in the induction and maintenance of pluripotency in mouse cells. The *Terf1* gene is highly expressed in stem cells through direct regulation by the pluripotency factor oct3/4 (Schneider et al. 2013). Accordingly, TRF1 depletion reduces in vivo reprogramming efficiency (Marion et al. 2017) and causes loss of the naïve state of mouse ES (embryonic stem) cells through increased binding of the repressive PRC2 complex on some of its targets, leading to activation of differentiation programs. The mechanism was reported to involve the binding of the telomeric RNA TERRA (overexpressed when downregulating TRF1) to these TRF1-sensitive PRC2 binding sites and consequential recruitment of PRC2 (Marion et al. 2019). TRF1 is also involved in cancer stem cells maintenance, therefore explaining why its deletion impairs initiation and progression of tumors in a glioblastoma mouse model (Bejarano et al. 2017). This led the authors to propose TRF1 as a possible target for cancer therapy and to launch screens and tests for TRF1 chemical inhibitors in mice (Mendez-Pertuz et al. 2017; Bejarano et al. 2019). Future studies should shed light on the usefulness of this strategy for humans.

3.1.2.1.5. Telomere spatial organization within the nucleus

Perhaps connected to its role in stemness, links between TRF1 and nucleus organization have been reported. TRF1 interacts with SA1 to allow sister chromatids cohesion at telomeres after replication and before their separation at mitosis (Canudas et al. 2007). This cohesin subunit is known to be involved in stabilizing topologically associating domain boundaries together with CCCTC-Binding Factor (Kojic et al. 2018). How this interaction influences nucleus topology is still unknown. TRF1 also mediates telomere attachment to the nuclear envelop during meiosis (Dunce et al. 2018 and references therein).

3.1.2.2. *TRF2*

As for *TERF1*, *TERF2* deletion causes embryonic lethality in mice, albeit at a later stage (embryonic day 13.5 (Celli and de Lange 2005) for *TERF2* deletion versus around 4 for *TERF1* (Karlseder et al. 1999)) (Figure 3.4). It is also a negative regulator of telomeric DNA lengthening in a telomerase-independent manner (Ancelin et al. 2002). The increase in telomeric DNA length observed after TRF2 inhibition and the decrease due to its overexpression can be explained by the deregulation of degradation and recombination pathways that are controlled by TRF2 (detailed below) (Ancelin et al. 2002; Wang et al. 2004; Kim et al. 2017a).

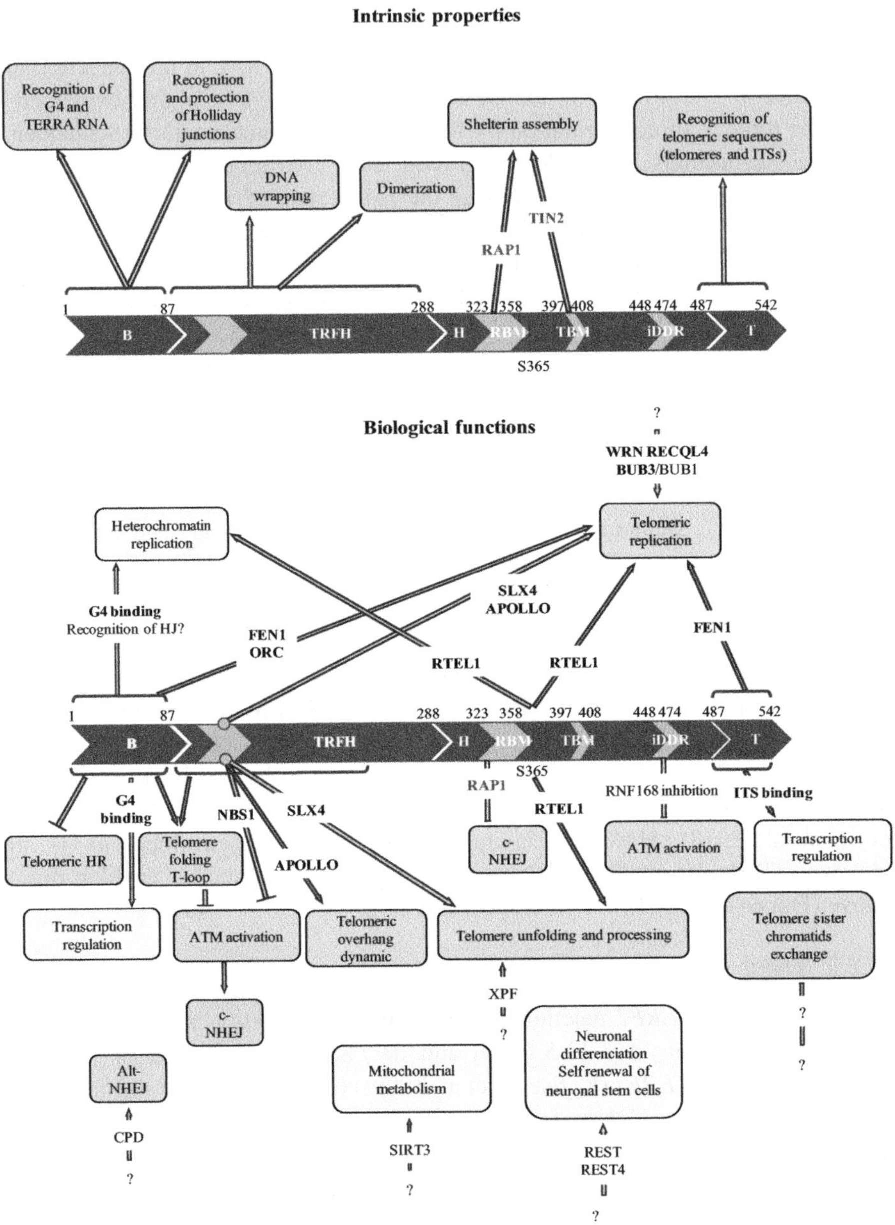

Figure 3.4. *Intrinsic properties and biological functions of TRF2*

COMMENTS ON FIGURE 3.4.– *The four domains of TRF2 are labelled B (Basic domain), TRFH (TRF Homology domain), H (Hinge domain) and T (Telobox). Orange boxes, green boxes and yellow boxes correspond to intrinsic properties, telomeric functions and extra-telomeric functions respectively. Arrows indicate the domains involved for each property or function and the shape of the arrow stimulatory or inhibiting effect of TRF2. When known, TRF2 partners for the corresponding property or function are noted on the arrow in bold for direct interaction or plain text if indirect. Shelterin partners are labelled in red. Question marks signal an unknown partner or domain involved. The light blue circle signals the protein hub (or docking site) in the TRFH domain. CPD stands for Claspin/PCNA/DONSON complex.*

3.1.2.2.1. End protection

Protecting telomeres termini from DNA damage signaling and repair is a major role of TRF2. This allows telomeres to be ignored by the repair machinery that would otherwise consider them as dsDNA break (DSB) and repair them accordingly. Of note, this essential function is also shared by POT1-TPP1 (although through different mechanisms, discussed below). TRF2 protects telomeres by at least two mechanisms: one is to avoid detection by inhibiting the activation of the Ataxia Telangiectasia Mutated PI3K-like kinase (Karlseder et al. 1999, 2004) (ATM, a major kinase in signaling and repair of DSBs; Clouaire et al. 2017) and its downstream kinase CHK2 (Celli and de Lange 2005; Buscemi et al. 2009); the other is by blocking DNA repair pathways such as non-homologous end joining (NHEJ) and homologous recombination (HR). Indeed, depletion of TRF2 in mouse or human cells causes telomeric recruitment of DDR proteins such as γH2AX and 53BP1, forming foci called TIFs (telomere dysfunction induced foci; Takai et al. 2003), leading to activation of ATM, phosphorylation of CHK2 and illicit repair mainly by classical-NHEJ (c-NHEJ, ligase 4 dependent process), therefore fusing telomeres together (van Steensel et al. 1998; Celli and de Lange 2005). For obvious reasons, TRF2-mediated control of the DDR and repair needs to be as tight as possible. This is achieved through a certain amount of redundancy (or additive effects) in the mechanisms involved: TRF2 blocks ATM activation by interaction of its TRFH domain with an unphosphorylated form of NBS1 (NBS1 phosphorylation by CDK2 detaches it from TRF2 which allows NBS1 to associate with Mre11-RAD50 and to activate ATM-dependent repair) (Rai et al. 2017) and by promoting the formation of the lasso-like T-loop structure, which hides the 3′ single strand tail (Griffith et al. 1999; Doksani et al. 2013; Bower and Griffith 2014) (a process also involving the TRFH domain and DNA wrapping; Amiard et al. 2007; Benarroch-Popivker et al. 2016; Van Ly et al. 2018); it inhibits CHK2 activation by direct interaction, of note, even without prior DNA damage (Buscemi et al. 2009); it blocks c-NHEJ through its interaction with RAP1 (discussed below), by interacting with Ku70 and inhibiting its action (Ribes-Zamora et al. 2013) and by suppressing RNF168 activation thanks to its iDDR motif (see Figure 3.4) (Okamoto et al. 2013); it limits HR thanks to its basic N-terminal domain, which recognizes (Fouche et al. 2006) and protects branched DNA structures from resolution (Wang et al. 2004; Poulet et al. 2009; Saint-Leger et al. 2014; Schmutz et al. 2017) and dissolution by the Werner syndrome helicase (Nora et al. 2010). This is perceived as the mechanism behind T-loop stability since a Holliday junction (or double-Holliday junction) is thought to be formed at the foot the T-loop. Interestingly, TRF2 blocks the telomere sister-chromatid exchange independently of the presence of its N-terminal basic domain (Celli et al. 2006), suggesting a different mechanism for the inhibition of this specific recombination event.

This plethora of protective activities, being for the most part mechanistically independent, allows fail-safe pathways to be engaged if needed. For example, while a wrapping-deficient mutant of TRF2 (Top-less, see below) causes a significant decrease in T-loops in human cells, no fusion phenotype could be observed in human cells due to an anti-fusion back-up pathway mediated by RAP1 (Benarroch-Popivker et al. 2016; Lototska et al. 2020) (the biological significance of this will be discussed more extensively in the RAP1 section).

3.1.2.2.2. Telomere folding and topology

Intimately linked to its protective functions, TRF2 takes charge of the folding of telomeres. As mentioned above, it is responsible for T-loop formation (Stansel et al. 2001; Doksani et al. 2013), a process that requires the wrapping of DNA around the TRFH domain in a right handed orientation (Benarroch-Popivker et al. 2016). This is achieved through non-sequence specific binding of DNA on several lysines and arginines (nine were identified) located at strategic positions on the outer surface of the domain. Mutating these residues to alanines causes a dramatic decrease in DNA wrapping and in TRF2 ability to control telomeric DNA topology (hence the name of Top-less for this wrapping-deficient mutant). Indeed, in a constrained DNA (as telomeres might be, considering their attachment to the nuclear matrix; de Lange 1992; Luderus et al. 1996), this right-handed wrapping causes consequential unwinding of the surrounding unbound DNA therefore favoring the invasion of single stranded (ss) DNA (or RNA). Hence, by wrapping DNA, TRF2 stimulates the invasion of ss DNA in telomeric DNA (Verdun et al. 2005; Amiard et al. 2007) to create a D-loop and form the T-loop (or the invasion of the TERRA RNA to create R-loops; Lee et al. 2018). As a result, the Top-less mutant of TRF2 causes deficiency in T-loop formation and leads to relaxation of telomeric DNA (Benarroch-Popivker et al. 2016). TRF2 may also participate in telomeres conformation through its interaction with the telomeric RNA TERRA (Deng et al. 2009) (more about TERRA in other chapter of this book). TRF2 is also capable of auto-associating in vitro (Baker et al. 2011; Poulet et al. 2012; Benarroch-Popivker et al. 2016), which may explain the condensation observed by Kaur et al. (2016) on long telomeric tracts. Conflicting results on the effect of removal of TRF2 on telomeric compaction in cells using high-resolution microscopy (Bandaria et al. 2016; Timashev et al. 2017; Vancevska et al. 2017) do not allow us to conclude on the biological relevance of this condensation as yet. Interestingly, TRF2 is also capable to foldback telomere to internal subtelomeric sequences, thus forming large terminal chromatin loops involved in the regulation of subtelomeric gene expression (Robin et al. 2020).

As much as it controls telomeric DNA folding, TRF2 is also responsible for its unfolding and processing.

3.1.2.2.3. Telomeric DNA processing

Studying TRF2 partners outside the shelterin context revealed interactions with many different DNA processing enzymes, helicases and even nucleases. This could seem rather puzzling, knowing the protective role of TRF2. Of course, the obvious answer is that these DNA processing activities are required at some point or another for telomere functioning and TRF2 plays the maestro by orchestrating them. This is definitively the case for the following enzymes: the exonuclease Apollo (SNM1B) (Lenain et al. 2006; van Overbeek and de Lange 2006), which interacts with the TRFH domain of TRF2 (Chen et al. 2008) and its activity is enhanced by this interaction in vitro (Lenain et al. 2006), is recruited to allow resection of the 5′ end of the leading strand after synthesis to recreate the 3′ overhang necessary for its protection probably through T-loop formation (Lam et al. 2010; Wu et al. 2010). It also assists TRF2 for efficient replication of telomere tracks (van Overbeek and de Lange 2006; Ye et al. 2010); TRF2 interacts with FEN1 through its Telobox and N-terminal domain in vitro (Muftuoglu et al. 2006) and recruits it on telomeres to aid replication of the lagging strand in cells (Saharia et al. 2008). It also interacts with the Werner (WRN) and RECQL4 helicases to assist in DNA synthesis (Opresko et al. 2002; Machwe et al. 2004; Ghosh et al. 2012). It binds the SLX4 protein platform (again through interaction with the TRFH domain; Chen et al. 2008) and thus recruits on telomeres and regulates the nuclease partners of SLX4, ERCC1-XPF, MUS81-EME1/2 and SLX1 (for review, see Guervilly and Gaillard (2018)). TRF2-mediated recruitment of SLX4 is particularly visible on long telomeres and peaks during the S phase. Indeed, SLX1 activity is necessary for efficient replication of telomeres in ALT (telomerase negative cells that replenish their telomeres through recombination) U2OS human cells (XPF and MUS81 are not). However, all three nucleases are important for telomere length regulation in these ALT cells. Of note, the SLX4 TRF2 binding site is absent outside primates. The reason behind this late addition is unknown. Lastly, TRF2 interacts with the RTEL1 helicase (regulator of telomere length 1) (Ding et al. 2004), a helicase involved in the dismantling of T-loops during S phase (Sarek et al. 2015) which is also recruited by TRF2 to satellite 2/3 sequences (pericentromeres) probably to remove aberrant structures that would impede replication of these regions (Mendez-Bermudez et al. 2018).

One can readily notice that many of these processing activities are connected to replication. Thus, it is no surprise that TRF2 assists DNA synthesis on telomeres.

3.1.2.2.4. Telomeric DNA replication

One of the first reports showing a replicative role of TRF2 does not concern telomeres, but, rather surprisingly, the maintenance of the Epstein–Bahr virus (EBV). The EBV episome is maintained in B lymphocytes thanks to its replication origin OriP. This origin is flanked by TTAGGG repeats and TRF2 is necessary for efficient maintenance of the virus during latency (Deng et al. 2002; Atanasiu et al. 2006). Knowing that the firing of this origin is regulated by a sequence highly sensitive to negative topological constraints (Polonskaya et al. 2004), it is tempting to assume that TRF2 provides the required negative supercoils thanks to DNA wrapping. Further studies will be necessary to really ascertain that point. Another way in which TRF2 could assist EBV replication initiation is the recruitment of the ORC complex on the origin since TRF2 has been shown to interact with ORC in the EBV context as well as on telomeres (Atanasiu et al. 2006; Tatsumi et al. 2008). Besides this possible role in replication initiation, TRF2 has been shown to be important for avoiding fork stalling on telomeric repeats. TRF2 recruits the BUB3-BUB1 complex of the spindle assembly checkpoint on telomeres during S phase, the BUB1 kinase activity phosphorylates TRF1 which promotes BLM binding and thus facilitates DNA synthesis (Li et al. 2018). Furthermore, interactions between TRF2 and the replisome (Claspin PCNA DONSON (CPD) complex) have been observed in the context of TPP1-POT1 deficiency (Rai et al. 2019). Although this may seem a rather particular situation, we can wonder if this may happen in a different context such as stalled forks.

TRF2 does not affect the ATR-dependent response, which is the classical pathway in response to an increased amount of single-stranded (ss) DNA upon fork stalling, but rather inhibits an ATM-dependent pathway of replicative damage (Ye et al. 2010). This suggests that the replication problem requiring TRF2 is of a different nature than the classical one based on ss DNA accumulation and RPA recruitment. One possible lead to answer this question is the concerted action between TRF2 and topoisomerase 2α during the S phase. Indeed, telomeres of HT1080 cells are exquisitely sensitive to the topoisomerase 2 inhibitor ICRF-193 (Chen et al. 2015), exhibiting telomeric DNA damage in its presence in conditions where bulk DNA is virtually untouched. Importantly, this phenotype is rescued by increasing the amount of TRF2 in cells (Ye et al. 2010). This strongly suggests that TRF2 is involved in solving topological problems on telomeric tracts as a failsafe mechanism in the event of topoisomerase 2 failure or overloading. Knowing that TRF2 can recognize positive supercoils (Ye et al. 2010), a topology that can arise ahead of the replication fork in case of blockade and which is dealt with by topoisomerase 2α, one can readily hypothesize that TRF2 could be recruited on "topologically" blocked forks to help resume replication. Importantly, the Apollo catalytic domain is required for this activity, drawing an interesting parallel with the rescue of replication forks on fragile sites after Aphidicolin treatment (Mason et al. 2013). Another way TRF2 can help telomeric replication is through its capacity to recognize non-B DNA structures such as G4 (Pedroso et al. 2009; Purohit et al. 2018; Mukherjee et al. 2019) and or Holliday junctions (Fouche et al. 2006; Poulet et al. 2009) and to recruit enzymes that can unfold or resolve these structures (WRN, RTEL1, SLX4 interacting enzymes, etc., see above). This activity would be particularly important on a telomeric sequence due to its propensity to form these kinds of structures. Importantly, either recognition of G4, positive supercoiling or Holliday junctions do not strictly depend on DNA sequence but rather on the said propensity of these sequences to form these conformations. Hence, with insight, it is not surprising that TRF2 was found to bind and protect the regions which are hard to replicate outside telomeric sequences (Mendez-Bermudez et al. 2018) and again, unsurprisingly, to be slightly but statistically enriched on HU-dependent stalled forks genome-wide (Dungrawala et al. 2015).

3.1.2.2.5. TRF2 a player outside telomeres

Reports of extra-telomeric localization of TRF2 were published as early as 2005 thanks to the study of Nijar et al. (2005) using breast tumor cell lines imaging. But, direct evidence of this non-telomeric binding only came out in 2011 after ChiP-seq (where ChiP is chromatin immuno precipitation) experiments performed in Zhou Songyang's and Eric Gilson's laboratories (Simonet et al. 2011; Yang et al. 2011). Further ChIP experiments performed by Mukherjee et al. (2019) gave additional support to these extra-telomeric interactions. Several interesting points can be drawn from these studies: some binding sites are shared by TRF1/TRF2 or TRF2/RAP1 but they can also bind on their own; interstitial telomeric sequences (ITS) are over-represented in the sequencing reads but they correspond to a subset of the total genomic ITSs, the selection appearing to be based on length and homogeneity of sequence. Consequently, sequences able to form G4 structures (defined as a minimal run of four GGG containing repeats) are also over-represented; localization of these binding sites is not random, but rather concentrates close to the genes, suggesting a regulation of genes expression (Ye et al. 2014). Surprisingly, besides ITSs another prominent binding of TRF2 lies on pericentromeric satellite 2/3 repeats, (ATTCC)n (Simonet et al. 2011). These last two findings opened far-reaching avenues of research. TRF2-dependent gene regulation has been documented for several genes such as *SIRT3* in postmitotic human and mouse skeletal tissues as well as in vitro differentiated human myoblasts, therefore linking TRF2 to mitochondrial metabolism (Robin et al. 2020); *PDGFRβ* in human vein endothelial cells, a gene encoding the platelet derived growth factor receptor beta and involved in angiogenesis (El Mai et al. 2014); positive regulation by TRF2 of *HS3ST4*, *GPC6* and *VCAN*, three genes involved in the biosynthesis of heparin sulfate proteoglycans of the glycocalyx, participates in the immunosuppressive effect of tumours through regulation of myeloid-derived suppressor cells, thus inhibiting the NK- and T-cell responses (Biroccio et al. 2013; Cherfils-Vicini et al. 2019); TRF2-dependent up- or downregulation of G4 containing promoters of nine genes was also reported (*ANXA2*, *CHRM2*, *INHA*, *KCNH2*, *OBSL1*, *OPN4*, *SAMD14*, *SMAD7* and *THRA*) (Mukherjee et al. 2019), five of which are related to neuronal development; the other four are connected with cell growth, differentiation and cancer. In addition to neuronal development, TRF2 was shown, in zebrafishes, to regulate the expression of the *PPP2R2C* gene, a neuronal isoform of the regulatory subunit of the PP2A phosphatase (Wagner et al. 2017) and to interact with the RE1-silencing transcription factors REST and REST4, participating either in self-renewal of neuronal stem cells (REST) or neuronal differentiation (REST4) (for a review, see Ye et al. (2014)). Again through REST, binding of TRF2 on the G4-containing promoter of the p21 coding gene causes downregulation of the expression of this protein (Hussain et al. 2017). Overall, these data point out to direct or indirect roles of TRF2 in mitochondrial metabolism, neuronal differentiation (Lobanova et al. 2017) and oncogenesis. Indeed, cellular amounts of TRF2 are often elevated in tumours of different origins (Cherfils-Vicini et al. 2019 and references therein) and are controlled by several cancer-associated proteins including p53 (Fujita et al. 2010), Wnt/β-Catenin (Diala et al. 2013) and WT1 (Wagner et al. 2014). Importantly, for the cases where it has been analyzed, this TRF2-dependent transcriptional response has been shown to be disconnected to telomeric damage (Biroccio et al. 2013; Cherfils-Vicini et al. 2019) and to be dependent on TRF2 dosage (Biroccio et al. 2013; Cherfils-Vicini et al. 2019; Mukherjee et al. 2019).

Another interesting point to note from ChIP-seq studies is the presence of TRF2 on pericentromeric satellite 2/3 DNA repeats (Simonet et al. 2011). In fact, TRF2 is required for an efficient replication of these hard to replicate DNA sequences (Mendez-Bermudez et al. 2018). Accordingly, depletion of TRF2 causes pericentromeric DNA damage (again through ATM and not ATR) and fork stalling on these sequences. The recruitment of TRF2 does not rely on the C-terminal Telobox domain but rather on the N-terminal and TRFH domains, suggesting recognition of non-B DNA structures (G4 and/or branched structures for the B domain, positive supercoils for the TRFH). Indeed, TRF2 binding is increased by a treatment of the cells with a G4-stabilizing agent, by topological stress (downregulation or inhibition of topoisomerases) and Top-less the TRF2 mutant deficient in supercoil recognition is not recruited to these sequences. Another argument in favor of complex DNA structures being at the source of the replicative problem is the recruitment by TRF2 of RTEL1, a helicase known to dismantle G4 and D-loops (see above). Downregulating G9a or Suv39H1 and treating cells with TSA decreased TRF2 binding, suggesting an important role of the heterochromatic status in TRF2 recruitment. Even more strikingly, the sequence of satellite 2/3 DNA is not strictly required since it is possible to recruit TRF2 to forks blocked by LacI repressor/Lac operator complexes and, impressively, to augment this recruitment by increasing the heterochromatic status of this tracts through fusion of the LacI repressor to HP1α (heterochromatin protein 1 alpha). It suggests that TRF2 could well be an accessory protein for rescuing forks blocked by the build-up of complex DNA structures and supercoils that can arise in heterochromatin. Indeed, TRF2 protects many regions in the genome, most of which correspond to heterochromatin. Interestingly, we could draw a parallel with the study of Gilley et al. (2012) showing recruitment of TRF2 at sites of high energy laser DNA damage. Perhaps, aberrant structures are also part of this process.

Overall, the fact that TRF2 is involved in telomeric protection as well as in transcriptional regulation and heterochromatin stability genome-wide suggests a coordination between the telomeric state and various cellular processes controlling cell fate (Ye et al. 2014; Mendez-Bermudez et al. 2020). These functional relationships could provide an explanation for the broad contribution of telomeres to development as well as normal and pathological ageing.

3.2. RAP1

Discovered in the late 1990s (Li et al. 2000) because of a yeast two-hybrid screen using TRF2 as bait performed in Titia de Lange's laboratory, hRAP1 shares three domains with its budding yeast ortholog ScRap1 (BRCT, RCT and Myb-like domains) but contains only one copy of the Myb-like domain (Figure 3.5). Early experiments did not reveal DNA binding using Electrophoretic Mobility Shift Assay (EMSA) (Li et al. 2000), a fact in part explained by structural differences in the Myb domain (Hanaoka et al. 2001), but a more recent publication reports a significant binding on ds-ss DNA junctions albeit with no sequence preference (Arat and

Griffith 2012). RAP1 is recruited on telomeres by its interaction with TRF2. Several teams have analyzed the RAP1-TRF2 complex by different biophysical or biochemical means. In terms of stoichiometry, two monomers of RAP1 bind one dimer of TRF2 (or four monomers to two dimers) (Chen et al. 2011; Arat and Griffith 2012; Gaullier et al. 2016). This interaction involves several domains of each protein: A first interaction involves the RCT domain of RAP1 located in the C-terminal region and two helices located in the hinge/linker domain of TRF2 (RAP1 binding motif (RBM)) (Li et al. 2000; Chen et al. 2011). This is the main and stronger interaction; the second interaction concerns a large area of the RAP1 N-terminal moiety and TRF2 TRFH domain and includes the Apollo/SLX4 docking site (Gaullier et al. 2016). An interesting dynamic occurs between both proteins. The binding of the two molecules of RAP1 cause structural modifications on TRF2, in which consequential length (290 Å, to be compared to the 90 Å of TRF1; Boskovic et al. 2016) increases even further to 345 Å. On the RAP1 moiety, this secondary binding causes dimerization of the BRCT domain. The biological consequences of this dimerization have not been explored so far but would certainly be of interest knowing the biological role of BRCT dimerization in several other proteins involved in DDR and DNA repair (e.g. in BRCA1 where the BRCT dimer recognizes the ATM-phosphorylated Abraxas protein, thus allowing BRCA1 accumulation on site of DNA damage; Wu et al. 2016). Analyzing the respective affinities between the TRFH and peptides from Apollo, TIN2 and RAP1, and Marie-Hélène Le Du et al. proposed that this secondary interaction, being in the same range of that of TIN2 but lower than that of Apollo, would allow us to block illegitimate TIN2 binding on the TRFH but would be competed out by Apollo if needed. In vitro, the RAP1-TRF2 complex has a higher affinity for ds-ss DNA junctions than TRF2 alone (10-fold increase) (Arat and Griffith 2012) without altering the capacity of TRF2 to wrap DNA (Gaullier et al. 2016). This suggests that RAP1 could help TRF2 for the formation of the T-loop. However, protection of Holliday junctions against cleavage by several resolving enzymes seems affected (Gaullier et al. 2016). Again in vitro, the RAP1 has been shown to cooperate with TRF2 to protect telomeric DNA against NHEJ (Bae and Baumann 2007; Sarthy et al. 2009; Bombarde et al. 2010). If in vitro experiments clearly suggested a role of RAP1 in assisting TRF2 for telomere protection against NHEJ, cellular and in vivo experiments provide a somewhat different explanation for this protein. Depleting human or mouse telomeres of RAP1 caused activation of homology-directed recombination repair (Sfeir et al. 2010; Rai et al. 2016), although the mechanism involved is still controversial (Schmutz et al. 2017). Importantly, this depletion had, at best, very limited effect on DDR pathways, did not modify telomere protection against NHEJ and mice lacking RAP1 developed metabolic and inflammatory but not telomeric disorders (Jiang et al. 2007; Sfeir et al. 2010; Teo et al. 2010; Martinez et al. 2013; Yeung et al. 2013; Kim et al. 2017a). The non-telomeric phenotypes of RAP1-compromised mice are linked to the capacity of RAP1 to bind internally in the genome (Yang et al. 2011) to regulate metabolic genes (Martinez et al. 2010) and NFkB activation (Teo et al. 2010).

We can explain the apparent discrepancy between the in vitro role of hRAP1 to protect against NHEJ and the apparent lack of telomere fusion in RAP1-depleted human cells by redundancies in the mechanisms protecting telomeres from NHEJ. On the one hand, TRF2, independently of RAP1, ensures protection of long telomeres against NHEJ by forming the t-loop. On the other hand, RAP1, by itself can prevent NHEJ once it is targeted to telomeres via TRF2. Indeed, human telomeres unable to form t-loops through expression of the Top-less form of TRF2 are still protected against NHEJ thanks to the ability of Top-less to recruit RAP1 (Benarroch-Popivker et al. 2016). Accordingly, telomeres thought to be too short to form t-loops, such as those appearing during replicative senescence, rely on RAP1 to be protected against NHEJ (Lototska et al. 2020). This anti-NHEJ mechanism of the TRF2-RAP1 complex has been documented in vivo by Maria Blasco's laboratory showing that the short telomeres of telomerase negative mice required RAP1 to be protected against uncapping (Martinez et al. 2016). Hence, as was predicted from in vitro experiment, mammalian RAP1 does protect the telomere, as a failsafe pathway to protect telomeres when they become too short to be efficiently protected by TRF2-mediated t-loops. Of note, a very recent publication from Titia de Lange's group suggests that RAP1 may not act as an anti-NHEJ factor in MEFs cells harboring long telomeres, another fact pointing to the importance of the differences between mouse and human cells (Timashev and de Lange 2020). Interestingly, mouse RAP1 has also been recently shown to promote rather than inhibit c-NHEJ in the genome through interactions with several proteins of this pathway (Ku70, Ku80, DNA-PK, MRE11, Rad50 among others) in an independent manner from TRF2 (Khattar et al. 2019). Much earlier, Songyang and colleagues (2004) had reported similar TRF2-independent interactions with hRAP1 (involving Ku70, Ku86, MRE11, Rad50) that involved several domains of RAP1 spanning from the N-terminal BRCT to the coil domains included. This suggests that RAP1 has an inherent pro-NHEJ activity that is inhibited when it is recruited on telomeres by TRF2. One possible mechanism could involve the interaction observed in Allison Bertuch's laboratory between TRF2 and Ku that inhibits Ku heterotetramerization and thus c-NHEJ (Ribes-Zamora et al. 2013). Furthermore, structural studies performed by Marie-Hélène Le Du et al. have shown the involvement of the whole N-terminal part of hRAP1 in a secondary interaction with TRF2, which may alter the way RAP1 interacts with these NHEJ proteins. Overall, it emerges that outside telomeres RAP1 would activate DSB repair while, on telomeres, its interaction with TRF2 would convert it to an anti-NHEJ factor.

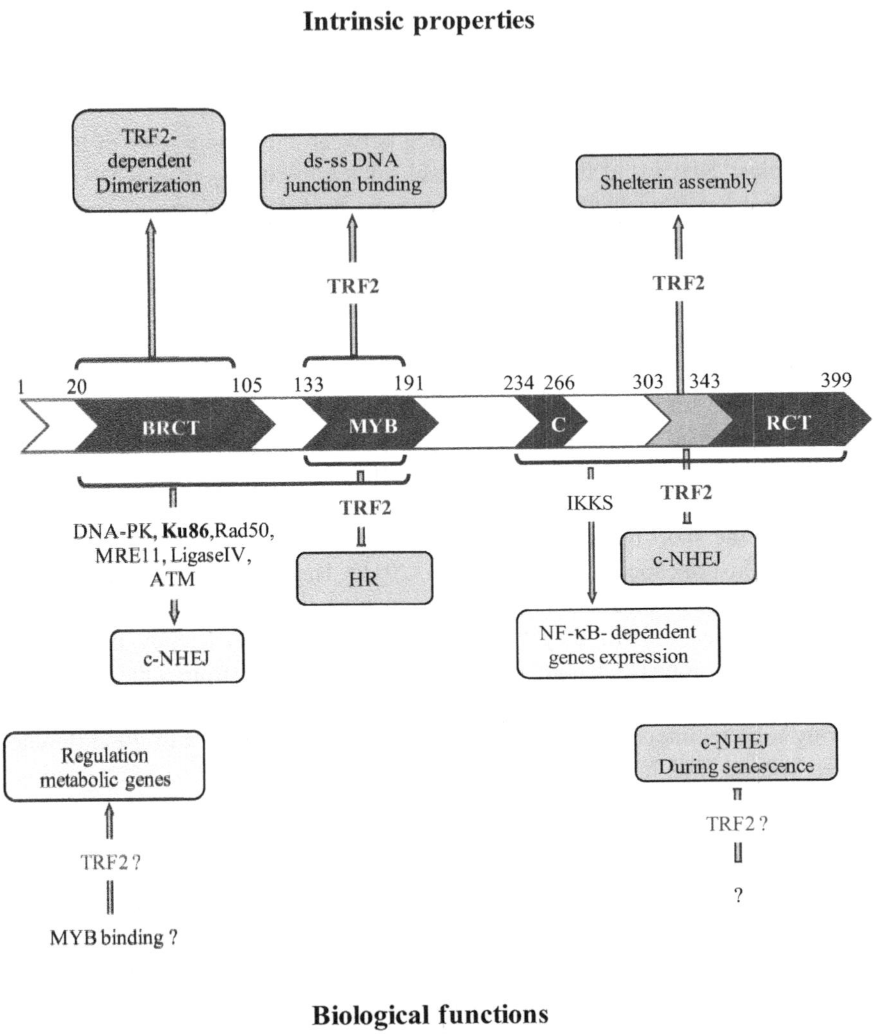

Figure 3.5. *Intrinsic properties and biological functions of RAP1. The four domains of RAP1 are labelled BRCT (BRCA1 C Terminal domain), MYB (Myb domain), C (Coil domain) and RCT (RAP1 C Terminal domain). Orange boxes, green boxes and yellow boxes correspond to intrinsic properties, telomeric functions and extra-telomeric functions respectively. Arrows indicate the domains involved for each property or function and the shape of the arrow stimulatory or inhibiting effect of RAP1. When known, RAP1 partners for the corresponding property or function are noted on the arrow in bold for direct interaction or plain text if indirect. Shelterin partners are labelled in red. Question marks signal an unknown partner or domain involved*

3.3. TIN2

TIN2 (UNIPROT-Q9BSI4) was the third shelterin member to be identified, because of a yeast two-hybrid screen using TRF1 as bait and performed in Judith Campisi's laboratory in the late nineties (Kim et al. 1999) (Figure 3.6). Encoded by the *TINF2* gene (chromosome 14, ENSG00000092330), the TIN2 protein exists as three isoforms in human cells (TIN2S, TIN2L and TIN2M) (Kaminker et al. 2009; Pike et al. 2019) that are expressed roughly at similar level, at least in 293T cells (Pike et al. 2019). Compared to TIN2S, TIN2M harbors 62 additional residues at the C-terminus and TIN2L, 97. These isoforms all originate from alternative splicing of *TINF2* and seem to all participate in telomere functions (Pike et al. 2019). Of note, only the longer isoform has been reported in mouse cells. As for TRF1 and TRF2, TIN2 is essential for mouse development, its deletion causing embryonic lethality before day 7.5 (Chiang et al. 2004). It harbors two structural domains present in all isoforms (Figure 3.2). At the N-terminus, a module allows for interaction with TRF2 and TPP1. Surprisingly, this domain is structurally similar to the TRFH domains of TRF1 and TRF2 (Hu et al. 2017). The telomeric Poz1 and Taz1 proteins from *Schizosaccharomyces pombe* also harbor a TRFH domain (Xue et al. 2017), suggesting that this module could be at the root of the evolution tree of the telomeric proteins (Giraud-Panis et al. 2018). Positioned at the center of the protein, the TBM motif mediates interaction with the TRFH domain of TRF1 and, with lower affinity, with the one of TRF2 (Chen et al. 2008). Interactions with TRF1 and TRF2 are crucial to TIN2 loading on telomeres (Frescas and de Lange 2014c; Hu et al. 2017). Juxtaposed to the TBM on its C-terminal side, a sequence called DC cluster contains most of the common mutations found in the TIN2-mediated forms of DC, a telomeropathy characterized by bone marrow failure and abnormal shortening of telomeres amongst several other clinical features (for more information on this, please see Chapter 12). The role of the DC cluster in the etiology of the disease is still elusive, but a link has been found between the C-terminal extension in TIN2L (in the form of a phosphorylation event on S396) and the most commonly mutated residue in DC, R282 (Nelson et al. 2018). The DC cluster also contains a binding site for the heterochromatin protein 1γ (HP1γ), a binding important to establish and/or maintain telomere cohesion during S phase (Canudas et al. 2011). TIN2 is also involved in telomere cohesion through its interaction with the cohesin subunit SA1 (Canudas et al. 2007). It is unknown if this interaction with SA1 and cohesion regulation, as well as HP1γ binding have any bearing on the disease, although DC patients derived cell lines have shown defects in telomere cohesion (Canudas et al. 2011).

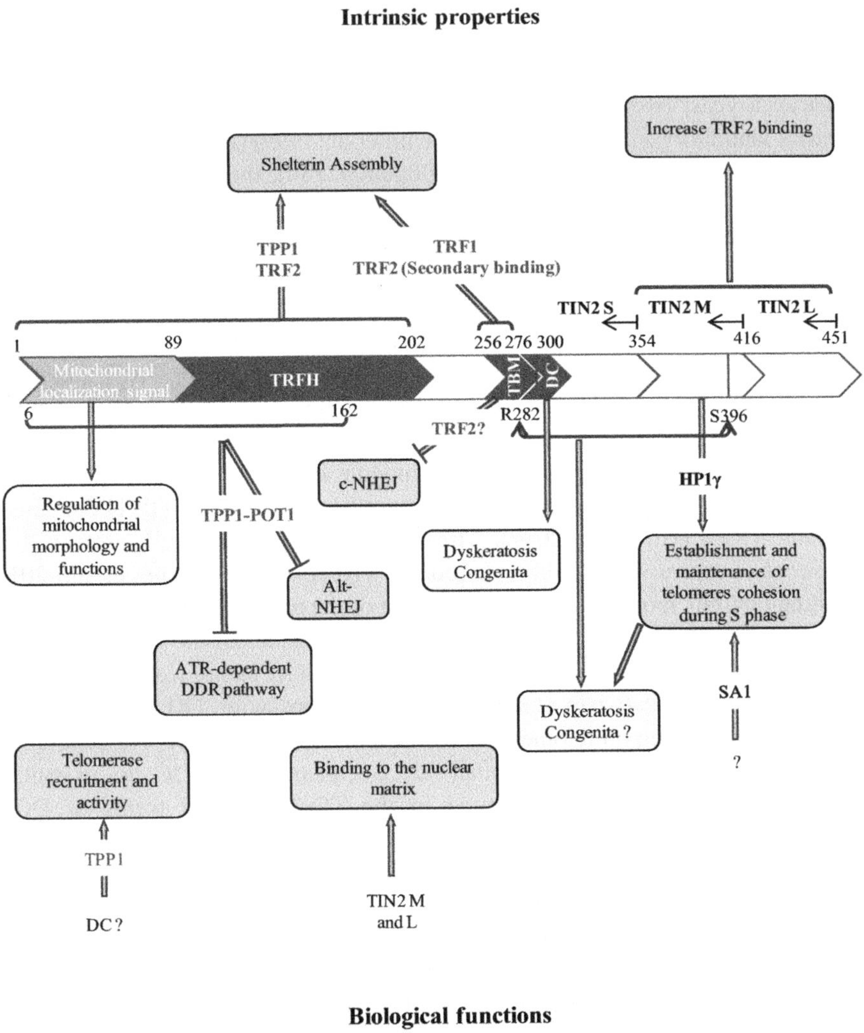

Figure 3.6. *Intrinsic properties and biological functions of TIN2. The three domains of TIN2 are labelled TRFH (TRF Homology domain), TBM (TRFH Binding Motif), DC (Dyskeratosis Congenita hotspot). Orange boxes, green boxes and yellow boxes correspond to intrinsic properties, telomeric functions and extra-telomeric functions respectively. Arrows indicate the domains involved for each property or function and the shape of the arrow stimulatory or inhibiting effect of TIN2. When known, TIN2 partners for the corresponding property or function are noted on the arrow in bold for direct interaction or plain text if indirect. Shelterin partners are labelled in red. Question marks signal an unknown partner or domain involved*

In the shelterin complex, TIN2 has the crucial role to bridge the double strand DNA binders (TRF1 and TRF2-RAP1) to the single strand DNA binding complex TPP1-POT1. Its presence on the telomere is required for the tethering of TPP1-POT1 on telomeres (Takai et al. 2011; Hu et al. 2017). As a result, TIN2 KO in MEFs and HeLa cells causes a marked decrease of telomeric localization of TPP1 and POT1, thus leading to phenotypes that are associated with this loss (Takai et al. 2011; Frescas and de Lange 2014a; Kim et al. 2017a): activation of the ATR-dependent DDR pathway and increase in telomeric fusions by the alternative NHEJ pathway (Alt-NHEJ, see section 3.4). These abnormalities were recapitulated in human and mouse cells by abrogating TIN2-TPP1 interaction (Frescas and de Lange 2014a; Hu et al. 2017), underlining the importance of the TIN2-TPP1 bridge in telomeres functions. Of note, differences were observed in the outcome of TIN2 KO in human and mouse cells, suggesting that the interdependence of the subunits and their role within the shelterin complex was somewhat different between these two species (Kim et al. 2017a). Linked to the TIN2-TPP1 interaction, TIN2 participates in telomerase recruitment through TPP1 (Abreu et al. 2010) and increases telomerase activity in a biochemical assay based on coimmunoprecipitated complexes (Pike et al. 2019). However, it is still debated whether this involvement of TIN2 in telomerase recruitment and activity is at the origin of DC. Indeed, the most common TIN2 mutations in DC did not affect TIN2-mediated telomerase stimulation and the telomere shortening observed in a DC mouse model was independent of the presence of telomerase (Frescas and de Lange 2014b). Conversely, a reduction of the frequency and extension rate of telomerase activity was observed in HCT116 in the context of the DC TIN2-R282H mutant (Frank et al. 2015). More work on DC and on the role of the different isoforms of TIN2 should shed some light on this controversial topic. Indeed, as mentioned above, a functional link has been proposed between the C-terminal extension found in TIN2L and the DC cluster (Nelson et al. 2018), an extension that is also responsible for increasing TIN2 interactions with TRF2 (Nelson et al. 2018) and for binding of TIN2 to the nuclear matrix (Kaminker et al. 2009).

In addition to its binding to TPP1, TIN2 interaction with TRF1 and TRF2 is essential for shelterin assembly. Indeed, TIN2 stabilizes both proteins on telomeres (Ye and de Lange 2004; Ye et al. 2004b; Zeng et al. 2010). Interestingly, it could also regulate TRF1 nuclear localization (Bhanot and Smith 2012). Hence, TIN2 deletion and point mutations that abrogate its interaction with TRF1 and TRF2 cause complex phenotypes such as robust DDR activation and telomeric fusions though c- and Alt-NHEJ (Frescas and de Lange 2014c; Hu et al. 2017; Kim et al. 2017a). Surprisingly, mutations in TIN2 TBM show that abrogating its link with TRF1 in mouse cells does not lead to a TRF1 KO-like or TPP1-POT1 deletion phenotypes but rather causes defects similar to those observed when deleting TRF2 (activation of ATM and CHK2, fusions by c-NHEJ) (Frescas and de Lange 2014c; Hu et al. 2017). This apparent conundrum can perhaps be explained by the loss of the weak interaction between TIN2 TBM (mutated in these studies) and TRF2 TRFH that has been observed in vitro (Chen et al. 2008) and which partially rescues the effects of mutations in TIN2 TRFH targeting its stronger link with TRF2 (Hu et al. 2017). The role of this secondary interaction and its consequence on the binding of the other TRF2 TRFH interactors would be interesting to investigate in the future. Finally, another aspect of TIN2 biological function that is still an opened question to this day is its localization to mitochondria which has been observed by Songyang et al. (2012). Thanks to a mitochondrial localization signal located in the first 89 residues of the protein, TIN2 shuttles to the mitochondria, a shuttling inhibited by TPP1 binding on TIN2, where it regulates mitochondrial morphology and function through modulation of mitochondrial oxidative phosphorylation and regulation of metabolic pathways (Chen et al. 2012; Kim et al. 2017a).

3.4. TPP1-POT1

Although discovered separately, TPP1 and POT1 form a heterodimeric complex that is now considered as one functional unit because both of them are strictly dependent on one another and the similarities they present to the TEBPαβ complex that protects telomeres of several ciliates (Baumann et al. 2002) (Figure 3.7). Human POT1 (UniProtKB-Q9NUX5) was identified in early 2000 in Tom Cech's laboratory as a putative telomeric ss DNA binding protein because of weak sequence similarities with the Pot1 protein from *Schizosaccharomyces pombe* and with the TEBPα protein from the ciliate *Oxytricha nova* (Baumann and Cech 2001). Of note, there are two Pot1 proteins in mice (Pot1a and Pot1b) with distinct roles in telomere (discussed later in the text). hPOT1 is encoded by the *Pot1* gene (chromosome 7, ENSG00000128513). It is also known for its several variants identified in various types of cancer such as melanomas, gliomas or colorectal cancer (Gong et al. 2020).

hTPP1 (UniProtKB-Q96AP0), briefly named PTOP/PIP1/TINT1 since its identification was reported by three different groups at the same time in 2004 (Titia de Lange's, Susan Smith's and Zhou Songyang's laboratories) (Colgin and Reddel 2004), is the last shelterin protein discovered. It is encoded by the *ACD* gene for Adrenocortical dysplasia protein homolog (chromosome 16, ENSG00000102977). Two isoforms have been identified in human cells (TPP1-S and TPP1-L, the latter being 86 residues longer at the N-terminus). TPP1-S is the major isoform found in somatic cells, while TPP1-L is predominant in differentiated male germ cells where it correlates with the persistence of telomerase activity (Grill et al. 2019).

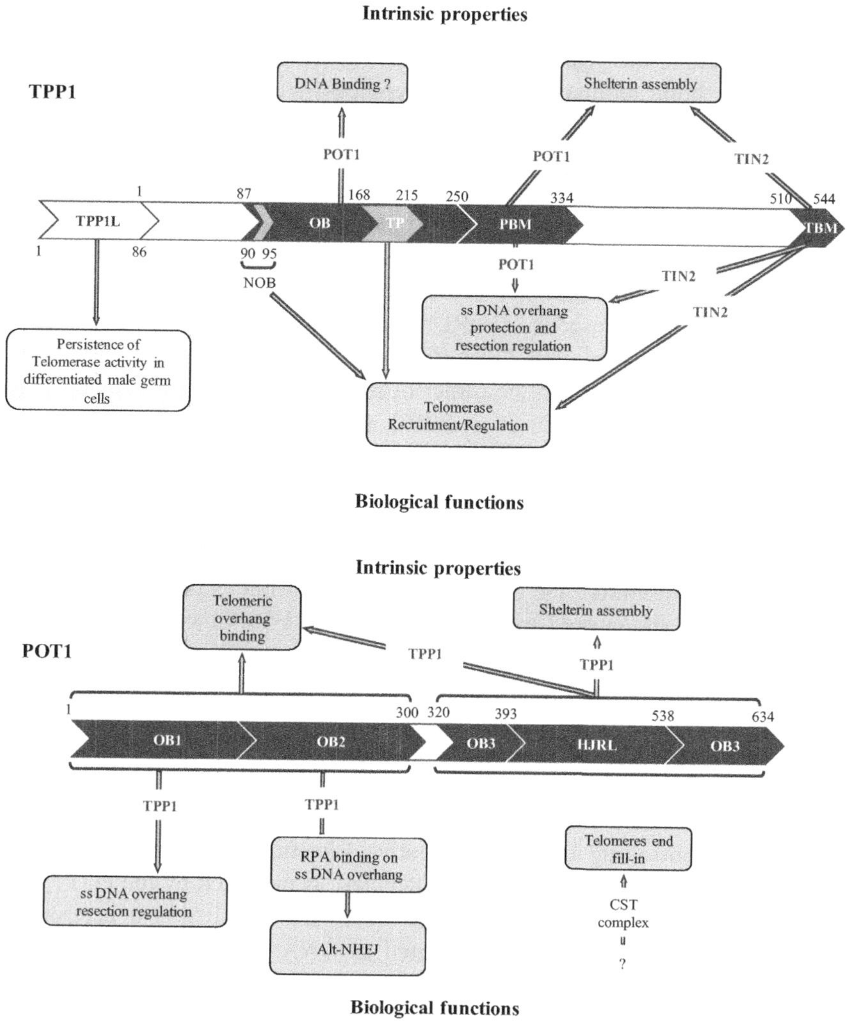

Figure 3.7. *Intrinsic properties and biological functions of TPP1 and POT1. The four domains of TPP1 are labelled NOB (N terminus of the OB domain), OB (Oligosaccharide Binding domain), PBM (POT1 Binding Motif), TBM (TIN2 Binding Motif). The four domains of POT1 are labelled OB1,2 and 3 (Oligosaccharide Binding domain), HJRL (Holliday Junction Resolvase Like domain) and TP (Tel Patch). Orange boxes, green boxes and yellow boxes correspond to intrinsic properties, telomeric functions and extra-telomeric functions respectively. Arrows indicate the domains involved for each property or function and the shape of the arrow stimulatory or inhibiting effects. When known, partners for the corresponding property or function are noted on the arrow in bold for direct interaction or plain text if indirect. Shelterin partners are labelled in red. Question marks signal an unknown partner or domain involved*

The 3D structure of several elements of the heterodimer (Figure 3.2) has been published throughout the years (Lei et al. 2004; Chen et al. 2017; Rice et al. 2017), which showed both similarities and differences with the TEBPαβ complex, particularly in the organization of the domains (Chen et al. 2017). However, a common structural theme is clear, the OB-fold, a versatile conformation that can serve as an ss DNA binding motif as well as a protein–protein interaction hub (Bochkarev and Bochkareva 2004). hPOT1 contains three OB-folds (Figure 3.2): the first two (OB1 and 2) are located at the N-terminus and are used to bind the telomeric overhang; the third one (OB3) at the C-terminus is interrupted by a Holliday Junction Resolvase like (HJRL) motif and both OB3 and HJRL provide residues for the interaction with TPP1 (Chen et al. 2017; Rice et al. 2017). TPP1 contains one OB-fold located in the center of the protein; prediction models place it in position to bind DNA (Chen et al. 2017), although no DNA binding has been observed for this protein on its own (Wang et al. 2007; Xin et al. 2007), very likely because it requires POT1 to do so. Hence, TPP1 increases POT1 affinity for ssDNA (Xin et al. 2007). Immediately following this OB-fold lies the POT1 interaction domain (PBM) required for recruiting POT1 to telomeres (Liu et al. 2004; Ye et al. 2004a). At the C-terminus, the TBM domain allows interactions with TIN2 (Hu et al. 2017). Thus, the current model is that the TPP1-POT1 complex is primarily recruited to telomeres via TIN2, while its binding to the 3′ DNA overhang contributes to its positioning and stabilization at telomere ends. The TPP1-POT1 complex has a crucial role in protecting/regulating the dynamics of the ss overhang and in recruiting/regulating telomerase at telomeres.

Protection of the ss overhang is mediated by inhibiting the activation of the ATR kinase and by controlling the post-replicative resection of the ends necessary to reform the overhang. Inhibition of the ATR kinase is mediated by the POT1 subunit by preventing the recognition of the ssDNA by RPA. RPA is a heterotrimeric complex that binds ssDNA when unwanted ss DNA accumulation occurs in the genome. It forms a platform that activates the ATR kinase and engages repair pathways such as HR, BIR or Alt-NHEJ (as opposed to c-NHEJ which relies on Ku, DNA_PKcs and Ligase 4, Alt-NHEJ is based on MRN/CtIP, PARP1 and Ligase 3 among others; Brambati et al. 2020). On telomeres, RPA eviction is mediated through the coating by TPP1-POT1 of the ss DNA present at telomeric ends and, potentially, on the displaced strand of t-loop (Taylor et al. 2011). Although the absolute quantity of the POT-TPP1 complex is far lower than that of RPA, its recruitment by TIN2 is thought to compensate this unfavorable ratio by increasing the local concentration of the POT-TPP1 dimer (Kratz and de Lange 2018). The eviction of RPA impedes ATR activation and thus ultimately inhibits the Alt-NHEJ pathway (Badie et al. 2015). Of note, in mouse cells, both Pot1a and Potb can inhibit

ATR, but this function is mainly promoted by Pot1a, Pot1b is more dedicated to the formation of the ss overhang (He et al. 2006; Palm et al. 2009).

Resection of telomeric ends is an important process crucial for telomere protection as well as length homeostasis. In most mouse and human cells, all telomeres end with a single strand G-rich overhang (McElligott et al. 1997). Of note, this structure is not universal, for example, *C. elegans* telomeres harbor 5′ C- and 3′ G-rich overhangs (Raices et al. 2008), and even in human cells C-rich overhangs have been detected in cancer cells such as ALT cells (Zhang et al. 2019). The overhang is necessary to allow for the formation of the T-loop, thus enabling T-loop mediated telomere protection, and for elongation of telomeres, since it serves as a substrate for telomerase. Although an overhang is produced naturally at the end of replication on the lagging strand, the leading strand replication machinery creates a blunt end that needs to be resected (Lingner et al. 1995). This resection is initiated by the Apollo exonuclease that catalyzes a limited resection and is followed by the action of ExoI and BLM (McElligott et al. 1997; Wu et al. 2012; Kibe et al. 2016). A fill-in step mediated by the CST complex (CC1-STN1-TEN1 complex that also negatively regulates telomerase and assists telomeres replication; Stewart et al. 2018) through interaction with POT1 (Pot1b in mice; Wu et al. 2012) creates the final overhang. POT1 regulates this resection that can lead to a precise determination of the terminal sequence of the ds part of telomeric DNA (CCAATC-5'/ GGTTAG-3'), at least in Hela cells (more variability is observed in BJ fibroblasts) (Sfeir et al. 2005a, 2005b). Thus, in both mice and humans, deleting POT1 (or Pot1a/b) leads to extension of the ss overhang, binding of RPA and fusions by Alt-NHEJ. One notable difference between mice and humans was reported by Songyang et al. (2017a) and concerned the functional redundancy in ss DNA length regulation between natural variants of POT1 linked to cancer that was observed in human cells. Several of these variants were able to protect the telomeric overhang independently of TIN2 and TPP1. The biological significance of this independent pathway is still unknown, but it might have some bearing in POT1-mediated diseases (Gong et al. 2020). Another noticeable difference is a mild activation of the ATM kinase in hPOT1 KO (Kim et al. 2017a) which is perhaps caused by the decrease in TRF2 telomere occupancy observed in this condition. Interestingly, the Alt-NHEJ fusions observed when deleting TPP1-POT1 rely on TRF2 through interaction with the Claspin-PCNA-DONSON replisome complex that recruits ExoI-CtIP, ExoI catalyzing the long resection required for Alt-NHEJ (Rai et al. 2019). Thus, in a rather paradoxical way, TRF2 appears to both inhibit (c-NHEJ) and promote (Alt-NHEJ) fusions.

TPP1 is the subunit that connects the TPP1-POT1 heterodimer to telomerase through direct interaction with dedicated domains: TEL patch and NOB have been described on TPP1 (Nandakumar et al. 2012; Sexton et al. 2014; Grill et al. 2018) and several domains of telomerase (TEN, IFD, CTE) have been put forward (Grill et al. 2018). TPP1-POT1 stimulates the processivity of the telomerase enzyme, with the aid of TIN2 (Pike et al. 2019) and because of its TEL patch domain (Nandakumar et al. 2012; Sexton et al. 2012; Zhong et al. 2012; Schmidt et al. 2014), a stimulatory effect is missing for mouse TPP1 (Zaug et al. 2010). This activity of human TPP1 probably explains the telomere shortening observed in KO of TPP1 (Kim et al. 2017a). It can also explain the severe phenotype caused by a germline mutation in the TEL patch sequence linked to a Hoyeraal–Hreidarsson syndrome (severe form of DC) (Kocak et al. 2004).

3.5. Quaternary structure of the shelterin

Biochemical and structural analyses of several shelterin sub-complexes have revealed a high level of interdependency between several shelterin members. Indeed, the binding of POT1-TPP1 on human or mouse telomeres depends upon the presence of TIN2 (Takai et al. 2011; Frescas and de Lange 2014a; Hu et al. 2017). In vitro and in 293T cells, the binding of TPP1 on TIN2 increases the affinity of TIN2 to TRF2 (Kim et al. 2017b), suggesting that shelterin formation is a hierarchical process. However, there is clear evidence that sub-complexes exist and, judging by the stoichiometry of the different proteins, probably constitute the main complexes on telomeres. Indeed, TPP1-POT1 are roughly 10 times less abundant than TRF1, TRF2, RAP1 and TIN2 in the chromatin bound fraction of telomeric proteins in human cells (Takai et al. 2020). Furthermore, studies on the residence time of TRF1, TRF2 and POT1 in human cells by FRAP show two types of dynamics: TRF1 shares a residence time of ~44 seconds with 70% of the pool of TRF2 proteins. The remaining 30% behave as POT1 with a much longer residence time of ~11 min. This suggests an absence of TRF1 in some complexes containing TRF2. Accordingly, all co-immunoprecipitation studies show several types of complexes besides the six proteins containing complex (Ye et al. 2004b; O'Connor et al. 2006; Giannone et al. 2007; Kim et al. 2008). Moreover, reconstitution performed in Tom R. Cech's laboratory revealed that a $(TRF2\text{-}RAP1)_2$-TIN2-TPP1-POT1 complex can be readily formed in vitro and exhibits preferential binding to ss-ds DNA junctions as would be expected for a terminal, T-loop forming complex (Lim et al. 2017).

3.6. Concluding remarks

We are now approaching the 30 years benchmark since the discovery of the first telomeric protein. Many data have been collected, complexes and sub-complexes have been discovered, their properties unraveled and many of their functions unveiled. Now is perhaps the time to deconstruct the somewhat rigid image that was built around the shelterin and to deepen our knowledge on the crucial notion of sub-complex dynamics. We need also to take a step backwards, as it were, and investigate the overall organization of telomeric chromatin. How the different shelterin subcomplexes, together with TERRA, nucleosomes, CST and shelterin interactors, shape a functional telomere? Also, it is now clear that telomeres are neither autonomous regions of the genome nor are they disconnected to cellular homeostasis. Indeed, changes in telomere state signal a diversity of pathways determining cell fate (Ye et al. 2014). Deciphering these interconnections will have a great impact on our understanding and treatment of age-associated diseases and will probably be the aim of many telomeric studies in the future.

3.7. References

Abreu, E., Aritonovska, E., Reichenbach, P., Cristofari, G., Culp, B., Terns, R.M., Lingner, J., Terns, M.P. (2010). TIN2-tethered TPP1 recruits human telomerase to telomeres in vivo. *Mol. Cell. Biol.*, 30, 2971–2982.

d'Alcontres, M.S., Palacios, J.A., Mejias, D., Blasco, M.A. (2014). TopoIIalpha prevents telomere fragility and formation of ultra thin DNA bridges during mitosis through TRF1-dependent binding to telomeres. *Cell Cycle*, 13, 1463–1481.

Amiard, S., Doudeau, M., Pinte, S., Poulet, A., Lenain, C., Faivre-Moskalenko, C., Angelov, D., Hug, N., Vindigni, A., Bouvet, P. et al. (2007). A topological mechanism for TRF2-enhanced strand invasion. *Nature Structural & Molecular Biology*, 14, 147–154.

Ancelin, K., Brunori, M., Bauwens, S., Koering, C.E., Brun, C., Ricoul, M., Pommier, J.P., Sabatier, L., Gilson, E. (2002). Targeting assay to study the cis functions of human telomeric proteins: Evidence for inhibition of telomerase by TRF1 and for activation of telomere degradation by TRF2. *Mol. Cell. Biol.*, 22, 3474–3487.

Arat, N.O. and Griffith, J.D. (2012). Human Rap1 interacts directly with telomeric DNA and regulates TRF2 localization at the telomere. *J. Biol. Chem.*, 287, 41583–41594.

Atanasiu, C., Deng, Z., Wiedmer, A., Norseen, J., Lieberman, P.M. (2006). ORC binding to TRF2 stimulates OriP replication. *EMBO Rep.*, 7, 716–721.

Badie, S., Carlos, A.R., Folio, C., Okamoto, K., Bouwman, P., Jonkers, J., Tarsounas, M. (2015). BRCA1 and CtIP promote alternative non-homologous end-joining at uncapped telomeres. *EMBO J.*, 34, 410–424.

Bae, N.S. and Baumann, P.A. (2007). RAP1/TRF2 complex inhibits nonhomologous end-joining at human telomeric DNA ends. *Mol. Cell.*, 26, 323–334.

Baker, A.M., Fu, Q., Hayward, W., Victoria, S., Pedroso, I.M., Lindsay, S.M., Fletcher, T.M. (2011). The telomere binding protein TRF2 induces chromatin compaction. *PLoS One*, 6, e19124.

Bandaria, J.N., Qin, P., Berk, V., Chu, S., Yildiz, A. (2016). Shelterin protects chromosome ends by compacting telomeric chromatin. *Cell*, 164, 735–746.

Baumann, P. and Cech, T.R. (2001). Pot1, the putative telomere end-binding protein in fission yeast and humans. *Science*, 292, 1171–1175.

Baumann, P., Podell, E., Cech, T.R. (2002). Human Pot1 (protection of telomeres) protein: Cytolocalization, gene structure, and alternative splicing. *Mol. Cell. Biol.*, 22, 8079–8087.

Bejarano, L., Schuhmacher, A.J., Méndez, M., Megías, D., Blanco-Aparicio, C., Martínez, S., Pastor, J., Squatrito, M., Blasco, M.A. (2017). Inhibition of TRF1 telomere protein impairs tumor initiation and progression in glioblastoma mouse models and patient-derived xenografts. *Cancer Cell*, 32, 590–607, e594.

Bejarano, L., Louzame, J., Montero, J.J., Megías, D., Flores, J.M., Blasco, M.A. (2019). Safety of whole-body abrogation of the TRF1 shelterin protein in wild-type and cancer-prone mouse models. *iScience*, 19, 572–585.

Benarroch-Popivker, D., Pisano, S., Mendez-Bermudez, A., Lototska, L., Kaur, P., Bauwens, S., Djerbi, N., Latrick, C.M., Fraisier, V., Pei, B. et al. (2016). TRF2-mediated control of telomere DNA topology as a mechanism for chromosome-end protection. *Mol. Cell.*, 61, 274–286.

Bhanot, M. and Smith, S. (2012). TIN2 stability is regulated by the E3 ligase Siah2. *Mol. Cell. Biol.*, 32, 376–384.

Bianchi, A., Smith, S., Chong, L., Elias, P., de Lange, T. (1997). TRF1 is a dimer and bends telomeric DNA. *EMBO J.*, 16, 1785–1794.

Bianchi, A., Stansel, R.M., Fairall, L., Griffith, J.D., Rhodes, D., de Lange, T. (1999). TRF1 binds a bipartite telomeric site with extreme spatial flexibility. *EMBO J.*, 18, 5735–5744.

Bilaud, T., Koering, C.E., Binet-Brasselet, E., Ancelin, K., Pollice, A., Gasser, S.M., Gilson, E. (1996). The telobox, a Myb-related telomeric DNA binding motif found in proteins from yeast, plants and human. *Nucleic Acids Res.*, 24, 1294–1303.

Bilaud, T., Brun, C., Ancelin, K., Koering, C.E., Laroche, T., Gilson, E. (1997). Telomeric localization of TRF2, a novel human telobox protein. *Nat. Genet.*, 17, 236–239.

Biroccio, A., Cherfils-Vicini, J., Augereau, A., Pinte, S., Bauwens, S., Ye, J., Simonet, T., Horard, B., Jamet, K., Cervera, L. et al. (2013). TRF2 inhibits a cell-extrinsic pathway through which natural killer cells eliminate cancer cells. *Nat. Cell. Biol.*, 15, 818–828.

Bochkarev, A. and Bochkareva, E. (2004). From RPA to BRCA2: Lessons from single-stranded DNA binding by the OB-fold. *Curr. Opin. Struct. Biol.*, 14, 36–42.

Bombarde, O., Boby, C., Gomez, D., Frit, P., Giraud-Panis, M.J., Gilson, E., Salles, B., Calsou, P. (2010). TRF2/RAP1 and DNA-PK mediate a double protection against joining at telomeric ends. *EMBO J.*, 29, 1573–1584.

Bosco, N. and de Lange, T. (2012). TRF1-controlled common fragile site containing interstitial telomeric sequences. *Chromosoma*, 121, 465–474.

Boskovic, J., Martinez-Gago, J., Mendez-Pertuz, M., Buscato, A., Martinez-Torrecuadrada, J.L., Blasco, M.A. (2016). Molecular architecture of full-length TRF1 favors its interaction with DNA. *J. Biol. Chem.*, 291, 21829–21835.

Bower, B.D. and Griffith, J.D. (2014). TRF1 and TRF2 differentially modulate Rad51-mediated telomeric and nontelomeric displacement loop formation in vitro. *Biochemistry*, 53, 5485–5495.

Brambati, A., Barry, R.M., Sfeir, A.J. (2020). DNA polymerase theta (Poltheta) – An error-prone polymerase necessary for genome stability. *Curr. Opin. Genet. Dev.*, 60, 119–126.

Broccoli, D., Smogorzewska, A., Chong, L., de Lange, T. (1997). Human telomeres contain two distinct Myb-related proteins, TRF1 and TRF2. *Nat. Genet.*, 17, 231–235.

Buscemi, G., Zannini, L., Fontanella, E., Lecis, D., Lisanti, S., Delia, D. (2009). The shelterin protein TRF2 inhibits Chk2 activity at telomeres in the absence of DNA damage. *Current Biology: CB*, 19, 874–879.

Canudas, S., Houghtaling, B.R., Kim, J.Y., Dynek, J.N., Chang, W.G., Smith, S. (2007). Protein requirements for sister telomere association in human cells. *EMBO J.*, 26, 4867–4878.

Canudas, S., Houghtaling, B.R., Bhanot, M., Sasa, G., Savage, S.A., Bertuch, A.A., Smith, S. (2011). A role for heterochromatin protein 1γ at human telomeres. *Genes Dev.*, 25, 1807–1819.

Celli, G.B. and de Lange, T. (2005). DNA processing is not required for ATM-mediated telomere damage response after TRF2 deletion. *Nat. Cell. Biol.*, 7, 712–718.

Celli, G.B., Denchi, E.L., de Lange, T. (2006). Ku70 stimulates fusion of dysfunctional telomeres yet protects chromosome ends from homologous recombination. *Nat. Cell. Biol.*, 8, 885–890.

Chen, Y., Yang, Y., van Overbeek, M., Donigian, J.R., Baciu, P., de Lange, T., Lei, M. (2008). A shared docking motif in TRF1 and TRF2 used for differential recruitment of telomeric proteins. *Science*, 319, 1092–1096.

Chen, Y., Rai, R., Zhou, Z.R., Kanoh, J., Ribeyre, C., Yang, Y., Zheng, H., Damay, P., Wang, F., Tsujii, H. et al. (2011). A conserved motif within RAP1 has diversified roles in telomere protection and regulation in different organisms. *Nature Structural & Molecular Biology*, 18, 213–221.

Chen, L.Y., Zhang, Y., Zhang, Q., Li, H., Luo, Z., Fang, H., Kim, S.H., Qin, L., Yotnda, P., Xu, J. et al. (2012). Mitochondrial localization of telomeric protein TIN2 links telomere regulation to metabolic control. *Mol. Cell*, 47, 839–850.

Chen, L., Zhu, X., Zou, Y., Xing, J., Gilson, E., Lu, Y., Ye, J. (2015). The topoisomerase II catalytic inhibitor ICRF-193 preferentially targets telomeres that are capped by TRF2. *Am. J. Physiol. Cell Physiol.*, 308, C372–C377.

Chen, C., Gu, P., Wu, J., Chen, X., Niu, S., Sun, H., Wu, L., Li, N., Peng, J., Shi, S. et al. (2017). Structural insights into POT1-TPP1 interaction and POT1 C-terminal mutations in human cancer. *Nat. Commun.*, 8, 14929.

Cherfils-Vicini, J., Iltis, C., Cervera, L., Pisano, S., Croce, O., Sadouni, N., Győrffy, B., Collet, R., Renault, V.M., Rey-Millet, M. et al. (2019). Cancer cells induce immune escape via glycocalyx changes controlled by the telomeric protein TRF2. *EMBO J.*, 38, e100012.

Chiang, Y.J., Kim, S.H., Tessarollo, L., Campisi, J., Hodes, R.J. (2004). Telomere-associated protein TIN2 is essential for early embryonic development through a telomerase-independent pathway. *Mol. Cell. Biol.*, 24, 6631–6634.

Chong, L., van Steensel, B., Broccoli, D., Erdjument-Bromage, H., Hanish, J., Tempst, P., de Lange, T. (1995). A human telomeric protein. *Science*, 270, 1663–1667.

Clouaire, T., Marnef, A., Legube, G. (2017). Taming tricky DSBs: ATM on duty. *DNA Repair*, 56, 84–91.

Colgin, L. and Reddel, R. (2004). Telomere biology: A new player in the end zone. *Current Biology: CB*, 14, R901–R902.

Deng, Z., Lezina, L., Chen, C.J., Shtivelband, S., So, W., Lieberman, P.M. (2002). Telomeric proteins regulate episomal maintenance of Epstein-Barr virus origin of plasmid replication. *Mol. Cell*, 9, 493–503.

Deng, Z., Norseen, J., Wiedmer, A., Riethman, H., Lieberman, P.M. (2009). TERRA RNA binding to TRF2 facilitates heterochromatin formation and ORC recruitment at telomeres. *Mol. Cell*, 35, 403–413.

Deng, Z., Glousker, G., Molczan, A., Fox, A.J., Lamm, N., Dheekollu, J., Weizman, O.E., Schertzer, M., Wang, Z., Vladimirova, O. et al. (2013). Inherited mutations in the helicase RTEL1 cause telomere dysfunction and Hoyeraal-Hreidarsson syndrome. *Proc. Natl. Acad. Sci. USA*, 110, E3408–E3416.

Diala, I., Wagner, N., Magdinier, F., Shkreli, M., Sirakov, M., Bauwens, S., Schluth-Bolard, C., Simonet, T., Renault, V.M., Ye, J. et al. (2013). Telomere protection and TRF2 expression are enhanced by the canonical Wnt signalling pathway. *EMBO Rep.*, 14, 356–363.

Ding, H., Schertzer, M., Wu, X., Gertsenstein, M., Selig, S., Kammori, M., Pourvali, R., Poon, S., Vulto, I., Chavez, E. et al. (2004). Regulation of murine telomere length by Rtel: An essential gene encoding a helicase-like protein. *Cell*, 117, 873–886.

Doksani, Y., Wu, J.Y., de Lange, T., Zhuang, X. (2013). Super-resolution fluorescence imaging of telomeres reveals TRF2-dependent T-loop formation. *Cell*, 155, 345–356.

Drosopoulos, W.C., Kosiyatrakul, S.T., Schildkraut, C.L. (2015). BLM helicase facilitates telomere replication during leading strand synthesis of telomeres. *The Journal of Cell Biology*, 210, 191–208.

Dunce, J.M., Milburn, A.E., Gurusaran, M., da Cruz, I., Sen, L.T., Benavente, R., Davies, O.R. (2018). Structural basis of meiotic telomere attachment to the nuclear envelope by MAJIN-TERB2-TERB1. *Nat. Commun.*, 9, 5355.

Dungrawala, H., Rose, K.L., Bhat, K.P., Mohni, K.N., Glick, G.G., Couch, F.B., Cortez, D. (2015). The replication checkpoint prevents two types of fork collapse without regulating replisome stability. *Mol. Cell*, 59, 998–1010.

El Maï, M., Wagner, K.D., Michiels, J.F., Ambrosetti, D., Borderie, A., Destree, S., Renault, V., Djerbi, N., Giraud-Panis, M.J., Gilson, E. et al. (2014). The telomeric protein TRF2 regulates angiogenesis by binding and activating the PDGFRbeta promoter. *Cell Reports*, 9, 1047–1060.

Fairall, L., Chapman, L., Moss, H., de Lange, T., Rhodes, D. (2001). Structure of the TRFH dimerization domain of the human telomeric proteins TRF1 and TRF2. *Mol. Cell.*, 8, 351–361.

Fouché, N., Cesare, A.J., Willcox, S., Ozgür, S., Compton, S.A., Griffith, J.D. (2006). The basic domain of TRF2 directs binding to DNA junctions irrespective of the presence of TTAGGG repeats. *J. Biol. Chem.*, 281, 37486–37495.

Frank, A.K., Tran, D.C., Qu, R.W., Stohr, B.A., Segal, D.J., Xu, L. (2015). The shelterin TIN2 subunit mediates recruitment of telomerase to telomeres. *PLoS Genet.*, 11, e1005410.

Frescas, D. and de Lange, T. (2014a). Binding of TPP1 protein to TIN2 protein is required for POT1a,b protein-mediated telomere protection. *J. Biol. Chem.*, 289, 24180–24187.

Frescas, D. and de Lange, T. (2014b). A TIN2 dyskeratosis congenita mutation causes telomerase-independent telomere shortening in mice. *Genes Dev.*, 28, 153–166.

Frescas, D. and de Lange, T. (2014c). TRF2-tethered TIN2 can mediate telomere protection by TPP1/POT1. *Mol. Cell. Biol.*, 34, 1349–1362.

Fujita, K., Horikawa, I., Mondal, A.M., Jenkins, L.M., Appella, E., Vojtesek, B., Bourdon, J.C., Lane, D.P., Harris, C.C. (2010). Positive feedback between p53 and TRF2 during telomere-damage signalling and cellular senescence. *Nat. Cell Biol.*, 12, 1205–1212.

Galati, A., Micheli, E., Alicata, C., Ingegnere, T., Cicconi, A., Pusch, M.C., Giraud-Panis, M.J., Gilson, E., Cacchione, S. (2015). TRF1 and TRF2 binding to telomeres is modulated by nucleosomal organization. *Nucleic Acids Res.*, 43, 5824–5837.

Gaullier, G., Miron, S., Pisano, S., Buisson, R., Le Bihan, Y.V., Tellier-Lebègue, C., Messaoud, W., Roblin, P., Guimarães, B.G., Thai, R. et al. (2016). A higher-order entity formed by the flexible assembly of RAP1 with TRF2. *Nucleic Acids Res.*, 44, 1962–1976.

Ghosh, A.K., Rossi, M.L., Singh, D.K., Dunn, C., Ramamoorthy, M., Croteau, D.L., Liu, Y., Bohr, V.L. (2012). RECQL4, the protein mutated in Rothmund-Thomson syndrome, functions in telomere maintenance. *J. Biol. Chem.*, 287, 196–209.

Giannone, R.J., McDonald, W.H., Hurst, G.B., Huang, Y., Wu, J., Liu, Y., Wang, Y. (2007). Dual-tagging system for the affinity purification of mammalian protein complexes. *Biotechniques*, 43, 296, 298, 300 passim.

Giraud-Panis, M.J., Pisano, S., Benarroch-Popivker, D., Pei, B., Le Du, M.H., Gilson, E. (2013). One identity or more for telomeres? *Frontiers in Oncology*, 3, 48.

Giraud-Panis, M.J., Ye, J., Gilson, E. (2018). TRFH domain: At the root of telomere protein evolution? *Cell Research*, 28, 7–8.

Gong, Y., Stock, A.J., Liu, Y. (2020). The enigma of excessively long telomeres in cancer: Lessons learned from rare human POT1 variants. *Curr. Opin. Genet. Dev.*, 60, 48–55.

Griffith, J.D., Comeau, L., Rosenfield, S., Stansel, R.M., Bianchi, A., Moss, H., de Lange, T. (1999). Mammalian telomeres end in a large duplex loop. *Cell*, 97, 503–514.

Grill, S., Tesmer, V.M., Nandakumar, J. (2018). The N terminus of the OB domain of telomere protein TPP1 Is critical for telomerase action. *Cell Reports*, 22, 1132–1140.

Grill, S., Bisht, K., Tesmer, V.M., Shami, A.N., Hammoud, S.S., Nandakumar, J. (2019). Two separation-of-function isoforms of human TPP1 dictate telomerase regulation in somatic and germ cells. *Cell Reports*, 27, 3511–3521, e3517.

Guervilly, J.H. and Gaillard, P.H. (2018). SLX4: Multitasking to maintain genome stability. *Critical Reviews in Biochemistry and Molecular Biology*, 53, 475–514.

Hanaoka, S., Nagadoi, A., Yoshimura, S., Aimoto, S., Li, B., de Lange, T., Nishimura, Y. (2001). NMR structure of the hRap1 Myb motif reveals a canonical three-helix bundle lacking the positive surface charge typical of Myb DNA-binding domains. *J. Mol. Biol.*, 312, 167–175.

He, H., Multani, A.S., Cosme-Blanco, W., Tahara, H., Ma, J., Pathak, S., Deng, Y., Chang, S. (2006). POT1b protects telomeres from end-to-end chromosomal fusions and aberrant homologous recombination. *EMBO J.*, 25, 5180–5190.

Hockemeyer, D., Daniels, J.P., Takai, H., de Lange, T. (2006). Recent expansion of the telomeric complex in rodents: Two distinct POT1 proteins protect mouse telomeres. *Cell*, 126, 63–77.

Hu, C., Rai, R., Huang, C., Broton, C., Long, J., Xu, Y., Xue, J., Lei, M., Chang, S., Chen, Y. (2017). Structural and functional analyses of the mammalian TIN2-TPP1-TRF2 telomeric complex. *Cell Research*, 27, 1485–1502.

Huda, N., Abe, S., Gu, L., Mendonca, M.S., Mohanty, S., Gilley, D. (2012). Recruitment of TRF2 to laser-induced DNA damage sites. *Free Radic. Biol. Med.*, 53, 1192–1197.

Hussain, T., Saha, D., Purohit, G., Kar, A., Kishore Mukherjee, A., Sharma, S., Sengupta, S., Dhapola, P., Maji, B., Vedagopuram, S. et al. (2017). Transcription regulation of CDKN1A (p21/CIP1/WAF1) by TRF2 is epigenetically controlled through the REST repressor complex. *Sci. Rep.*, 7, 11541.

Jiang, W.Q., Zhong, Z.H., Henson, J.D., Reddel, R.R. (2007). Identification of candidate alternative lengthening of telomeres genes by methionine restriction and RNA interference. *Oncogene*, 26, 4635–4647.

Kaminker, P.G., Kim, S.H., Desprez, P.Y., Campisi, J. (2009). A novel form of the telomere-associated protein TIN2 localizes to the nuclear matrix. *Cell Cycle*, 8, 931–939.

Karlseder, J., Broccoli, D., Dai, Y., Hardy, S., de Lange, T. (1999). p53- and ATM-dependent apoptosis induced by telomeres lacking TRF2. *Science*, 283, 1321–1325.

Karlseder, J., Kachatrian, L., Takai, H., Mercer, K., Hingorani, S., Jacks, T., de Lange, T. (2003). Targeted deletion reveals an essential function for the telomere length regulator Trf1. *Mol. Cell. Biol.,* 23, 6533–6541.

Karlseder, J., Hoke, K., Mirzoeva, O.K., Bakkenist, C., Kastan, M.B., Petrini, J.H., de Lange, T. (2004). The telomeric protein TRF2 binds the ATM kinase and can inhibit the ATM-dependent DNA damage response. *PLoS Biol.*, 2, E240.

Kaur, P., Wu, D., Lin, J., Countryman, P., Bradford, K.C., Erie, D.A., Riehn, R., Opresko, P.L., Wang, H. (2016). Enhanced electrostatic force microscopy reveals higher-order DNA looping mediated by the telomeric protein TRF2. *Sci Rep.*, 6, 20513.

Khattar, E., Maung, K.Z.Y., Chew, C.L., Ghosh, A., Mok, M.M.H., Lee, P., Zhang, J., Chor, W.H.J., Cildir, G., Wang, C.Q. et al. (2019). Rap1 regulates hematopoietic stem cell survival and affects oncogenesis and response to chemotherapy. *Nat. Commun.*, 10, 5349.

Kibe, T., Zimmermann, M., de Lange, T. (2016). TPP1 blocks an ATR-mediated resection mechanism at telomeres. *Mol. Cell*, 61, 236–246.

Kim, S.H., Kaminker, P., Campisi, J. (1999). TIN2, a new regulator of telomere length in human cells. *Nat. Genet.*, 23, 405–412.

Kim, S.H., Davalos, A.R., Heo, S.J., Rodier, F., Zou, Y., Beausejour, C., Kaminker, P., Yannone, S.M., Campisi, J. (2008). Telomere dysfunction and cell survival: Roles for distinct TIN2-containing complexes. *The Journal of Cell Biology*, 181, 447–460.

Kim, H., Lee, O.-H., Xin, H., Chen, L.-Y., Qin, J., Chae, H.K., Lin, S.-Y., Safari, A., Liu, D., Songyang, Z. (2009). TRF2 functions as a protein hub and regulates telomere maintenance by recognizing specific peptide motifs. *Nature Structural & Molecular Biology*, 16, 372–379.

Kim, H., Li, F., He, Q., Deng, T., Xu, J., Jin, F., Coarfa, C., Putluri, N., Liu, D., Songyang, Z. (2017a). Systematic analysis of human telomeric dysfunction using inducible telosome/shelterin CRISPR/Cas9 knockout cells. *Cell Discov.*, 3, 17034.

Kim, J.K., Liu, J., Hu, X., Yu, C., Roskamp, K., Sankaran, B., Huang, L., Komives, E.A., Qiao, F. (2017b). Structural basis for shelterin bridge assembly. *Mol. Cell*, 68, 698–714, e695.

Kocak, H., Ballew, B.J., Bisht, K., Eggebeen, R., Hicks, B.D., Suman, S., O'Neil, A., Giri, N., NCI DCEG Cancer Genomics Research Laboratory, NCI DCEG Cancer Sequencing Working Group et al. (2014). Hoyeraal-Hreidarsson syndrome caused by a germline mutation in the TEL patch of the telomere protein TPP1. *Genes Dev.*, 28, 2090–2102.

Kojic, A., Cuadrado, A., De Koninck, M., Giménez-Llorente, D., Rodríguez-Corsino, M., Gómez-López, G., Le Dily, F., Marti-Renom, M.A., Losada, A. (2018). Distinct roles of cohesin-SA1 and cohesin-SA2 in 3D chromosome organization. *Nature Structural & Molecular Biology*, 25, 496–504.

Kratz, K. and de Lange, T. (2018). Protection of telomeres 1 proteins POT1a and POT1b can repress ATR signaling by RPA exclusion, but binding to CST limits ATR repression by POT1b. *J. Biol. Chem.*, 293, 14384–14392.

de Lange, T. (1992). Human telomeres are attached to the nuclear matrix. *EMBO J.*, 11, 717–724.

de Lange, T. (2005). Shelterin: The protein complex that shapes and safeguards human telomeres. *Genes Dev.*, 19, 2100–2110.

de Lange, T. (2018). Shelterin-mediated telomere protection. *Annual Review of Genetics*, 52, 223–247.

Lam, Y.C., Akhter, S., Gu, P., Ye, J., Poulet, A., Giraud-Panis, M.J., Bailey, S.M., Gilson, E., Legerski, R.J., Chang, S. (2010). SNMIB/Apollo protects leading-strand telomeres against NHEJ-mediated repair. *EMBO J.*, 29, 2230–2241.

Lan, J., Zhu, Y., Xu, L., Yu, H., Yu, J., Liu, X., Fu, C., Wang, X., Ke, Y., Huang, H., Dou, Z. (2014). The 68-kDa telomeric repeat binding factor 1 (TRF1)-associated protein (TAP68) interacts with and recruits TRF1 to the spindle pole during mitosis. *J. Biol. Chem.*, 289, 14145–14156.

Lee, Y.W., Arora, R., Wischnewski, H., Azzalin, C.M. (2018). TRF1 participates in chromosome end protection by averting TRF2-dependent telomeric R loops. *Nature Structural & Molecular Biology*, 25, 147–153.

Lei, M., Podell, E.R., Baumann, P., Cech, T.R. (2003). DNA self-recognition in the structure of POT1 bound to telomeric single-stranded DNA. *Nature*, 426, 198–203.

Lei, M., Podell, E.R., Cech, T.R. (2004). Structure of human POT1 bound to telomeric single-stranded DNA provides a model for chromosome end-protection. *Nature Structural & Molecular Biology*, 11, 1223–1229.

Leman, A.R., Dheekollu, J., Deng, Z., Lee, S.W., Das, M.M., Lieberman, P.M., Noguchi, E. (2012). Timeless preserves telomere length by promoting efficient DNA replication through human telomeres. *Cell Cycle*, 11, 2337–2347.

Lenain, C., Bauwens, S., Amiard, S., Brunori, M., Giraud-Panis, M.J., Gilson, E. (2006). The Apollo 5' exonuclease functions together with TRF2 to protect telomeres from DNA repair. *Current Biology: CB*, 16, 1303–1310.

Levis, R.W., Ganesan, R., Houtchens, K., Tolar, L.A., Sheen, F.M. (1993). Transposons in place of telomeric repeats at a Drosophila telomere. *Cell*, 75, 1083–1093.

Li, B., Oestreich, S., de Lange, T. (2000). Identification of human Rap1: Implications for telomere evolution. *Cell*, 101, 471–483.

Li, B., Qiao, R., Wang, Z., Zhou, W., Li, X., Xu, W., Rao, Z. (2016). Crystal structure of a tankyrase 1-telomere repeat factor 1 complex. *Acta Crystallogr. F Struct. Biol. Commun.*, 72, 320–327.

Li, F., Kim, H., Ji, Z., Zhang, T., Chen, B., Ge, Y., Hu, Y., Feng, X., Han, X., Xu, H. et al. (2018). The BUB3-BUB1 complex promotes telomere DNA replication. *Mol. Cell*, 70, 395–407.

Lim, C.J., Zaug, A.J., Kim, H.J., Cech, T.R. (2017). Reconstitution of human shelterin complexes reveals unexpected stoichiometry and dual pathways to enhance telomerase processivity. *Nat. Commun.*, 8, 1075.

Lingner, J., Cooper, J.P., Cech, T.R. (1995). Telomerase and DNA end replication: No longer a lagging strand problem? *Science*, 269, 1533–1534.

Liu, D., Safari, A., O'Connor, M.S., Chan, D.W., Laegeler, A., Qin, J., Songyang, Z. (2004). PTOP interacts with POT1 and regulates its localization to telomeres. *Nat. Cell. Biol.*, 6, 673–680.

Lobanova, A., She, R., Pieraut, S., Clapp, C., Maximov, A., Denchi, E.L. (2017). Different requirements of functional telomeres in neural stem cells and terminally differentiated neurons. *Genes Dev.*, 31, 639–647.

Lototska, L., Yue, J.X., Li, J., Giraud-Panis, M.J., Songyang, Z., Royle, N.J., Liti, G., Ye, J., Gilson, E., Mendez-Bermudez, A. (2020). Human RAP1 specifically protects telomeres of senescent cells from DNA damage. *EMBO Rep.*, e49076.

Ludérus, M.E., van Steensel, B., Chong, L., Sibon, O.C., Cremers, F.F., de Lange, T. (1996). Structure, subnuclear distribution, and nuclear matrix association of the mammalian telomeric complex. *The Journal of Cell Biology*, 135, 867–881.

Machwe, A., Xiao, L., Orren, D.K. (2004). TRF2 recruits the Werner syndrome (WRN) exonuclease for processing of telomeric DNA. *Oncogene*, 23, 149–156.

Marión, R.M., López de Silanes, I., Mosteiro, L., Gamache, B., Abad, M., Guerra, C., Megías, D., Serrano, M., Blasco, M.A. (2017). Common telomere changes during in vivo reprogramming and early stages of tumorigenesis. *Stem Cell Reports*, 8, 460–475.

Marión, R.M., Montero, J.J., López de Silanes, I., Graña-Castro, O., Martínez, P., Schoeftner, S., Palacios-Fábrega, J.A., Blasco, M.A. (2019). TERRA regulate the transcriptional landscape of pluripotent cells through TRF1-dependent recruitment of PRC2. *Elife*, 8, e44656.

Martínez, P., Thanasoula, M., Muñoz, P., Liao, C., Tejera, A., McNees, C., Flores, J.M., Fernández-Capetillo, O., Tarsounas, M., Blasco, M.A. (2009). Increased telomere fragility and fusions resulting from TRF1 deficiency lead to degenerative pathologies and increased cancer in mice. *Genes Dev.*, 23, 2060–2075.

Martinez, P., Thanasoula, M., Carlos, A.R., Gómez-López, G., Tejera, A.M., Schoeftner, S., Dominguez, O., Pisano, D.G., Tarsounas, M., Blasco, M.A. (2010). Mammalian Rap1 controls telomere function and gene expression through binding to telomeric and extratelomeric sites. *Nat. Cell Biol.*, 12, 768–780.

Martínez, P., Gómez-López, G., García, F., Mercken, E., Mitchell, S., Flores, J.M., de Cabo, R., Blasco, M.A. (2013). RAP1 protects from obesity through its extratelomeric role regulating gene expression. *Cell Reports*, 3, 2059–2074.

Martinez, P., Gomez-Lopez, G., Pisano, D.G., Flores, J.M., Blasco, M.A. (2016). A genetic interaction between RAP1 and telomerase reveals an unanticipated role for RAP1 in telomere maintenance. *Aging Cell*, 15, 1113–1125.

Mason, J.M., Das, I., Arlt, M., Patel, N., Kraftson, S., Glover, T.W., Sekiguchi, J.M. (2013). The SNM1B/Apollo DNA nuclease functions in resolution of replication stress and maintenance of common fragile site stability. *Human Molecular Genetics*, 22, 4901–4913.

Mason, J.M., Randall, T.A., Capkova Frydrychova, R. (2016). Telomerase lost? *Chromosoma*, 125, 65–73.

McElligott, R. and Wellinger, R.J. (1997). The terminal DNA structure of mammalian chromosomes. *EMBO J.*, 16, 3705–3714.

Mendez-Bermudez, A., Lototska, L., Bauwens, S., Giraud-Panis, M.J., Croce, O., Jamet, K., Irizar, A., Mowinckel, M., Koundrioukoff, S., Nottet, N. et al. (2018). Genome-wide control of heterochromatin replication by the telomere capping protein TRF2. *Mol. Cell*, 70, 449–461, e445.

Mendez-Bermudez, A., Giraud-Panis, M.J., Ye, J., Gilson, E. (2020). Heterochromatin replication goes hand in hand with telomere protection. *Nature Structural & Molecular Biology*, 27, 313–318.

Méndez-Pertuz, M., Martínez, P., Blanco-Aparicio, C., Gómez-Casero, E., Belen García, A., Martínez-Torrecuadrada, J., Palafox, M., Cortés, J., Serra, V., Pastor, J. et al. (2017). Modulation of telomere protection by the PI3K/AKT pathway. *Nat. Commun.*, 8, 1278.

Muftuoglu, M., Wong, H.K., Imam, S.Z., Wilson 3rd, D.M., Bohr, V.A., Opresko, P.L. (2006). Telomere repeat binding factor 2 interacts with base excision repair proteins and stimulates DNA synthesis by DNA polymerase beta. *Cancer Res.*, 66, 113–124.

Mukherjee, A.K., Sharma, S., Bagri, S., Kutum, R., Kumar, P., Hussain, A., Singh, P., Saha, D., Kar, A., Dash, D. et al. (2019). Telomere repeat-binding factor 2 binds extensively to extra-telomeric G-quadruplexes and regulates the epigenetic status of several gene promoters. *J. Biol. Chem.*, 294, 17709–17722.

Muñoz, P., Blanco, R., de Carcer, G., Schoeftner, S., Benetti, R., Flores, J.M., Malumbres, M., Blasco, M.A. (2009). TRF1 controls telomere length and mitotic fidelity in epithelial homeostasis. *Mol. Cell. Biol.*, 29, 1608–1625.

Nakamura, M., Zhou, X.Z., Kishi, S., Kosugi, I., Tsutsui, Y., Lu, K.P. (2001). A specific interaction between the telomeric protein Pin2/TRF1 and the mitotic spindle. *Current Biology: CB*, 11, 1512–1516.

Nandakumar, J., Bell, C.F., Weidenfeld, I., Zaug, A.J., Leinwand, L.A., Cech, T.R. (2012). The TEL patch of telomere protein TPP1 mediates telomerase recruitment and processivity. *Nature*, 492, 285–289.

Neidle, S. (2016). Quadruplex nucleic acids as novel therapeutic targets. *J. Med. Chem.*, 59, 5987–6011.

Nelson, N.D., Dodson, L.M., Escudero, L., Sukumar, A.T., Williams, C.L., Mihalek, I., Baldan, A., Baird, D.M., Bertuch, A.A. (2018). The C-Terminal extension unique to the long isoform of the shelterin component TIN2 enhances its interaction with TRF2 in a phosphorylation- and dyskeratosis congenita cluster-dependent fashion. *Mol. Cell. Biol.*, 38, 18–34.

Nijjar, T., Bassett, E., Garbe, J., Takenaka, Y., Stampfer, M.R., Gilley, D., Yaswen, P. (2005). Accumulation and altered localization of telomere-associated protein TRF2 in immortally transformed and tumor-derived human breast cells. *Oncogene*, 24, 3369–3376.

Nishikawa, T., Nagadoi, A., Yoshimura, S., Aimoto, S., Nishimura, Y. (1998). Solution structure of the DNA-binding domain of human telomeric protein, hTRF1. *Structure*, 6, 1057–1065.

Nishikawa, T., Okamura, H., Nagadoi, A., König, P., Rhodes, D., Nishimura, Y. (2001). Solution structure of a telomeric DNA complex of human TRF1. *Structure*, 9, 1237–1251.

Nora, G.J., Buncher, N.A., Opresko, P.L. (2010). Telomeric protein TRF2 protects Holliday junctions with telomeric arms from displacement by the Werner syndrome helicase. *Nucleic Acids Res.*, 38, 3984–3998.

O'Connor, M.S., Safari, A., Liu, D., Qin, J., Songyang, Z. (2004). The human Rap1 protein complex and modulation of telomere length. *J. Biol. Chem.*, 279, 28585–28591.

O'Connor, M.S., Safari, A., Xin, H., Liu, D., Songyang, Z. (2006). A critical role for TPP1 and TIN2 interaction in high-order telomeric complex assembly. *Proc. Natl. Acad. Sci. USA*, 103, 11874–11879.

Ohishi, T., Muramatsu, Y., Yoshida, H., Seimiya, H. (2014). TRF1 ensures the centromeric function of Aurora-B and proper chromosome segregation. *Mol. Cell. Biol.*, 34, 2464–2478.

Okamoto, K., Iwano, T., Tachibana, M., Shinkai, Y. (2008). Distinct roles of TRF1 in the regulation of telomere structure and lengthening. *J. Biol. Chem.*, 283, 23981–23988.

Okamoto, K., Bartocci, C., Ouzounov, I., Diedrich, J.K., Yates 3rd, J.R., Denchi, E.L. (2013). A two-step mechanism for TRF2-mediated chromosome-end protection. *Nature*, 494, 502–505.

Opresko, P.L., von Kobbe, C., Laine, J.P., Harrigan, J., Hickson, I.D., Bohr, V.A. (2002). Telomere-binding protein TRF2 binds to and stimulates the Werner and bloom syndrome helicases. *J. Biol. Chem.*, 277, 41110–41119.

van Overbeek, M. and de Lange, T. (2006). Apollo, an artemis-related nuclease, interacts with TRF2 and protects human telomeres in S phase. *Current Biology: CB*, 16, 1295–1302.

Palm, W. and de Lange, T. (2008). How shelterin protects mammalian telomeres. *Annual Review of Genetics*, 42, 301–334.

Palm, W., Hockemeyer, D., Kibe, T., de Lange, T. (2009). Functional dissection of human and mouse POT1 proteins. *Mol. Cell. Biol.*, 29, 471–482.

Pedroso, I.M., Hayward, W., Fletcher, T.M. (2009). The effect of the TRF2 N-terminal and TRFH regions on telomeric G-quadruplex structures. *Nucleic Acids Res.*, 37, 1541–1554.

Pike, A.M., Strong, M.A., Ouyang, J.P.T., Greider, C.W. (2019). TIN2 Functions with TPP1/POT1 To stimulate telomerase processivity. *Mol. Cell. Biol.*, 39, 1–15.

Polonskaya, Z., Benham, C.J., Hearing, J. (2004). Role for a region of helically unstable DNA within the Epstein-Barr virus latent cycle origin of DNA replication oriP in origin function. *Virology*, 328, 282–291.

Poulet, A., Buisson, R., Faivre-Moskalenko, C., Koelblen, M., Amiard, S., Montel, F., Cuesta-Lopez, S., Bornet, O., Guerlesquin, F., Godet, T. et al. (2009). TRF2 promotes, remodels and protects telomeric Holliday junctions. *EMBO J.*, 28, 641–651.

Poulet, A., Pisano, S., Faivre-Moskalenko, C., Pei, B., Tauran, Y., Haftek-Terreau, Z., Brunet, F., Le Bihan, Y.V., Ledu, M.H., Montel, F. et al. (2012). The N-terminal domains of TRF1 and TRF2 regulate their ability to condense telomeric DNA. *Nucleic Acids Res.*, 40, 2566–2576.

Prime, G. and Markie, D. (2005). The telomere repeat binding protein Trf1 interacts with the spindle checkpoint protein Mad1 and Nek2 mitotic kinase. *Cell Cycle*, 4, 121–124.

Purohit, G., Mukherjee, A.K., Sharma, S., Chowdhury, S. (2018). Extratelomeric binding of the telomere binding protein TRF2 at the PCGF3 promoter is G-quadruplex motif-dependent. *Biochemistry*, 57, 2317–2324.

Raffa, G.D., Ciapponi, L., Cenci, G., Gatti, M. (2011). Terminin: A protein complex that mediates epigenetic maintenance of Drosophila telomeres. *Nucleus*, 2, 383–391.

Rai, R., Chen, Y., Lei, M., Chang, S. (2016). TRF2-RAP1 is required to protect telomeres from engaging in homologous recombination-mediated deletions and fusions. *Nat. Commun.*, 7, 10881.

Rai, R., Hu, C., Broton, C., Chen, Y., Lei, M., Chang, S. (2017). NBS1 phosphorylation status dictates repair choice of dysfunctional telomeres. *Mol. Cell*, 65, 801–817, e804.

Rai, R., Gu, P., Broton, C., Kumar-Sinha, C., Chen, Y., Chang, S. (2019). The replisome mediates A-NHEJ repair of telomeres lacking POT1-TPP1 independently of MRN function. *Cell Reports*, 29, 3708–3725, e3705.

Raices, M., Verdun, R.E., Compton, S.A., Haggblom, C.I., Griffith, J.D., Dillin, A., Karlseder, J. (2008). *C. elegans* telomeres contain G-strand and C-strand overhangs that are bound by distinct proteins. *Cell*, 132, 745–757.

Ribes-Zamora, A., Indiviglio, S.M., Mihalek, I., Williams, C.L., Bertuch, A.A. (2013). TRF2 interaction with Ku heterotetramerization interface gives insight into c-NHEJ prevention at human telomeres. *Cell Reports*, 5, 194–206.

Rice, C., Shastrula, P.K., Kossenkov, A.V., Hills, R., Baird, D.M., Showe, L.C., Doukov, T., Janicki, S., Skordalakes, E. (2017). Structural and functional analysis of the human POT1-TPP1 telomeric complex. *Nat. Commun.*, 8, 14928.

Robin, J.D., Jacome Burbano, M.S., Peng, H., Croce, O., Thomas, J.L., Laberthonniere, C., Renault, V., Lototska, L., Pousse, M., Tessier, F. et al. (2020). Mitochondrial function in skeletal myofibers is controlled by a TRF2-SIRT3 axis over lifetime. *Aging Cell*, e13097.

Saharia, A., Guittat, L., Crocker, S., Lim, A., Steffen, M., Kulkarni, S., Stewart, S.A. (2008). Flap endonuclease 1 contributes to telomere stability. *Current Biology: CB*, 18, 496–500.

Saint-Léger, A., Koelblen, M., Civitelli, L., Bah, A., Djerbi, N., Giraud-Panis, M.J., Londoño-Vallejo, A., Ascenzioni, F., Gilson, E. (2014). The basic N-terminal domain of TRF2 limits recombination endonuclease action at human telomeres. *Cell Cycle*, 13, 2469–2474.

Sarek, G., Vannier, J.B., Panier, S., Petrini, J.H.J., Boulton, S.J. (2015). TRF2 recruits RTEL1 to telomeres in S phase to promote t-loop unwinding. *Mol. Cell*, 57, 622–635.

Sarthy, J., Bae, N.S., Scrafford, J., Baumann, P. (2009). Human RAP1 inhibits non-homologous end joining at telomeres. *EMBO J.*, 28, 3390–3399.

Schmidt, J.C., Dalby, A.B., Cech, T.R. (2014). Identification of human TERT elements necessary for telomerase recruitment to telomeres. *Elife*, 3, e03563.

Schmutz, I., Timashev, L., Xie, W., Patel, D.J., de Lange, T. (2017). TRF2 binds branched DNA to safeguard telomere integrity. *Nature Structural & Molecular Biology*, 24, 734–742.

Schneider, R.P., Garrobo, I., Foronda, M., Palacios, J.A., Marión, R.M., Flores, I., Ortega, S., Blasco, M.A. (2013). TRF1 is a stem cell marker and is essential for the generation of induced pluripotent stem cells. *Nat. Commun.*, 4, 1946.

Sexton, A.N., Youmans, D.T., Collins, K. (2012). Specificity requirements for human telomere protein interaction with telomerase holoenzyme. *J. Biol. Chem.*, 287, 34455–34464.

Sexton, A.N., Regalado, S.G., Lai, C.S., Cost, G.J., O'Neil, C.M., Urnov, F.D., Gregory, P.D., Jaenisch, R., Collins, K., Hockemeyer, D. (2014). Genetic and molecular identification of three human TPP1 functions in telomerase action: Recruitment, activation, and homeostasis set point regulation. *Genes Dev.*, 28, 1885–1899.

Sfeir, A.J., Chai, W., Shay, J.W., Wright, W.E. (2005a). Telomere-end processing the terminal nucleotides of human chromosomes. *Mol. Cell*, 18, 131–138.

Sfeir, A.J., Shay, J.W., Wright, W.E. (2005b). Fine-tuning the chromosome ends: The last base of human telomeres. *Cell Cycle*, 4, 1467–1470.

Sfeir, A.J., Kosiyatrakul, S.T., Hockemeyer, D., MacRae, S.L., Karlseder, J., Schildkraut, C.L., de Lange, T. (2009). Mammalian telomeres resemble fragile sites and require TRF1 for efficient replication. *Cell*, 138, 90–103.

Sfeir, A.J., Kabir, S., van Overbeek, M., Celli, G.B., de Lange, T. (2010). Loss of Rap1 induces telomere recombination in the absence of NHEJ or a DNA damage signal. *Science*, 327, 1657–1661.

Shen, M., Haggblom, C., Vogt, M., Hunter, T., Lu, K.P. (1997). Characterization and cell cycle regulation of the related human telomeric proteins Pin2 and TRF1 suggest a role in mitosis. *Proc. Natl. Acad. Sci. USA*, 94, 13618–13623.

Simonet, T., Zaragosi, L.E., Philippe, C., Lebrigand, K., Schouteden, C., Augereau, A., Bauwens, S., Ye, J., Santagostino, M., Giulotto, E. et al. (2011). The human TTAGGG repeat factors 1 and 2 bind to a subset of interstitial telomeric sequences and satellite repeats. *Cell Research*, 21, 1028–1038.

Smogorzewska, A., van Steensel, B., Bianchi, A., Oelmann, S., Schaefer, M.R., Schnapp, G., de Lange, T. (2000). Control of human telomere length by TRF1 and TRF2. *Mol. Cell. Biol.*, 20, 1659–1668.

Stansel, R.M., de Lange, T., Griffith, J.D. (2001). T-loop assembly in vitro involves binding of TRF2 near the 3' telomeric overhang. *EMBO J.*, 20, 5532–5540.

van Steensel, B. and de Lange, T. (1997). Control of telomere length by the human telomeric protein TRF1. *Nature*, 385, 740–743.

van Steensel, B., Smogorzewska, A., de Lange, T. (1998). TRF2 protects human telomeres from end-to-end fusions. *Cell*, 92, 401–413.

Stewart, J.A., Wang, Y., Ackerson, S.M., Schuck, P.L. (2018). Emerging roles of CST in maintaining genome stability and human disease. *Front. Biosci. (Landmark Ed.)*, 23, 1564–1586.

Takai, H., Smogorzewska, A., de Lange, T. (2003). DNA damage foci at dysfunctional telomeres. *Current Biology: CB*, 13, 1549–1556.

Takai, K.K., Hooper, S., Blackwood, S., Gandhi, R., de Lange, T. (2010). In vivo stoichiometry of shelterin components. *J. Biol. Chem.*, 285, 1457–1467.

Takai, K.K., Kibe, T., Donigian, J.R., Frescas, D., de Lange, T. (2011). Telomere protection by TPP1/POT1 requires tethering to TIN2. *Mol. Cell*, 44, 647–659.

Tatsumi, Y., Ezura, K., Yoshida, K., Yugawa, T., Narisawa-Saito, M., Kiyono, T., Ohta, S., Obuse, C., Fujita, M. (2008). Involvement of human ORC and TRF2 in pre-replication complex assembly at telomeres. *Genes to Cells: Devoted to Molecular & Cellular Mechanisms*, 13, 1045–1059.

Taylor, D.J., Podell, E.R., Taatjes, D.J., Cech, T.R. (2011). Multiple POT1-TPP1 proteins coat and compact long telomeric single-stranded DNA. *J. Mol. Biol.*, 410, 10–17.

Teo, H., Ghosh, S., Luesch, H., Ghosh, A., Wong, E.T., Malik, N., Orth, A., de Jesus, P., Perry, A.S., Oliver, J.D. et al. (2010). Telomere-independent Rap1 is an IKK adaptor and regulates NF-kappaB-dependent gene expression. *Nat. Cell Biol.*, 12, 758–767.

Timashev, L.A. and de Lange, T. (2020). Characterization of t-loop formation by TRF2. *Nucleus*, 11, 164–177.

Timashev, L.A., Babcock, H., Zhuang, X., de Lange, T. (2017). The DDR at telomeres lacking intact shelterin does not require substantial chromatin decompaction. *Genes Dev.*, 31, 578–589.

Van Ly, D., Low, R.R.J., Frölich, S., Bartolec, T.K., Kafer, G.R., Pickett, H.A., Gaus, K., Cesare, A.J. (2018). Telomere loop dynamics in chromosome end protection. *Mol. Cell*, 71, 510–525, e516.

Vancevska, A., Douglass, K.M., Pfeiffer, V., Manley, S., Lingner, J. (2017). The telomeric DNA damage response occurs in the absence of chromatin decompaction. *Genes Dev.*, 31, 567–577.

Verdun, R.E., Crabbe, L., Haggblom, C., Karlseder, J. (2005). Functional human telomeres are recognized as DNA damage in G2 of the cell cycle. *Mol. Cell*, 20, 551–561.

Wagner, K.D., Cherfils-Vicini, J., Hosen, N., Hohenstein, P., Gilson, E., Hastie, N.D., Michiels, J.F., Wagner, N. (2014). The Wilms' tumour suppressor Wt1 is a major regulator of tumour angiogenesis and progression. *Nat. Commun.*, 5, 5852.

Wagner, K.D., Ying, Y., Leong, W., Jiang, J., Hu, X., Chen, Y., Michiels, J.F., Lu, Y., Gilson, E., Wagner, N., Ye, J. (2017). The differential spatiotemporal expression pattern of shelterin genes throughout lifespan. *Aging (Albany NY)*, 9, 1219–1232.

Wan, B., Yin, J., Horvath, K., Sarkar, J., Chen, Y., Wu, J., Wan, K., Lu, J., Gu, P., Yu, E.Y. et al. (2013). SLX4 assembles a telomere maintenance toolkit by bridging multiple endonucleases with telomeres. *Cell Reports*, 4, 861–869.

Wang, R.C., Smogorzewska, A., de Lange, T. (2004). Homologous recombination generates T-loop-sized deletions at human telomeres. *Cell*, 119, 355–368.

Wang, F., Podell, E.R., Zaug, A.J., Yang, Y., Baciu, P., Cech, T.R., Lei, M. (2007). The POT1-TPP1 telomere complex is a telomerase processivity factor. *Nature*, 445, 506–510.

Wu, P., van Overbeek, M., Rooney, S., de Lange, T. (2010). Apollo contributes to G-overhang maintenance and protects leading-end telomeres. *Mol. Cell*, 39, 606–617.

Wu, P., Takai, H., de Lange, T. (2012). Telomeric 3' overhangs derive from resection by Exo1 and Apollo and fill-in by POT1b-associated CST. *Cell*, 150, 39–52.

Wu, Q., Paul, A., Su, D., Mehmood, S., Foo, T.K., Ochi, T., Bunting, E.L., Xia, B., Robinson, C.V., Wang, B., Blundell, T.L. (2016). Structure of BRCA1-BRCT/Abraxas complex reveals phosphorylation-dependent BRCT dimerization at DNA damage sites. *Mol. Cell*, 61, 434–448.

Xin, H., Liu, D., Wan, M., Safari, A., Kim, H., Sun, W., O'Connor, M.S., Songyang, Z. (2007). TPP1 is a homologue of ciliate TEBP-beta and interacts with POT1 to recruit telomerase. *Nature*, 445, 559–562.

Xue, J., Chen, H., Wu, J., Takeuchi, M., Inoue, H., Liu, Y., Sun, H., Chen, Y., Kanoh, J., Lei, M. (2017). Structure of the fission yeast S. pombe telomeric Tpz1-Poz1-Rap1 complex. *Cell Research*, 27, 1503–1520.

Yang, D., Xiong, Y., Kim, H., He, Q., Li, Y., Chen, R., Songyang, Z. (2011). Human telomeric proteins occupy selective interstitial sites. *Cell Research*, 21, 1013–1027.

Yang, L., Sun, L., Teng, Y., Chen, H., Gao, Y., Levine, A.S., Nakajima, S., Lan, L. (2017). Tankyrase1-mediated poly(ADP-ribosyl)ation of TRF1 maintains cell survival after telomeric DNA damage. *Nucleic Acids Res.*, 45, 3906–3921.

Ye, J.Z. and de Lange, T. (2004). TIN2 is a tankyrase 1 PARP modulator in the TRF1 telomere length control complex. *Nat. Genet.*, 36, 618–623.

Ye, J.Z., Hockemeyer, D., Krutchinsky, A.N., Loayza, D., Hooper, S.M., Chait, B.T., de Lange, T. (2004a). POT1-interacting protein PIP1: A telomere length regulator that recruits POT1 to the TIN2/TRF1 complex. *Genes Dev.*, 18, 1649–1654.

Ye, J.Z., Donigian, J.R., van Overbeek, M., Loayza, D., Luo, Y., Krutchinsky, A.N., Chait, B.T., de Lange, T. (2004b). TIN2 binds TRF1 and TRF2 simultaneously and stabilizes the TRF2 complex on telomeres. *J. Biol. Chem.*, 279, 47264–47271.

Ye, J., Lenain, C., Bauwens, S., Rizzo, A., Saint-Léger, A., Poulet, A., Benarroch, D., Magdinier, F., Morere, J., Amiard, S. et al. (2010). TRF2 and Apollo cooperate with topoisomerase 2alpha to protect human telomeres from replicative damage. *Cell*, 142, 230–242.

Ye, J., Renault, V.M., Jamet, K., Gilson, E. (2014). Transcriptional outcome of telomere signalling. *Nat. Rev. Genet.*, 15, 491–503.

Yeung, F., Ramírez, C.M., Mateos-Gomez, P.A., Pinzaru, A., Ceccarini, G., Kabir, S., Fernández-Hernando, C., Sfeir, A.J. (2013). Nontelomeric role for Rap1 in regulating metabolism and protecting against obesity. *Cell Reports*, 3, 1847–1856.

Zaug, A.J., Podell, E.R., Nandakumar, J., Cech, T.R. (2010). Functional interaction between telomere protein TPP1 and telomerase. *Genes Dev.*, 24, 613–622.

Zeng, Z., Wang, W., Yang, Y., Chen, Y., Yang, X., Diehl, J.A., Liu, X., Lei, M. (2010). Structural basis of selective ubiquitination of TRF1 by SCFFbx4. *Dev. Cell*, 18, 214–225.

Zhang, T., Zhang, Z., Shengzhao, G., Li, X., Liu, H., Zhao, Y. (2019). Strand break-induced replication fork collapse leads to C-circles, C-overhangs and telomeric recombination. *PLoS Genet.*, 15, e1007925.

Zhong, Z., Shiue, L., Kaplan, S., de Lange, T. (1992). A mammalian factor that binds telomeric TTAGGG repeats in vitro. *Mol. Cell. Biol.*, 12, 4834–4843.

Zhong, F.L., Batista, L.F., Freund, A., Pech, M.F., Venteicher, A.S., Artandi, S.E. (2012). TPP1 OB-fold domain controls telomere maintenance by recruiting telomerase to chromosome ends. *Cell*, 150, 481–494.

Zhou, X.Z. and Lu, K.P. (2001). The Pin2/TRF1-interacting protein PinX1 is a potent telomerase inhibitor. *Cell*, 107, 347–359.

Zhu, Q., Meng, L., Hsu, J.K., Lin, T., Teishima, J., Tsai, R.Y. (2009). GNL3L stabilizes the TRF1 complex and promotes mitotic transition. *The Journal of Cell Biology*, 185, 827–839.

Zimmermann, M., Kibe, T., Kabir, S., de Lange, T. (2014). TRF1 negotiates TTAGGG repeat-associated replication problems by recruiting the BLM helicase and the TPP1/POT1 repressor of ATR signaling. *Genes Dev.*, 28, 2477–2491.

4

Mammalian Telomere Organization and Dynamics in the Nuclear Volume

Sonia STINUS and Laure CRABBE
Molecular, Cellular and Developmental Biology Department, Integrative Biology Center, University of Toulouse, CNRS, UPS, France

4.1. Introduction

Chromosomes occupy discrete regions of the nucleus called chromosome territories. The organization of chromosome territories within the nuclear volume influences major biological processes including regulation of gene expression, replication timing, genome stability, cancer progression and senescence. The nuclear envelope (NE), which defines the boundaries of the nuclear volume, is a key player in organizing nuclear architecture. In this chapter, we introduce the organization of the NE and review the evidence that supports its physical connection with telomeres, which is revealed to be important for nuclear architecture, replication timing, mitosis and telomere dynamics.

For a color version of all the figures in this chapter, see www.iste.co.uk/saintome/telomeres.zip.

Telomeres,
coordinated by Carole SAINTOMÉ.

4.2. Nuclear envelope and lamins: orchestrators of cell and genome architecture

4.2.1. *The nuclear envelope landmarks*

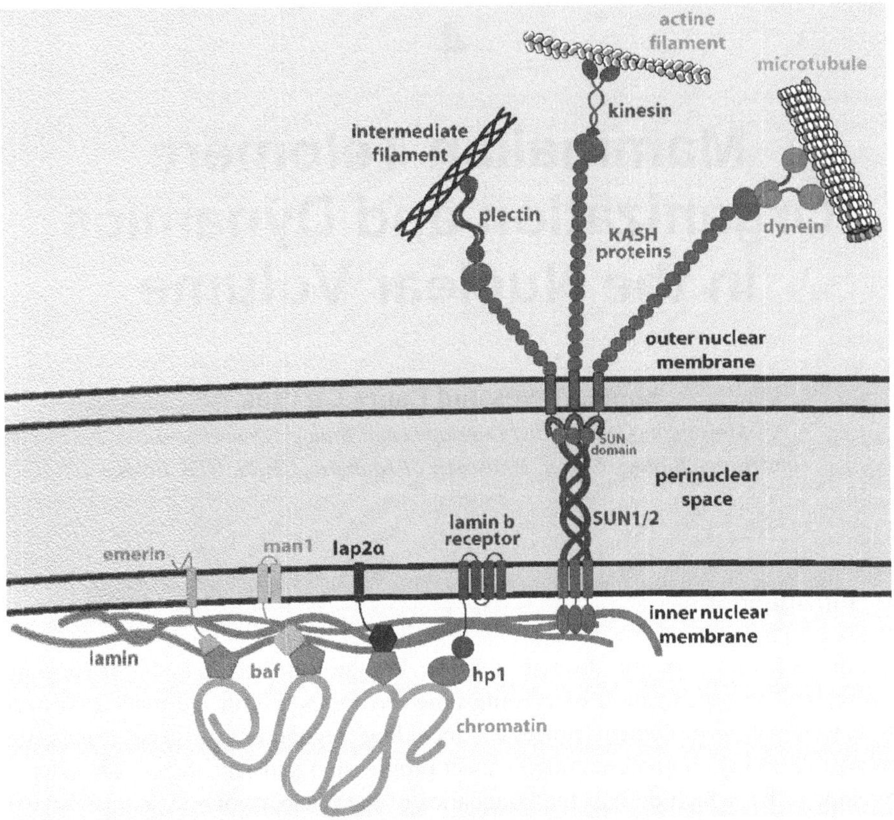

Figure 4.1. *Organization of the nuclear envelope*

The NE is the barrier that confines the nucleus, providing both architectural support and communication between the nucleus and the cytoplasm. This highly organized double membrane is composed of the outer nuclear membrane (ONM) on the cytoplasmic side, continuous with the endoplasmic reticulum, the inner nuclear membrane (INM) that faces the nucleoplasm and the perinuclear space comprised between the ONM and the INM (Figure 4.1). Both membranes are joined at sites of insertion of nuclear pore complexes (NPC), which are located all along the NE. NPCs are large channel structures mediating the transport of proteins and RNA from the nucleus to the cytoplasm. Remarkably, both INM and ONM harbor specific sets of proteins that confer the NE the ability to function way beyond a simple barrier. Particularly, the INM is lined with a meshwork of type V intermediate filament polymers forming the nuclear lamina, composed of A- and B-type lamins. Together with additional INM-associated factors, the lamina provides a docking site for chromatin and creates a regulation hub for several essential cellular functions. The NE landscape is also composed of the LINC complex (linker of nucleoskeleton and cytoskeleton), which is associated with both inner and outer membranes of the nucleus and coordinates chromatin dynamics using the cytoskeleton network. The LINC complex serves as force transmitter between the two cell compartments thanks to SUN-domain proteins (SUN1 and SUN2), which span from the nucleoplasm to the perinuclear space, and KASH proteins, that interact with SUN proteins in the perinuclear space and with the cytoskeleton in the cytoplasm (Cain et al. 2018).

4.2.2. *The nuclear envelope as a scaffold to organize nuclear architecture*

The NE maintains the structural integrity of the nucleus during interphase but, together with nuclear lamins, also organizes nuclear architecture through chromatin tethering. Organization of chromatin in the nuclear volume influences essential processes such as replication timing, transcription and DNA repair. Generally, chromatin located in the periphery of the nucleus is transcriptionally repressed and late replicated, while chromatin positioned toward the nuclear interior is transcriptionally active and early replicated. The association between chromatin and the nuclear lamina occurs via multiple NE-associated proteins (Figure 4.1). On the one hand, Emerin, Man1 and Lap2 connect chromatin to the NE through the DNA-bridging protein BAF (barrier-to-autointegration factor) (Cai et al. 2001; Lee et al. 2001). On the other hand, lamin B receptor, which interacts with heterochromatin protein 1 (HP1) (Ye and Worman 1996), and lamin A/C are known to tether heterochromatin to the NE. When lamin B receptor or lamin A/C are absent, the peripheral localization of heterochromatin shifts toward the nuclear interior (Solovei et al. 2013). Remarkably, damaged DNA, such as eroded telomeres, unrepaired DNA double-stranded breaks and stalled replication forks are known to be repositioned to the NPCs in yeast and mammals for repair (Nagai et al., 2008; Khadaroo et al. 2009; Oza et al., 2009; Churikov et al. 2016). Therefore, the scaffolding function of the NE directly influences the organization of the nuclear content and the maintenance of genomic integrity.

4.2.3. *Nuclear envelope disassembly during mitosis*

The 3D organization of the interphase nuclei is reset during mitotic cell division to allow chromosome segregation and distribution of a single and complete copy of the mother genome to each daughter cell. Two types of mitosis exist, known as closed mitosis and open mitosis. Several organisms, including budding and fission yeast, encounter closed mitosis, where the NE remains intact throughout the entire mitotic process. However, most higher eukaryotes experience an open mitosis, which comprises full NE breakdown during cell division (Boettcher and Barral 2014). Open mitosis induces a dramatic change in nuclear organization, with NPC disassembly, depolymerization of the nuclear lamina and detachment of nuclear membrane from chromatin (reviewed in Kutay and Hetzer 2008). Most of these events are regulated by mitotic kinases, able to phosphorylate NE proteins, lamins, INM proteins and chromatin-associated factors. At the end of mitosis, the NE is reformed around the two sets of condensed daughter chromosomes to re-built fully functional and organized nuclei. NE reformation implies the coordination of essential events such as membrane reformation, NPC insertion, re-addressing of the lamina meshwork and repositioning of chromatin domains within the 3D nuclear space.

4.3. Interplay between telomeres and nuclear structures

The search for potential connections of mammalian telomeres with nuclear structures was influenced by findings in budding and fission yeast. In budding yeast *S. cerevisiae*, telomeres are organized as clusters tethered to the INM via the SUN-domain protein Mps3 (Conrad et al. 2007). The peripheral localization of telomere clusters is critical for telomere silencing, likely by creating a nuclear sub-compartment enriched for silencing factors (Hediger et al. 2002). Moreover, binding of budding yeast telomeres to the NE also protects telomeric repeats from recombination, adding a new dimension to telomere protection (Schober et al. 2009). Telomere clustering and telomere movement have also been described in depth in fission yeast. Upon meiotic induction, the telomeric protein Taz1 recruits Rap1 to interact with members of the LINC complex and meiosis-specific proteins Bqt1 and Bqt2 (Cooper et al. 1998). The interaction of the LINC complex with nucleating microtubules in the cytoplasm drives telomere movement toward the spindle pole body to compose the telomere bouquet and to promote meiotic recombination (Chikashige et al. 1994, 2007, 2006, 2009). These early findings clearly drew a functional link between 3D nuclear positioning and telomere regulation that paved the way to studies in mammals.

In the following section, we wish to highlight the physical interactions described between mammalian telomeres and NE proteins, focusing on the resulting functional connection.

4.4. Telomere tethering to the NE to promote meiotic pairing

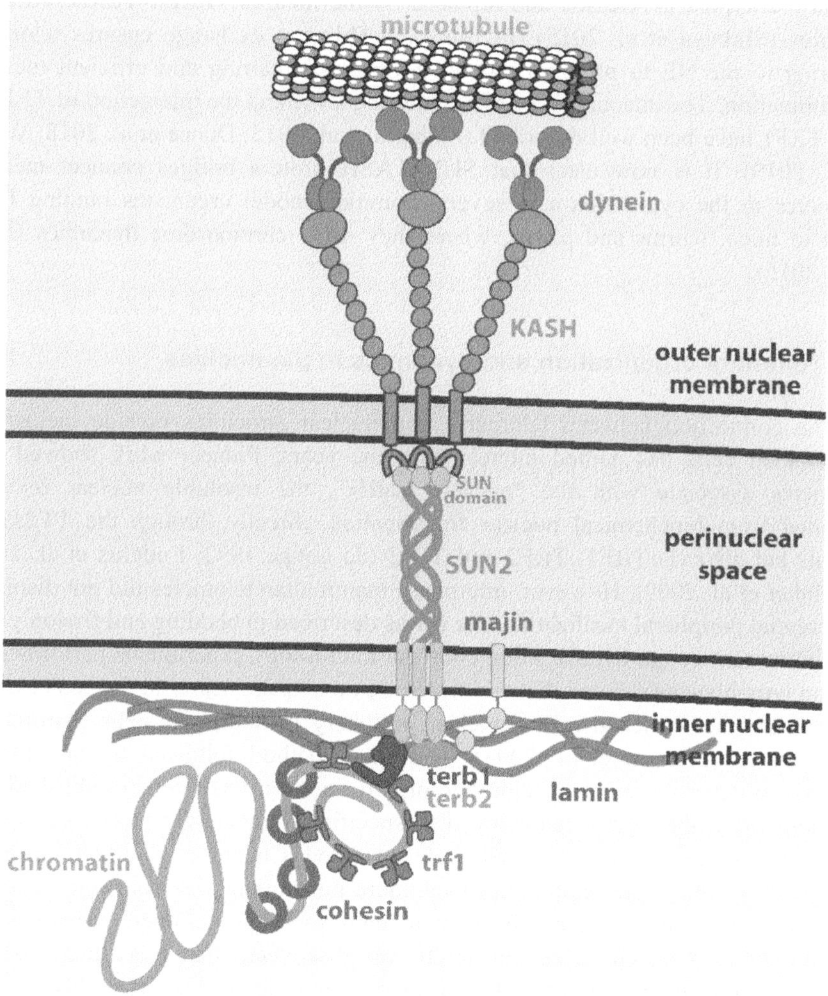

Figure 4.2. *Telomere "cap exchange" during meiosis in mouse cells. The Shelterin complex is released and replaced by the meiotic TERB1-TERB2-MAJIN complex*

After the discovery of SUN-domain proteins and their role in bouquet formation via telomere tethering in yeast, studies performed in mouse cells showed that a similar mechanism was also taking place. The protein SUN2 is involved in telomere tethering for bouquet formation in mammalian telomeres but, unlike in yeast telomeres, Rap1 is dispensable for this process (Schmitt et al. 2007; Scherthan et al. 2011). Instead, mammalian telomeres experience a cap exchange process where the shelterin complex is released and replaced by the meiotic TERB1-TERB2-MAJIN complex (Shibuya et al. 2015) (Figure 4.2). This cap exchange ensures telomere tethering to the NE to promote proper chromosome pairing and efficient meiotic recombination. The attachment of MAJIN to the INM and the interaction of TERB1 with TRF1 have been well described (Shibuya et al. 2015; Dunce et al. 2018; Wang et al. 2019). It is now clear that SUN-KASH protein bridges connect meiotic telomeres to the cytoskeleton in several common model organisms ranging from yeast to mice, worms and plants, where they drive chromosome dynamics (Link et al. 2015).

4.5. Telomere organization and dynamics in the nucleus

The connection between telomeres and nuclear structures outside meiosis in mammalian cells has gained interest over the years. Pioneer work showed that telomeres associate with the “nuclear matrix”, the insoluble nuclear remnant obtained from biochemical nuclear fractionation, directly through the TTAGGG repeats but also via TRF1, TRF2 and TIN2 (de Lange 1992; Ludérus et al. 1996; Kaminker et al. 2009). However, interphase mammalian telomeres did not display a preferential peripheral localization like it was described in budding and fission yeast. 3D-fluorescent in situ hybridization confocal microscopy experiments performed on human lymphocytes showed that telomeres are on average nearer to the center of the nucleus and not enriched at the nuclear periphery in interphase cells (Amrichová et al. 2003; Weierich et al. 2003). A similar method followed by quantitative analysis to determine nuclear telomeric organization on fixed cells established that telomeres assemble into a telomeric disk specifically in the G2 phase, suggesting that telomere position might be cell cycle regulated (Chuang et al. 2004; Vermolen et al. 2005). Cell cycle regulation of telomere positioning was confirmed later on using advanced time-lapse confocal microscopy to follow telomere dynamics and positioning over an entire cell cycle. It was discovered that more than 40% of telomeres colocalize with the NE during post mitotic nuclear assembly in both primary fibroblasts and HeLa cells (Crabbe et al. 2012). These results indicate that telomere tethering to the NE is a conserved mechanism which, unlike in yeast, where it takes place throughout the entire cell cycle, is restricted to a specific time-window in mitotic human cells. The organization and dynamics of telomeres are also influenced by the mechanism in charge of telomere length maintenance. Visualization of telomere dynamics using peptide nucleic acid probes in living U2OS cells, known to maintain telomere length via the recombination-based ALT (alternative lengthening of telomeres) pathway, showed that the majority of telomeres undergo a diffusive constrained movement in interphase cells (Molenaar et al. 2003), which is just about twice the estimated size of the nuclear lamina. Moreover, U2OS telomeres were shown to associate with PML (promyelocytic leukemia) bodies, hub spots for ALT-dependent recombination events. Interestingly, it was later on reported that the shortening of telomeres in ALT cells increases their mobility to facilitate the association between telomeres and PML bodies (Jegou et al. 2009).

Nuclear lamins are also involved in telomere chromatin dynamics. The anomalous diffusion of interphase telomeres depends on the lamin A network that creates a crosslink to restrict chromatin movements (De Vos et al. 2010; Bronshtein et al. 2015). These important findings highlight a major role of the nuclear matrix in chromatin movement and distribution, which also affects the natural ends of linear chromosomes.

4.6. Interaction between telomere chromatin and lamins during post-mitotic nuclear assembly

Several instances of interactions between the nuclear lamina and telomeres suggest that telomeres in mammalian cells are anchored to the nuclear lamina similarly to lamina-associated domains (LADs), which are regions composed of gene-poor and transcriptionally repressed chromatin located in the periphery of the nucleus. Unlike LADs, however, these interactions only occur during a very specific moment of the cell cycle. Telomeres decorate the NE during its reformation starting in late mitosis (late anaphase/early telophase), and are released later in G1, once the daughter nuclei have regained their proper size. During late anaphase and early telophase, telomeres interact with the NE-associated proteins lamin B1, Lap2α and Emerin (Crabbe et al. 2012). The interaction between the NE and telomeres has been further supported by Rap1 and SUN1 co-immunoprecipitation experiments (Crabbe et al. 2012). Moreover, Lap2α, which interacts both with lamin A and telomeres (Chojnowski et al. 2015), is located mainly in the cytoplasm until metaphase but is deposited on telomeres during anaphase and telophase, and then translocates to the nucleoplasm in G1 phase (Dechat et al. 2004). The telomere--NE interaction, occurring specifically during the late mitotic stages and the early times of the newly formed daughter cell, suggests that this interaction is important to organize the new born cell nucleus (Figure 4.3).

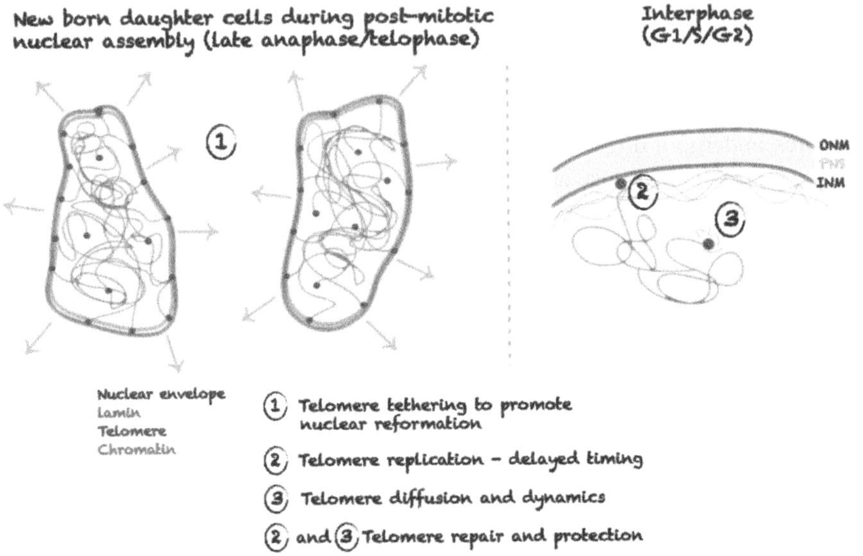

Figure 4.3. *Functional implication of telomere position relative to the nuclear envelope during the cell cycle*

4.7. Interaction between telomere chromatin and lamins for telomere maintenance

The interaction between telomeric chromatin and lamins has functional implications in telomere maintenance, including telomere replication, protection and repair (Figure 4.3).

While telomeres in human cells are known to replicate throughout S phase, the presence of a DNA sequence known as D4Z4 array in the subtelomeric region is sufficient to confer both peripheral localization and late-replicating property to the adjacent telomere (Ottaviani et al. 2009; Arnoult et al. 2010). The D4Z4 element is an 80 bp sequence that interacts with A-type lamins and confers the ability to address chromatin to the periphery. In fact, though the large majority of telomeres localize away from the nuclear periphery during interphase, a few telomeres show a peripheral localization that correlates with late replication (Arnoult et al. 2010). Moreover, the influence of lamins on telomere replication is also highlighted by AKTIP, a protein that interacts both with the shelterin complex through TRF1 and TRF2 and with A- and B-type lamins. Loss of AKTIP perturbs replication fork progression at telomeres, leading to defective telomere replication and an accumulation of DNA repair factors at chromosome ends (Burla et al. 2015, 2016).

Unlike B-type lamins, A-type lamins are also found throughout the nucleoplasm, especially in interphase (Dechat et al. 2010). This mobile pool of nucleoplasmic lamin A/C can interact with telomeres and promote telomere protection within the nuclear interior. Indeed, TRF2 is able to interact with interstitial telomeric sequences, which are telomeric TTAGGG repeats located elsewhere in the genome than at telomeres. The interaction of TRF2 with those interstitial telomeric sequences is stabilized by lamin A/C and has been proposed to promote t-loop formation (Wood et al. 2014). Telomeric integrity and lamins are also tightly bound in the context of laminopathies, which comprise several human premature ageing disorders that originate from mutations in the *LMNA* gene. Perhaps one of the best-studied laminopathies is Hutchinson Gilford progeria syndrome (HGPS) due to the expression of a mutant form of lamin A called progerin. One key feature of HGPS cells is their abnormal nuclear architecture, with lobulation of the NE, thickening of the nuclear lamina, and defects in 3D chromatin organization (Goldman et al. 2004). Progerin expression also results in telomere attrition that alters the proliferative capacity of the cells and entails premature senescence and premature organismal aging (Allsopp et al. 1992; Decker et al. 2009). An unquestionable crosstalk exists between progerin expression and telomere dysfunction, as elongation of telomeres by telomerase expression rescues the proliferative defects of lamina-mutant fibroblasts (Kudlow et al. 2008).

The link between human telomeres and lamins is also important with regard to the emerging role of nuclear architecture in DNA repair and genome maintenance. Recent work suggests that nuclear position dictates the choice of DNA repair, with the nuclear membrane being a compartment preventive of repair by homologous recombination (Lemaitre et al. 2014). Non-random spatial organization has also been shown to affect translocation events by influencing how broken chromosomes move and how they find their partners for homology-directed repair (Lemaitre et al. 2014; Roukos and Misteli 2014). Most importantly, the LINC complex is now shown to be involved in increasing the mobility of chromatin surrounding DNA double-stranded breaks or dysfunctional telomeres to promote repair (Lottersberger et al. 2015). The role of the nuclear membrane in DNA repair was also demonstrated in mice, for which A-type lamins are necessary for the processing of dysfunctional telomeres (Gonzalez-Suarez et al. 2009).

4.8. Conclusion

The nucleus is a highly organized organelle with specific microenvironments, which confer particular properties to chromatin that is distributed in its volume. The chromatin docking site provided by the NE, the nuclear lamina and associated proteins, combined with nucleoplasmic lamins and the nuclear matrix are key regulators of chromatin function and stability. We are only just starting to unravel the exciting implications these elements represent for telomere organization and function, opening a new era of research.

4.9. References

Allsopp, R.C., Vaziri, H., Patterson, C., Goldstein, S., Younglai, E.V., Futcher, A.B., Greider, C.W., Harley, C.B. (1992). Telomere length predicts replicative capacity of human fibroblasts. *Proc. Natl. Acad. Sci. USA*, 89, 10114–10118.

Amrichová, J., Lukásová, E., Kozubek, S., Kozubek, M. (2003). Nuclear and territorial topography of chromosome telomeres in human lymphocytes. *Exp. Cell Res.*, 289, 11–26.

Arnoult, N., Schluth-Bolard, C., Letessier, A., Drascovic, I., Bouarich-Bourimi, R., Campisi, J., Kim, S.-H., Boussouar, A., Ottaviani, A., Magdinier, F. et al. (2010). Replication timing of human telomeres is chromosome arm-specific, influenced by subtelomeric structures and connected to nuclear localization. *PLoS Genet*, 6, e1000920.

Boettcher, B. and Barral, Y. (2014). The cell biology of open and closed mitosis. *Nucleus*, 4, 160–165.

Bronshtein, I., Kepten, E., Kanter, I., Berezin, S., Lindner, M., Redwood, A.B., Mai, S., Gonzalo, S., Foisner, R., Shav-Tal, Y. et al. (2015). Loss of lamin A function increases chromatin dynamics in the nuclear interior. *Nat. Commun.*, 6, 8044.

Burla, R., Carcuro, M., Raffa, G.D., Galati, A., Raimondo, D., Rizzo, A., La Torre, M., Micheli, E., Ciapponi, L., Cenci, G. et al. (2015). AKTIP/Ft1, a new shelterin-interacting factor required for telomere maintenance. *PLoS Genet.*, 11, e1005167.

Burla, R., Carcuro, M., Torre, M.L., Fratini, F., Crescenzi, M., D'Apice, M.R., Spitalieri, P., Raffa, G.D., Astrologo, L., Lattanzi, G. et al. (2016). The telomeric protein AKTIP interacts with A- and B-type lamins and is involved in regulation of cellular senescence. *Open Biol.*, 6, 160103–160111.

Cai, M., Huang, Y., Ghirlando, R., Wilson, K.L., Craigie, R., Clore, G.M. (2001). Solution structure of the constant region of nuclear envelope protein LAP2 reveals two LEM-domain structures: One binds BAF and the other binds DNA. *Embo J.*, 20, 4399–4407.

Cain, N.E., Jahed, Z., Schoenhofen, A., Valdez, V.A., Elkin, B., Hao, H., Harris, N.J., Herrera, L.A., Woolums, B.M., Mofrad, M.R.K. et al. (2018). Conserved SUN-KASH interfaces mediate LINC complex-dependent nuclear movement and positioning. *Current Biology*, 28, 3086–3097.e4.

Chikashige, Y., Ding, D.Q., Funabiki, H., Haraguchi, T., Mashiko, S., Yanagida, M., Hiraoka, Y. (1994). Telomere-led premeiotic chromosome movement in fission yeast. *Science*, 264, 270–273.

Chikashige, Y., Tsutsumi, C., Yamane, M., Okamasa, K., Haraguchi, T., Hiraoka, Y. (2006). Meiotic proteins bqt1 and bqt2 tether telomeres to form the bouquet arrangement of chromosomes. *Cell*, 125, 59–69.

Chikashige, Y., Haraguchi, T., Hiraoka, Y. (2007). Another way to move chromosomes. *Chromosoma*, 116, 497–505.

Chikashige, Y., Yamane, M., Okamasa, K., Tsutsumi, C., Kojidani, T., Sato, M., Haraguchi, T., Hiraoka, Y. (2009). Membrane proteins Bqt3 and -4 anchor telomeres to the nuclear envelope to ensure chromosomal bouquet formation. *J. Cell Biol.*, 187, 413–427.

Chojnowski, A., Ong, P.F., Wong, E.S., Lim, J.S., Mutalif, R.A., Navasankari, R., Dutta, B., Yang, H., Liow, Y.Y., Sze, S.K., et al. (2015). Progerin reduces LAP2α-telomere association in Hutchinson-Gilford progeria. *Elife*, 4, e07759, doi: 10.7554/elife.07759.

Chuang, T.C.Y., Moshir, S., Garini, Y., Chuang, A.Y.-C., Young, I.T., Vermolen, B., van den Doel, R., Mougey, V., Perrin, M., Braun, M. et al. (2004). The three-dimensional organization of telomeres in the nucleus of mammalian cells. *BMC Biol.*, 2, 12.

Churikov, D., Charifi, F., Eckert-Boulet, N., Silva, S., Simon, M.N., Lisby, M., Géli, V. (2016). SUMO-dependent relocalization of eroded telomeres to nuclear pore complexes controls telomere recombination. *Cell Rep.*, 15, 1242–1253.

Conrad, M.N., Lee, C.-Y., Wilkerson, J.L., Dresser, M.E. (2007). MPS3 mediates meiotic bouquet formation in *Saccharomyces cerevisiae*. *Proc. Natl. Acad. Sci. USA*, 104, 8863–8868.

Cooper, J., Watanabe, Y., Nurse, P. (1998). Fission yeast Taz1 protein is required for meiotic telomere clustering and recombination. *Nature*, 392, 828–831.

Crabbe, L., Cesare, A.J., Kasuboski, J.M., Fitzpatrick, J.A.J., Karlseder, J. (2012). Human telomeres are tethered to the nuclear envelope during postmitotic nuclear assembly. *Cell Rep.*, 2, 1521–1529.

De Vos, W.H., Houben, F., Hoebe, R.A., Hennekam, R., van Engelen, B., Manders, E.M.M., Ramaekers, F.C.S., Broers, J.L.V., Van Oostveldt, P. (2010). Increased plasticity of the nuclear envelope and hypermobility of telomeres due to the loss of A-type lamins. Biochim. Biophys. *Acta*, 1800, 448–458.

Dechat, T., Gajewski, A., Korbei, B., Gerlich, D., Daigle, N., Haraguchi, T., Furukawa, K., Ellenberg, J., Foisner, R. (2004). LAP2alpha and BAF transiently localize to telomeres and specific regions on chromatin during nuclear assembly. *J. Cell Sci.*, 117, 6117–6128.

Dechat, T., Adam, S.A., Taimen, P., Shimi, T., Goldman, R.D. (2010). Nuclear lamins. *Cold Spring Harb. Perspect. Biol.*, 2, a000547.

Decker, M.L., Chavez, E., Vulto, I., Lansdorp, P.M. (2009). Telomere length in Hutchinson-Gilford progeria syndrome. *Mech. Ageing Dev.*, 130, 377–383.

Dunce, J.M., Milburn, A.E., Gurusaran, M., da Cruz, I., Sen, L.T., Benavente, R., Davies, O.R. (2018). Structural basis of meiotic telomere attachment to the nuclear envelope by MAJIN-TERB2-TERB1. *Nat. Commun.*, 9, 5355.

Goldman, R.D., Shumaker, D.K., Erdos, M.R., Eriksson, M., Goldman, A.E., Gordon, L.B., Gruenbaum, Y., Khuon, S., Mendez, M., Varga, R. et al. (2004). Accumulation of mutant lamin A causes progressive changes in nuclear architecture in Hutchinson-Gilford progeria syndrome. *Proc. Natl. Acad. Sci., USA*, 101, 8963–8968.

Gonzalez-Suarez, I., Redwood, A.B., Perkins, S.M., Vermolen, B., Lichtensztejin, D., Grotsky, D.A., Morgado-Palacin, L., Gapud, E.J., Sleckman, B.P., Sullivan, T. et al. (2009). Novel roles for A-type lamins in telomere biology and the DNA damage response pathway. *Embo J.*, 28, 2414–2427.

Hediger, F., Neumann, F.R., Van Houwe, G., Dubrana, K., Gasser, S.M. (2002). Live imaging of telomeres: yKu and Sir proteins define redundant telomere-anchoring pathways in yeast. *Curr. Biol.*, 12, 2076–2089.

Jegou, T., Chung, I., Heuvelman, G., Wachsmuth, M., Görisch, S.M., Greulich-Bode, K.M., Boukamp, P., Lichter, P., Rippe, K. (2009). Dynamics of telomeres and promyelocytic leukemia nuclear bodies in a telomerase-negative human cell line. *Mol. Biol. Cell.*, 20, 2070–2082.

Kaminker, P.G., Kim, S.-H., Desprez, P.-Y., Campisi, J. (2009). A novel form of the telomere-associated protein TIN2 localizes to the nuclear matrix. *Cell Cycle*, 8, 931–939.

Khadaroo, B., Teixeira, M.T., Luciano, P., Eckert-Boulet, N., Germann, S.M., Simon, M.N., Gallina, I., Abdallah, P., Gilson, E., Géli, V. et al. (2009). The DNA damage response at eroded telomeres and tethering to the nuclear pore complex. *Nat. Cell. Biol.*, 11, 980–987.

Kudlow, B.A., Stanfel, M.N., Burtner, C.R., Johnston, E.D., Kennedy, B.K. (2008). Suppression of proliferative defects associated with processing-defective lamin A mutants by hTERT or inactivation of p53. *Mol. Biol. Cell*, 19, 5238–5248.

Kutay, U. and Hetzer, M.W. (2008). Reorganization of the nuclear envelope during open mitosis. *Curr. Opin. Cell Biol.*, 20, 669–677.

de Lange, T. (1992). Human telomeres are attached to the nuclear matrix. *Embo J.*, 11, 717–724.

Lee, K.K., Haraguchi, T., Lee, R.S., Koujin, T., Hiraoka, Y., Wilson, K.L. (2001). Distinct functional domains in emerin bind lamin A and DNA-bridging protein BAF. *J. Cell. Sci.*, 114, 4567–4573.

Lemaitre, C., Grabarz, A., Tsouroula, K., Andronov, L., Furst, A., Pankotai, T., Heyer, V., Rogier, M., Attwood, K.M., Kessler, P. et al. (2014). Nuclear position dictates DNA repair pathway choice. *Genes Dev.*, 28, 2450–2463.

Link, J., Jahn, D., Alsheimer, M. (2015). Structural and functional adaptations of the mammalian nuclear envelope to meet the meiotic requirements. *Nucleus*, 6, 93–101.

Lottersberger, F., Karssemeijer, R.A., Dimitrova, N., de Lange, T. (2015). 53BP1 and the LINC complex promote microtubule-dependent DSB mobility and DNA repair. *Cell*, 163, 880–893.

Ludérus, M.E., van Steensel, B., Chong, L., Sibon, O.C., Cremers, F.F., de Lange, T. (1996). Structure, subnuclear distribution, and nuclear matrix association of the mammalian telomeric complex. *J. Cell Biol.*, 135, 867–881.

Molenaar, C., Wiesmeijer, K., Verwoerd, N.P., Khazen, S., Eils, R., Tanke, H.J., Dirks, R.W. (2003). Visualizing telomere dynamics in living mammalian cells using PNA probes. *Embo J.*, 22, 6631–6641.

Nagai, S., Dubrana, K., Tsai-Pflugfelder, M., Davidson, M.B., Roberts, T.M., Brown, G.W., Varela, E., Hediger, F., Gasser, S.M., Krogan, N.J. (2008). Functional targeting of DNA damage to a nuclear pore-associated SUMO-dependent ubiquitin ligase. *Science*, 322, 597–602.

Ottaviani, A., Schluth-Bolard, C., Rival-Gervier, S., Boussouar, A., Rondier, D., Foerster, A.M., Morere, J., Bauwens, S., Gazzo, S., Callet-Bauchu, E. et al. (2009). Identification of a perinuclear positioning element in human subtelomeres that requires A-type lamins and CTCF. *Embo J.*, 28, 2428–2436.

Oza, P., Jaspersen, S.L., Miele, A., Dekker, J., Peterson, C.L. (2009). Mechanisms that regulate localization of a DNA double-strand break to the nuclear periphery. *Genes Dev.*, 23, 912–927.

Roukos, V. and Misteli, T. (2014). The biogenesis of chromosome translocations. *Nat. Cell Biol.*, 16, 293–300.

Scherthan, H., Sfeir, A.J., de Lange, T. (2011). Rap1-independent telomere attachment and bouquet formation in mammalian meiosis. *Chromosoma*, 120, 151–157.

Schmitt, J., Benavente, R., Hodzic, D., Höög, C., Stewart, C.L., Alsheimer, M. (2007). Transmembrane protein Sun2 is involved in tethering mammalian meiotic telomeres to the nuclear envelope. *Proc. Natl. Acad. Sci. USA*, 104, 7426–7431.

Schober, H., Ferreira, H., Kalck, V., Gehlen, L.R., Gasser, S.M. (2009). Yeast telomerase and the SUN domain protein Mps3 anchor telomeres and repress subtelomeric recombination. *Genes Dev.*, 23, 928–938.

Shibuya, H., Hernández-Hernández, A., Morimoto, A., Negishi, L., Höög, C., Watanabe, Y. (2015). MAJIN links telomeric DNA to the nuclear membrane by exchanging telomere cap. *Cell*, 163, 1252–1266, doi: 10.1016/j.cell.2015.10.030.

Solovei, I., Wang, A.S., Thanisch, K., Schmidt, C.S., Krebs, S., Zwerger, M., Cohen, T.V., Devys, D., Foisner, R., Peichl, L. et al. (2013). LBR and lamin A/C sequentially tether peripheral heterochromatin and inversely regulate differentiation. *Cell*, 152, 584–598.

Vermolen, B.J., Garini, Y., Mai, S., Mougey, V., Fest, T., Chuang, T.C.-Y., Chuang, A.Y.-C., Wark, L., Young, I.T. (2005). Characterizing the three-dimensional organization of telomeres. *Cytometry A*, 67, 144–150.

Wang, Y., Chen, Y., Chen, J., Wang, L., Nie, L., Long, J., Chang, H., Wu, J., Huang, C., Lei, M. (2019). The meiotic TERB1-TERB2-MAJIN complex tethers telomeres to the nuclear envelope. *Nat. Commun.*, 10, 564.

Weierich, C., Brero, A., Stein, S., von Hase, J., Cremer, C., Cremer, T., Solovei, I. (2003). Three-dimensional arrangements of centromeres and telomeres in nuclei of human and murine lymphocytes. *Chromosome Res.*, 11, 485–502.

Wood, A.M., Rendtlew Danielsen, J.M., Lucas, C.A., Rice, E.L., Scalzo, D., Shimi, T., Goldman, R.D., Smith, E.D., Le Beau, M.M., Kosak, S.T. (2014). TRF2 and lamin A/C interact to facilitate the functional organization of chromosome ends. *Nat. Commun.*, 5, 5467.

Ye, Q. and Worman, H.J. (1996). Interaction between an integral protein of the nuclear envelope inner membrane and human chromodomain proteins homologous to Drosophila HP1. *J. Biol. Chem.*, 271, 14653–14656.

5

Saccharomyces cerevisiae, a Model at the Forefront of Telomere Biology

Aurélia BARASCU and Maria Teresa TEIXEIRA
Institute of Physico-Chemical Biology, PSL, CNRS, Sorbonne University, Paris, France

Telomeres are specialized structures that play a central role in maintaining genome stability in eukaryotic cells. *Saccharomyces cerevisiae*, the baker's yeast, was one of the first model organisms used to study telomeres. Over the last four decades, this single-cell species has made it possible to discover many aspects of telomere biology using the sophisticated genetic tools available in this model organism. In this chapter, we will discuss past and present findings in *S. cerevisiae* concerning telomeres and telomerase, most of which have revealed evolutionarily conserved biological mechanisms.

5.1. Introduction

Chromosomes are subjected to different genotoxic stresses daily, which can be of endogenous (replicative stress, oxidative stress, etc.) or exogenous origin (irradiation, exposure to chemical agents, etc.). Among the different types of DNA

For a color version of all the figures in this chapter, see www.iste.co.uk/saintome/telomeres.zip.

Telomeres,
coordinated by Carole SAINTOMÉ.

damage, double strand breaks are the most toxic because they can lead to the loss of large pieces of chromosomes. DNA damage repair mechanisms ensure genome stability through the orchestration of a set of enzymatic activities, which specifically recognize these accidental double-stranded ends, and repair the damaged ends, thus limiting the loss or alteration of genetic information for future generations. However, all eukaryotic cells have natural structures that resemble double-strand breaks. In fact, since eukaryotic chromosomes are linear, they have natural ends called telomeres. The major function of telomeres is to maintain the integrity of chromosomes by protecting them from the repair machinery from its normal action. Thus, telomeres are not recognized as accidental double-strand breaks and do not get fused or degraded by repair processes.

To better understand this essential function of telomeres, we will focus on these particular sequences that are dynamic entities, whose length, protein composition and structure change considerably during cell divisions, and even during the cell cycle. We will also discuss one of the mechanisms that inevitably affects telomeres, namely the replication machinery. Indeed, the mode of action of DNA polymerases, which are at the heart of the replication machinery, does not allow the complete replication of telomeres, leading to the so-called end replication problem. This feature leads to a progressive shortening of telomeres with each cell cycle. Telomerase is a reverse transcriptase that serves to counteract the end-replication problem in most eukaryotes studied to date. The study of the regulation of this enzyme is as complex as the study of its substrate. Telomerase is constitutively expressed in yeast cells. However, in many multicellular organisms such as humans, only stem cells and germ cells express telomerase and maintain telomeres at an average size that results from a balance between shortening and lengthening. Conversely, for somatic cells, telomeres shorten with each cell division until they reach a critical size up to a point in which telomeres change state. It is important to understand how this change leads to an activation of cell cycle checkpoints, and thus a permanent arrest of proliferation called replicative senescence, which applies to most cells. In rare cases, the mechanisms of damage response and repair of DNA damage can then generate genetic rearrangements at the expense of genome stability. Thus, we will discuss the central role that telomeres play in controlling cell viability and how their proper functioning ensures genome stability.

Many of these concepts were born through experiments carried out on the yeast *Saccharomyces cerevisiae*. We therefore propose to review the history of the emergence of these discoveries through the prism of this model organism.

5.2. Telomere characterization: various experimental organisms for first historical experiments

Double strand breaks are one of the most deleterious events for eukaryotic chromosomes because the chromosomal piece not associated with the centromere is no longer segregated faithfully and may be irreversibly lost. Double strand breaks are efficiently corrected in organisms with circular genomes, as well as in those with linear chromosomes, such as some bacteria (Borrelia, *A. tumefaciens*), some phages (T7, λ) and all investigated eukaryotes. However, very early in the first cytogenetic experiments in the 1940s, it was observed that the natural ends of chromosomes are not repairable (Müller 1938; McClintock 1940).

Even though the nature of DNA in chromosomes was not yet known, the existence of special material located at chromosome ends was stated in 1938 on the basis of studies conducted by Müller on Drosophila. Müller observed an absence of fusions of chromosome ends after X-ray treatment. He concluded that these natural ends must be protected by a special structure that he named "telomeres", from the Greek τέλος, telos (end) and μέρος, meros (segment) (Müller 1938). At the same time, Barbara McClintock also observed a specific hallmark of the ends in maize and that there were two types of ens, stable or unstable, corresponding to telomeres and double-strand breaks (McClintock 1939; Blackburn et al. 2006).

So how do cells distinguish telomeres from accidental breaks?

In 1978, Blackburn became interested in the distinctive structure and dynamics of the genome of *Tetrahymena thermophila*, a ciliated protozoans. This organism, as well as other ciliated protozoa, actually have a polyploid macronucleus, which contains hundreds of mini-chromosomes carrying the ribosomal gene, which proved to be a huge experimental advantage for the study of telomeres. Because the macronuclei were naturally enriched in DNA ends, Blackburn was able to sequence telomeric DNA for the first time and demonstrate the repetitive nature of these sequences. She also showed the nucleotide composition bias, with a strand oriented from the 5' to the 3' terminal containing between 20 and 70 GGGGTT repeats and presenting a single-stranded region terminated by a 3'-OH (Blackburn and Gall 1978).

Shortly thereafter, in collaboration with Szostak, Blackburn used these *Tetrahymena* terminal sequences to experimentally demonstrate that they were sufficient to protect and maintain the ends. In this experiment, the authors took advantage of recent advances in molecular biology, in particular the ease of manipulating the genetic content of the yeast *S. cerevisiae*, a pioneering eukaryotic experimental model. They cloned the telomeric sequences of the ciliate at the ends of a linear plasmid carrying sequences allowing autonomous replication in yeast cells. After the transformation of this hybrid plasmid, they observed that the yeast cells had recognized the heterologous telomeric sequences as functional and thus sufficient to maintain the integrity of the linear plasmid.

This historical experiment demonstrates that there is a high degree of conservation of telomere sequences, as well as the telomere maintenance mechanism, between these two species that are far apart in the phylogenetic tree of eukaryotes (Szostak and Blackburn 1982). Today, telomeric sequences are characterized in many organisms. They are T- and G-rich sequences on the strand containing the 3' terminal, and the yeast *S. cerevisiae* has degenerate TG1-3 repeats (a T followed by 1-3 Gs). Many eukaryotes also have a single-stranded region at the 3' end.

It was not until years later that an experiment in the yeast *S. cerevisiae* demonstrated that telomeres are also necessary for chromosome protection. In this experiment, a yeast strain was constructed to excise a single telomere placed on a dispensable supernumerary chromosome (Sandell and Zakian 1993; Haber 1995). This led to an immediate cycle arrest in G2/M, caused by the activation of the newly described DNA damage checkpoints (Weinert and Hartwell 1988), followed by the loss of the entire damaged chromosome. The cells thus behaved as if they had suffered an accidental break, and degraded the chromosome, demonstrating the necessity of these structures to distinguish between a natural and an accidental end, and, finally, to maintain chromosome integrity. Conversely, if telomeric repeats are artificially placed at the excision site, the cells do not activate the checkpoints and do not stop at G2/M (Michelson et al. 2005). Thus, the identity provided by the telomeric nucleoprotein system allows the cell to differentiate a telomere from a double-strand break and provides special protection to this natural end.

5.3. Protection and proliferation: the telomere dilemma

The study of the special features of telomeres in terms of sequence and binding proteins has led to a better understanding of how the cell is able to distinguish telomeres from double-strand breaks.

5.3.1. *Telomeric sequences: pattern conservation and length variations*

The nucleotide sequence of telomeres is highly conserved in eukaryotes. Telomeric DNA is formed by tandem repeats of a simple guanine-rich motif of 5–25 base pairs that terminates in a single-stranded 3' end in most organisms studied.

These telomeric repeats are regular (TTAGGG, as in humans) (Moyzis et al. 1988), or irregular as in the yeast *Schizosaccharomyces pombe* or *S. cerevisiae* ($TG_{1\text{-}3}$) (Wang and Zakian 1990; McEachern and Blackburn 1994; Teixeira and Gilson 2005).

The average length of telomeric DNA varies from a few tens of base pairs to several kilobases depending on the organism, but also varies within each cell. Shampay and Blackburn (1988) again used baker's yeast to characterize the dynamics of this size variation. They demonstrated that there is a size heterogeneity, around 300 bp at each end of most of the 16 chromosomes. By analyzing each telomere individually, they observed that their size varies over time and that this variability increases with the number of cell divisions. This observation allowed them to propose a model of equilibrium between, on the one hand, the progressive shortening due to the problem of replication of the ends and, on the other hand, the elongation of each telomere by a mechanism that was still unknown at that time. In view of the heterogeneity of the observed sizes, the nature of telomeres would therefore contain intrinsic sources of variability in the mechanisms of shortening and lengthening.

The size of the telomeric single strand region is also dynamic and heterogeneous in nature. The 3' single strand overhang of 5–10 nucleotides in *S. cerevisiae* can increase to several tens of nucleotides during S-phase, as the replication fork passes through the telomeric sequences (Wellinger et al. 1993; Dionne and Wellinger 1998; Larrivee et al. 2004; Soudet et al. 2014). It has also been observed that the single-stranded DNA (ssDNA) region increases further on the shortest telomeres at the time of senescence (Fallet et al. 2014). In mammals, the single-stranded telomeric region is able to invade double-stranded regions of the same telomere via a D-loop promoted by shelterin proteins. The overall structure, which has been observed by electron microscopy, is termed the “T-loop” (telomere loop) and its function in end protection remains to be clarified (Griffith et al. 1999; Ruis and Boulton 2021). In yeast *S. cerevisiae* telomeres, this ability to form T-loops has not been shown.

Because of its guanine richness, G-G pairs can form highly stable intramolecular four-stranded folds (Sen and Gilbert 1988). *In vitro* studies have shown that the guanine-rich strand of the telomere is capable of folding in on itself into a four-stranded structure maintained by hydrogen interactions between four guanines and a monovalent cation (Na+, K+), forming a G-quadruplex structure (Henderson et al. 1990). This structure is thought to sequester telomere 3’ ssDNA extensions in order to protect and regulate telomere size, but also to facilitate telomere compaction (Parkinson et al. 2002) or even facilitate or prevent telomerase action

(Zahler et al. 1991; Oganesian et al. 2006). Yet, the evidence for G4 formation at telomeres *in vivo* remains indirect.

Upstream of the telomeres are the regions called subtelomeres. They are made up of longer, heterogeneous and variable repeated elements depending on the telomere, including multigene families coding for functions involved in interaction with the environment (Louis 1995; Bergstrom et al. 2014). Unlike telomeres, the subtelomeric regions of *S. cerevisiae* are covered by nucleosomes (Wright et al. 1992; Pasquier and Wellinger 2020). The Sir (silent information regulator) proteins play a key role in the compaction of these regions and consequently affect the localizations of telomeres in the nucleus (Gilson and Géli 2007; Moser and Nakamura 2009; Giraud-Panis et al. 2010).

In *S. cerevisiae*, all subtelomeric regions have an X element, followed by one to four repeats of a Y' element on more than half of the subtelomeres (Chan and Tye 1983a, 1983b; Zakian et al. 1986; Yue et al. 2017). These regions contain potential replication origins or ARS (autonomously replicating sequences), as well as small TTAGGG repeats, corresponding to the widespread and potentially ancestral telomeric motif (Teixeira and Gilson 2005). The Y' elements encode a helicase, Y'-Help1 (Yamada et al. 1998).

In *S. cerevisiae* as well as in *S. pombe* and in humans, telomeres can be transcribed from subtelomeric regions into coding or non-coding RNA (Mak et al. 2009; Feuerhahn et al. 2010; Bah and Azzalin 2012; Kwapisz and Morillon 2020; Lalonde and Chartrand 2020). Among the non-coding RNAs transcribed from the regions closer to the tip are ARRET, αARRET, subTERRA and TERRA (TElomeric Repeat-containing RNA). The function of TERRA has been the subject of many studies. This RNA can form three-stranded DNA:RNA hybrid structures, called R-loops, whose regulation is linked to telomere size and could promote homologous recombination at the level of the shortest telomeres, playing a role in telomere maintenance and protection (see section 5.3.3) (Azzalin et al. 2007; Luke et al. 2008; Iglesias et al. 2011; Graf et al. 2017).

Although present on all telomeres, subtelomeric regions are not essential for cell viability under rich growth conditions (Louis 1995). On the other hand, the extreme plasticity of these regions ensures rapid adaptation of yeast under stress conditions, because subtelomeric genes encode functions related to interaction with the environment (Brown et al. 2010; Bergstrom et al. 2014; Jolivet et al. 2019). A study based on PacBio sequencing of different domesticated strains of *S. cerevisiae* and their respective wild-type versions, *S. paradoxus*, revealed a higher frequency of chromosomal rearrangements at these subtelomeric regions compared to other regions of the genome (Yue et al. 2017). This suggests that these regions are chromosomal regions that strongly contribute to the generation of new genes by duplications and gene losses, and thus to evolution.

5.3.2. *Telomeric proteins*

Telomere proteins play a central role in the protection and maintenance of telomere ends. They cooperate to prevent recognition, signaling and repair of the telomere such as a double-strand break. Telomeric proteins are generally classified into two groups according to their ability to bind to the double-stranded region of the telomere or to the single-stranded region. Their interaction forms a complex called "shelterin" in humans, which is essential for the protection and maintenance of the telomere ends.

5.3.2.1. *Telomeric double strand proteins*

In the yeast *S. cerevisiae*, Rap1 (repressor activator protein 1) binds with high affinity to telomeric sites 5′-GGTGTGTGGGTGT-3′ through two tandemly arranged Myb domains (Konig and Rhodes 1997). Rap1 also binds to non-telomeric sites and plays a role in regulating the transcription of many genes encoding glucolytic enzymes, ribosomal proteins and several components of the transcriptional machinery, and in repressing the expression of yeast sex-type loci (Shore 1994; Morse 2000). Because of the degenerate nature of the telomeric repeats in *S. cerevisiae*, it is estimated that one molecule of Rap1 is bound every 18 base pairs, so there would be 15–20 Rap1 per average-sized telomere (Gilson et al. 1993), and this binding would exclude nucleosomes from the telomeric DNA (Gilson and Gasser 1995; Pasquier and Wellinger 2020). Rap1 plays an essential role in telomere function by inhibiting inter-telomere fusion by non-homologous end joining (NHEJ), recruiting factors involved in the control of telomere size (Marcand et al. 1997) and in the transcriptional repression effect via its interaction with Sir2, Sir3 and Sir4 proteins (telomere silencing or TPE) (Kyrion et al. 1993; Shore 1994; Marcand et al. 1997; Pardo and Marcand 2005).

To regulate telomere length, Rap1 recruits Rif1, Rif2, Sir3 and Sir4 to the telomere through its C terminus to form Rap1-Rif1-Rif2 and Rap1-Sir3-Sir4 complexes at the telomeres (Moretti et al. 1994; Grunstein 1997; Kaizer et al. 2015) (Figure 5.1). While the binding of Rap1 to each telomeric site is independent of neighboring sites, structural analysis of these proteins shows that the interaction of Rif1 and Rif2 with Rap1 forms a molecular velcro that links the double-stranded proteins of a single telomere together (Shi et al. 2013; Mattarocci et al. 2017). By interacting with proteins of the Sir2-Sir3-Sir4 complex, Rap1 promotes the compaction of telomeres and the subtelomeric region (Moretti et al. 1994; Moretti and Shore 2001). This heterochromatin directly impacts the TPE and leads to transcriptional repression of genes located in these regions (Gottschling et al. 1990; Gilson et al. 1993; Ottaviani et al. 2008).

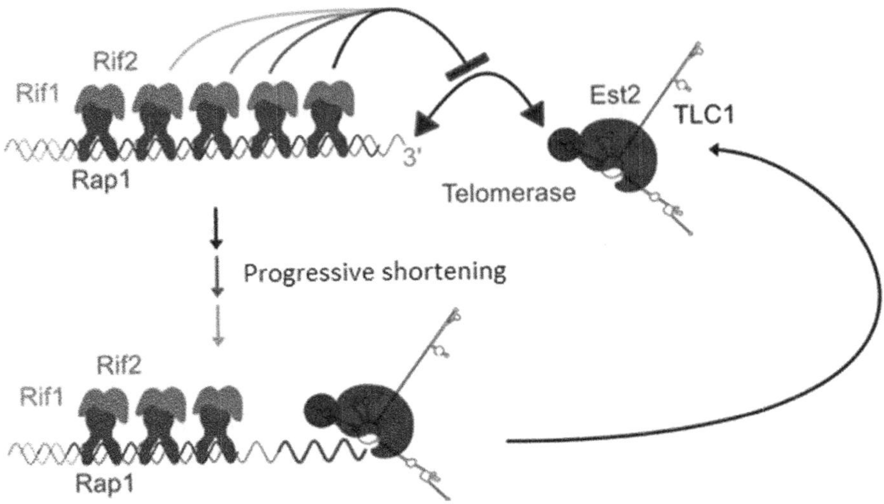

Figure 5.1. *Telomeric proteins present on the telomeres of the yeast Saccharomyces cerevisiae*

Rap1 binds to telomeric double-stranded repeats and allows the recruitment of Rif1 and Rif2 to telomeres. These telomeric proteins play a key role in controlling telomere length by inhibiting the action of telomerase. Telomerase is a reverse transcriptase composed of three protein subunits, Est1, Est2 and Est3 (only two are represented), responsible for catalytic activity and an RNA sequence, TLC1, used as a template for telomere lengthening.

The yeast Ku complex (yKu) plays a central role in the maintenance and repair by NHEJ of DNA ends generated during double-strand breaks (Emerson and Bertuch 2016). While telomeres protect themselves against this repair mechanism thanks to Rap1, Ku is found associated with telomeres and seems to play an essential role in the maintenance of telomere size, structure and function (Porter et al. 1996; Gravel et al. 1998). Indeed, a Ku deletion mutant has extremely short telomeres, particularly the strand containing the 5' end, but the mechanism involved is not yet well clarified. Through its DNA end binding (DEB) domain, Ku is able to bind directly to the telomere (Lopez et al. 2011), but also binds to and stabilizes TLC1, the RNA subunit of telomerase (Gallardo and Chartrand 2008; Mozdy et al.

2008; Zappulla et al. 2011; Williams et al. 2014). Ku also interacts with Sir4 and this interaction would be involved in the recruitment of telomerase (Hass and Zappulla 2015; Chen et al. 2018). Finally, the role of Ku in the maintenance of telomere size would also be via its effects on the Est1 subunit of telomerase. Indeed, the ability of Est1 to bind to telomeres is strongly reduced in the *yku70-R456E* mutant showing a defect in association with telomeres and DNA ends (Lemon et al. 2019). Thus, the loss of Ku's ability to bind DNA ends is sufficient to cause the short telomere phenotype.

5.3.2.2. *Telomeric single strand proteins*

In *S. cerevisiae*, the single-stranded 3' terminal DNA exiting the telomere is bound by the Cdc13 protein, a component of the CST(Cdc13-Stn1-Ten1) complex (Lin and Zakian 1996; Hughes et al. 2000). It plays a role both in protecting the end from degradation and in controlling telomere size homeostasis through telomerase recruitment. A heat-sensitive mutant *cdc13-1* reveals the protective function of Cdc13 as the mutants show a rapid accumulation of several kilobases of 3' ssDNA. A study conducted in the laboratory of Lundblad pointed out that the loss of viability of the *cdc13-1* mutant was the consequence of a sensitivity of the mutant to high temperatures coupled with a Cdc13 defect. This study thus questioned the conclusions of the genetic analyses of this *cdc13-1* mutant (Paschini et al. 2012). However, a recent study, using new heat-sensitive cdc13 mutants, established by Lundblad's laboratory, confirms the protective role of Cdc13 against telomere 5′-3′ resection (Westmoreland et al. 2018). The use of Cdc13 coupled to an inducible degradation system (the degron system) and placed in cells synchronized in the cell cycle made it possible to show that protection by the Cdc13 complex was required only at the end of S phase and in G2/M, but not in G1 phase or at the beginning of the S phase (Vodenicharov and Wellinger 2006, 2007; Vodenicharov et al. 2010).

By forming a complex with Cdc13, the proteins Stn1 and Ten1 are important for telomere protection (Grandin et al. 1997, 2001; Martin et al. 2007). Stn1 and Ten1 are also able to bind themselves to telomeric ssDNA, since their overexpression allows the rescue of Cdc13-1 mutants. Homologues of Stn1 and Ten1 have been identified in *S. pombe* (Martin et al. 2007; Miyake et al. 2009), plants (Surovtseva et al. 2009) and mammals (Miyake et al. 2009). Currently, there is no homolog of Cdc13 in *S. pombe*, but in mammals and plants, a protein, CTC1, shares common properties with Cdc13 and the CST complex has roles in replication beyond telomeres in these organisms, particularly in protecting double-strand breaks from 5'-3' resection (Surovtseva et al. 2009; Giraud-Panis et al. 2010; Chen et al. 2012; Mirman et al. 2018).

The CST complex has similarities in sequence and mode of action to the RPA complex and is therefore also referred to as t-RPA: for telomeric RPA (Gao et al. 2007). Cdc13 and CTC1 also share the ability to bind to polymerase alpha-primase for synthesis of the initiating RNA fragment of Okazaki fragments. It is therefore possible that CST protection of the ends is exclusively through an activity of stimulation of DNA synthesis initiation (Giraud-Panis et al. 2010; Lue et al. 2014; Mirman et al. 2018).

It should be noted that *S. cerevisiae* does not have a homolog of POT1, which is present in most of the other organisms studied including *S. pombe* and mammals. POT1 is involved in the protection of the single telomere strand by participating in the shelterin complex, and thus would carry some of the functions of Cdc13.

5.3.3. *TERRA or telomere transcription*

Despite the heterochromatin marks found in these regions, the subtelomeric sequences are transcribed into different non-coding RNAs (Luke et al. 2008; Bah and Azzalin 2012). One of these RNAs, named TERRA, is composed in 5' of subtelomere, followed by the telomeric UG-rich repeats (Azzalin et al. 2007; Iglesias et al. 2011) and its expression is conserved in most studied organisms. As in mammals, in *S. cerevisiae,* TERRA is transcribed by RNA polymerase II (Luke et al. 2008; Schoeftner and Blasco 2010; Iglesias et al. 2011) and has a variable size that correlates with telomere size, suggesting that RNA polymerase is capable of transcribing all the way to the end of chromosomes. In yeast, TERRA is rapidly degraded by an exonuclease named Rat1, involved in transcription termination (Luke et al. 2008), but also by RNAse H2 recruited by Rif2 to limit the formation of RNA:DNA hybrid structures in R-loops (Graf et al. 2017). Thus, TERRA and TERRA R-loops levels increase with a decrease in Rif2 at the telomere, that is, with a decrease in telomere size (Balk et al. 2014). Using the in situ hybridization technique of a fluorescent RNA in yeast, work by Chartrand's team showed a co-localization of telomerase and TERRA at the beginning of S-phase, suggesting that TERRA would play a role in the recruitment of telomerase to the shortest telomere by a still unknown mechanism (Cusanelli et al. 2013). In the absence of telomerase, R-loops formed by TERRA at the shortest telomeres would promote activation of the recombinational DNA repair pathway via recruitment of Rad51 (Graf et al. 2017). These data are supported by studies in mammals. Indeed, alternative lengthening of telomeres (ALT) cancer cells that are telomerase negative and maintain telomeres through homologous recombination-dependent mechanisms have elevated levels of R-loops (Arora et al. 2014) and higher levels of TERRA than telomerase positive cancer cells (Ng et al. 2009). Although many questions remain, including how TERRA interacts with telomerase, TERRA clearly appears to be a key regulator in telomere maintenance.

5.4. Telomere maintenance

5.4.1. *The difficulties of moving replication fork through telomeres*

In yeast and other organisms, studies have shown that the rate of replication fork progression is slower in telomeric sequences than in the rest of the genome (Ivessa et al. 2002; Miller et al. 2006; Bah et al. 2011). Moreover, unlike the rest of the genome, where a stalled replication fork can be rescued by the arrival of a second fork from the opposite direction, this rescue cannot take place at telomeres that are thought to lack replication origins.

The exact sources of the difficulties of replication fork passage through telomeric repeats are not yet well established. Among the potential causes is the fact that telomeres are GC base pair-rich regions, which makes the double strand very stable (G and C bases are linked by three hydrogen bonds). Second, the asymmetry of guanine richness on one of the strands, the one containing the 3'-overhang, gives telomeres the possibility to form G-quadruplexes, which are secondary structures comprising multiple consecutive planar structures comprising bonds between 4 G and a cation. Due to their structural complexity, G-quadruplexes are difficult to replicate and are therefore considered a source of genetic instability (Lopes et al. 2011), but their existence at yeast telomeres has not been proven. In 1999, the first microscopic observations revealed that the ends of mammalian chromosomes could also fold back on themselves and form a large terminal loop, called the "T-loop" (Griffith et al. 1999). This loop is made possible by the insertion of the 3' single-stranded end of the telomere into its double-stranded portion and would provide additional protection to the telomeres, but would also constitute an additional obstacle to replication. However, T-loops have not been detected in *S. cerevisiae*. On the other hand, it is possible that the presence of R-loops formed by TERRA and other telomeric transcripts constitutes a barrier to replication. It has also been proposed that the strong binding between Rap1 and telomeric sequences poses a problem for the fork progression (Goto et al. 2015).

5.4.2. *The end replication problem*

The replication machinery poses a problem for the replication of the ends of DNA molecules. This problem was first postulated in the 1970s by Watson and Olovnikov following the discovery that the mechanisms of cellular DNA synthesis do not allow the complete replication of the ends (Watson 1972; Olovnikov 1973). These authors propose, on the one hand, that the distance separating the ends of a polymerase from its catalytic region generates a region of template DNA at the DNA ends that is not copied by the polymerase. On the other hand, all DNA polymerases synthesize DNA in the 5' to 3' direction and require an RNA primer of 8–12 nucleotides for eukaryotic cells, which is degraded during replication in the process of maturation of Okazaki fragments. Conventional replication then inevitably generates an uncopied 5' extremity region left by the potential degradation of this RNA primer on the newly synthesized lagging strand. Years later, the discovery of the presence of an 3'-overhang end on both leading and lagging telomeres, and thus not just on the lagging strand, was thus unexpected and showed that telomere replication is in fact even more complicated than originally thought (Wellinger et al. 1993). This has led to the idea that it is the presence of the overhangs that causes the shortening during the synthesis of the leading strand due to a lack of template, independently of the origin of the overhang (Figure 5.2). Thus, the size of the telomeric single strand end determines the gap left on the newly synthesized leading strand and thus the rate of shortening at each passage of the replication fork (Lingner et al. 1995).

When the replisome reaches the telomere, Cdc13 assists the replication of the lagging strand. The last Okazaki fragment is initiated near the 3' end, leaving a gap of 5–10 nts. Synthesis of the leading strand can proceed up to the 5' end of the template, generating a blunt-ended intermediate product. This blunt end is quickly taken over by the Tel1/MRX/Sae2-Sgs1-Exo dependent 5'-3' resection pathway, which generates a single-stranded 3' end of approximately 40 nucleotides. Then, the C strand is filled by again using the polymerases of the lagging strand, in particular the polymerase alpha/primase. This synthesis would be similar to the synthesis of the last Okazaki fragment of the lagging strand, leaving a single-stranded end of 5–10 nts. In this model, telomeres shorten at a rate of 2.5–5 bp/population doubling, which has been verified experimentally.

This set of theoretical problems has been progressively confirmed *in vivo*. The lagging strand synthesis phase is handled by DNA polymerase α/primase that is recruited to the 3' end of the chromosome by Cdc13, which, in addition to recruiting the complex, also stimulates the transition between the RNA synthesis by primase and the DNA synthesis by polymerase α (Qi and Zakian 2000; Lue et al. 2014). It is thought that the degradation of the RNA primer by a still unknown mechanism allows the re-establishment of the 3' tail of the guanine-rich strand (G strand) at the lagging strand and thus the natural structure of this telomere. At the leading strand, this mechanism cannot take place and in theory, DNA synthesis could take place up to the end of the 5' of the template. However, studies carried out on different species have shown that a single-stranded end is present on most strands, and therefore potentially also at the level of the strand synthesized by the leading synthesis machinery (Henderson and Blackburn 1989; Wellinger et al. 1993; Raices et al. 2008). Although some of the steps are now known for mammals (Wu et al. 2012), it is the extensive studies in *S. cerevisiae* that have elucidated some of the details of the mechanisms of "overhang" reconstitution on both telomeric ends. In fact, it has been shown that the leading strand alone undergoes further maturation in the form of 5'-3' resection followed by DNA resynthesis that leads to the restoration of the 3' overhang. Indeed, the size of this single-stranded region increases during the S phase (Wellinger et al. 1993). More precisely, Soudet et al. (2014) have shown, thanks to a method that allows us to analyze the leading and lagging strands resulting from replication independently, that this additional 5'-3' resection takes place specifically on the leading strand.

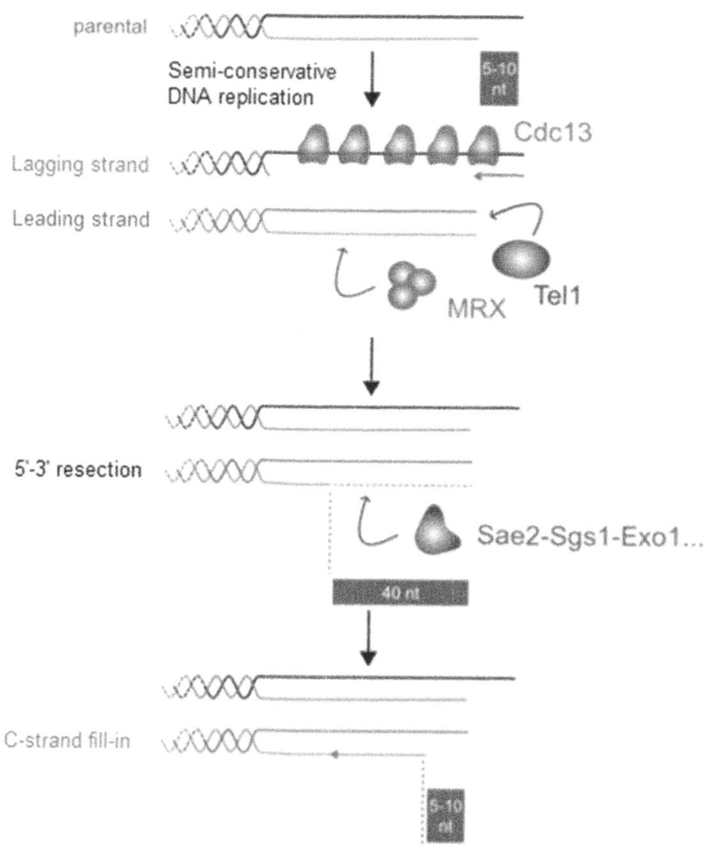

Figure 5.2. *Model of DNA end replication in Saccharomyces cerevisiae*

In *S. cerevisiae*, this 5'-3' resection is mediated by the Sae2/MRX protein complex (MRE11, Rad50, Xrs2), followed by the action of Sgs1 (RecQ family helicase) with Dna2 and Exo1 (Bonetti et al. 2009; Mimitou and Symington 2010) under the control of telomeric proteins that limit the extent of resection (Bonetti et al. 2010). The fact that resection occurs only on the leading strand is also consistent with a specific enrichment of Mre11 at the telomeric strand synthesized by the leading synthesis machinery (Faure et al. 2010).

Thus, a 3'-overhang end is reconstituted at each passage of the replication fork and this particular structure has a major consequence: the ssDNA portion cannot be replicated because of the lack of template during the synthesis of the leading strand and it is this step that will determine the shortening. Thus, without an adapted mechanism to compensate for this loss, chromosomes are bound to shorten at each passage of the replication fork and thus at each cell cycle. For *Saccharomyces cerevisiae*, this shortening is 3–5 base pairs per cell division (Singer and Gottschling 1994; Marcand et al. 1999; Larrivee et al. 2004).

The DNA end replication problem thus arises both from the absence of a terminal template on the leading strand, which leads to a loss of sequence on the daughter chromatid, and from the reconstitution after each replication of 3′-overhang at the origin of this absence of template. This problem is then recreated and transmitted from generation to generation.

Telomere replication is one of the last events of S phase in *S. cerevisiae* (Ferguson and Fangman 1992; Donaldson 2005). One of the causes of this late replication could be the presence of Sir3 and the heterodimer (Yku70/Yku80) involved in the positioning of telomeres at the periphery of the nucleus and the silencing of genes near telomeres (Stevenson and Gottschling 1999; Cosgrove et al. 2002). Another study also showed that Rif1 (Rap1-interacting factor 1) could be responsible for late telomere replication. Indeed, a mutation of Rif1 causes an earlier replication of telomeres (Bianchi and Shore 2007). Another study reveals the role of Tel1, which when recruited to a short telomere, phosphorylates Rif1 and thus allows earlier replication of the latter (Sridhar et al. 2014). More recent data reveal a major role for Rif1 in the control of replication initiation by directly binding to some origins of replication located at telomeres and elsewhere in the genome (Mattarocci et al. 2014; Hafner et al. 2018, 2019; Hiraga et al. 2018).

5.4.3. *Telomerase: a solution to the problem of telomere shortening*

For most eukaryotic cells, the problem of complete telomere replication is solved by the action of a specialized reverse transcriptase: telomerase (Blackburn et al. 2006).

Telomerase is a ribonucleoprotein complex (RNP) composed of an RNA that serves as a template for the synthesis of telomeric repeats. Telomerase activity was detected by Greider and Blackburn in extracts of the single-cell organism *Tetrahymena thermophila*. By transforming the yeast *S. cerevisiae* with a linear plasmid carrying telomeric ends from *Tetrahymena*, they observed that the latter were elongated following the addition of yeast telomeric sequences. This result served as evidence for the existence of an activity capable of specifically lengthening telomeres (Greider and Blackburn 1985, 1987). Furthermore, this activity depended on the presence of a telomerase-associated RNA, which contains a small sequence complementary to the telomeric repeats (Greider and Blackburn 1987, 1989).

In parallel, in order to identify the biochemical components of telomerase, Lundblad carried out a search for mutants with a defect in telomere maintenance in yeast. She showed that one of these mutants had telomeres that shortened with each cell division, which she named *est1*, for its "ever shorter telomeres" phenotype (Lundblad and Szostak 1989). This result highlighted the problem of end replication and the existence of an enzymatic activity essential for maintaining a constant telomere size.

Three other mutants, with a phenotype similar to that observed for *est1*, were then characterized, *est2*, *est3* and *est4*. Est2 encodes the only subunit capable of reverse transcriptase activity *in vitro* and corresponds to the catalytic subunit of telomerase. It was identified by sequence homology with the ciliate protein that co-purified with telomeric primer extension activity *in vitro* and had regions of homology with viral reverse transcriptases (Lingner et al. 1997). Like Est1, Est3 is a regulatory subunit of telomerase, and Est4 corresponds to Cdc13. *TLC1* encodes telomerase RNA in *S. cerevisiae* (Singer and Gottschling 1994). This RNA has a 16 nucleotides sequence complementary to the telomeric repeats. Thus, Est1-2 and TLC1 form a stable complex *in vitro* that associates with Est3 at the end of S phase, limiting telomerase activity at this phase of the cycle, before detachment of Est2 (Tucey and Lundblad 2014).

Telomerase activity is subject to other regulations. Its nuclear import is regulated by Mtr10 (mRNA transport defective protein 10) and Kap122 (karyopherin protein 122), and is stimulated by the interaction between TLC1 and Ku (Gallardo and Chartrand 2008). Once in the nucleus, telomerase can be recruited to telomeres. In *S. cerevisiae*, telomerase complexes cluster at telomeres that express TERRA, but it is not yet known to what extent this clustering is required for telomere lengthening (Gallardo et al. 2011; Cusanelli et al. 2013).

During telomere lengthening, telomerase uses its RNA template to associate by complementarity with the 3' end of the telomeric single strand (the G strand), and lengthen this strand. The length of the added sequences is very variable, but on average is 40 base pairs *in vivo* (Teixeira et al. 2004). This corresponds to reverse transcription of the telomeric template in an iterative fashion, whereas the enzyme extracted from yeast has very low activity *in vitro*. Telomerase lengthens only the G strand (the strand that ends at the telomere with a 3' single stranded end), but its activity is coupled to that of polymerases α and δ that fill-in the C strand (strand that ends at the telomere with a 5' end), thus synchronizing the elongation of the G and C strands (Diede and Gottschling 1999). In contrast, single-stranded telomeric DNA exists even in the absence of telomerase, indicating that single-stranded telomeric formation is independent of elongation by telomerase.

5.5. Telomeres: one size for one identity

Telomere length is extremely variable between organisms, ranging from a few hundred nucleotides in *S. cerevisiae* to several tens of kb in mice. Telomere length is also highly variable within a cell, each telomere having a different length. Despite this heterogeneity, telomerase manages to maintain telomeres at a size within a specific range for each organism, this is called telomere size homeostasis, described very early in *S. cerevisiae* (Shampay and Blackburn 1988). Since not all telomeres display the same size, telomerase acts on only a small proportion of the shortest telomeres to which it adds a variable number of repeats. The recruitment of telomerase to telomeres is dependent on both telomeric proteins and telomerase subunits. Indeed, a complex set of interactions between telomeric proteins and telomerase subunits allows its recruitment and its controlled activation, thus ensuring the maintenance of the cell's telomeres at a certain size homeostasis. The shorter a telomere is, the higher its elongation frequency, in agreement with a model of telomere length counting (Marcand et al. 1997; Teixeira et al. 2004). The STEX (Single Telomere EXtension assay) was one of the most relevant experiments to demonstrate the preferential activity of telomerase at shorter telomeres (Teixeira et al. 2004). Indeed, this technique allows us to follow the elongation of telomeres individually during an S phase. Thus, not all telomeres are elongated at each cell cycle, with a preferential action of telomerase at the shortest telomeres. This results in a dynamic equilibrium, which centers telomere size around an average length.

Other experimental systems set in yeast have served to characterize the telomerase activity. Telomerase-dependent elongation can take place in the G2/M phase at an artificial telomere created de novo by an HO cut (Diede and Gottschling 1999). A system allowing the generation of a single short telomere has shown that the recruitment of telomerase was only possible at the end of the S/G2 phase, but never at G1 or the beginning of S phase. Indeed, the association of telomerase with telomeres occurs preferentially in S phase (Marcand et al. 2000). This limitation of the telomerase action window can be explained by the abundance of telomerase-forming proteins at the end of S/G2 phase (Osterhage et al. 2006; Tuzon et al. 2011).

A study, published in 1997 by Shore's team, presents the number of molecules of the Rap1 protein as a key player in the regulation of telomere size; this is the "protein counting mechanism" (Figure 5.1). Thus, during incomplete telomere replication or telomere degradation, the loss of binding sites for Rap1 causes the telomere to switch to a new state which would then favor elongation. This elongation of the telomere allows the restoration of the Rap1 binding sites, which will ensure a return to the initial state of the telomere and repress the action of telomerase. The results of this study, supported by others, reinforce the idea of a threshold number of Rap1 proteins required to prevent telomere elongation (Marcand et al. 1997). In addition to Rap1, Rif proteins also appear to restrict the action of telomerase at the end of the S phase. Thus, in the absence of Rif proteins, short telomeres can be elongated in G1 (Gallardo et al. 2011). This recruitment is anticipated in the S phase when telomeres are short, consistent with early telomere replication (Bianchi and Shore 2007). Therefore, telomere elongation appears to be coordinated with their semi-conservative replication.

In *S. cerevisiae*, Est1 and Est2 are preferentially recruited to artificially shortened telomeres. This recruitment depends on the presence and recruitment to short telomeres of the DNA damage checkpoint kinase Tel1 (Tsukamoto et al. 2001; Goudsouzian et al. 2006; Sabourin et al. 2007; Keener et al. 2019). The stability of this recruitment depends on its interaction with the MRX complex, which is also preferential to short telomeres (McGee et al. 2010). Recruitment of telomerase to the telomere is also dependent on the Cdc13 protein (Lewis and Wuttke 2012), which interacts with the Est1 subunit of telomerase (Evans and Lundblad 1999; Wellinger and Zakian 2012). The Ku heterodimer also plays a role in telomerase recruitment by affecting its import and/or retention in the nucleus (Nandakumar and Cech 2013). Although shortened, a short telomere differs from longer telomeres in its protein composition and remains distinct from a double-strand break. Nevertheless, when further shortened, telomeres also expose subtelomeric single-stranded regions responsible for Mec1/ATR activation (Fallet et al. 2014). This shortest telomere also recruits DNA damage and replicative stress response factors. Nevertheless, a cell remains capable of distinguishing a double-stranded break from a telomere, in the presence of a limited number of telomeric repeats. Characterization of telomere-mediated activation of cell cycle checkpoints was addressed by Ribeyre and Shore (2012). By bordering a double-strand break of telomeric sequences of variable size in *S cerevisiae*, they observed that telomeric sequences have, via Rif1, an intrinsic ability to inhibit checkpoints.

Mutants in many proteins, telomeric or not, can affect telomere length. Environmental stresses can also affect telomere length. For example, after ethanol treatment telomeres are longer, while they shorten in response to increased temperature or oxidative stress (von Zglinicki 2002; Romano et al. 2013).

The telomeric ends of chromosomes are therefore highly organized nucleoprotein structures, and their identity changes as they become short. The maintenance of their size and structure ensures the protection of chromosomes and is therefore essential to maintain genome stability and cell viability.

5.6. Telomeres, replicative senescence, aging and genomic instability

5.6.1. *Aging and replicative senescence*

In the early 1960s, cellular senescence was described in human somatic cells by Hayflick and Moorhead (1961) as a progressive and irreversible arrest of the proliferative potential of a cell. This state is also accompanied by a set of drastic changes in cell morphology, gene expression, metabolism, epigenetics, etc. (van Deursen 2014). This phenotype is stable over time since senescent cells can be maintained in culture for several years after the first cycle arrest. Thus, the cell is only capable of a limited number of divisions known as the "Hayflick limit" (Hayflick and Moorhead 1961). This limit is set by the initial size of the telomeres which, in the absence of telomerase, progressively shorten until the arrest of the cell cycle that is replicative senescence (Harley et al. 1990). Indeed, ectopic expression of telomerase, which counteracts telomere shortening, circumvents senescence (Bodnar et al. 1998). This experiment thus demonstrates that telomere length is a limiting factor for cell proliferation and a causal factor for senescence.

While in unicellular organisms such as *S. cerevisiae*, telomeres are maintained at a constant length by the constitutive action of telomerase, this is not the case in many multicellular organisms. In humans, telomerase is active in germline tissues, but not in normal somatic tissues. Thus, the initial size of telomeres in a culture of primary fibroblasts, as well as other cell types, decreases progressively with increasing age of the individual, before decreasing further as the cells pass through (Harley et al. 1990; Hastie et al. 1990). The re-introduction of telomerase alone is sufficient to restore the proliferative capacity of the cells, thus demonstrating that these cells remain metabolically active (Bodnar et al. 1998).

In *S. cerevisiae*, replicative senescence is not a physiological event. Nevertheless, it can be experimentally achieved by blocking the expression of one of the telomerase genes, as described for the *EST* genes (Figure 5.3). Thus, after 40–60 population doublings of *est* mutant cell cultures, telomeres progressively shorten and a progressive loss of proliferative capacity is observed. Replicative senescence is characterized by an increased frequency in these populations of cells blocked in the G2/M phase of the cell cycle, corresponding to the peak of senescence. This evolution of cell populations is accompanied by morphological and gene expression changes and an increase in genomic instability (Hackett et al. 2001). Subsequently, a subpopulation with alternative telomere maintenance mechanisms emerges (Lundblad and Blackburn 1993). These cells, called "survivors", then invade the population and restore its viability.

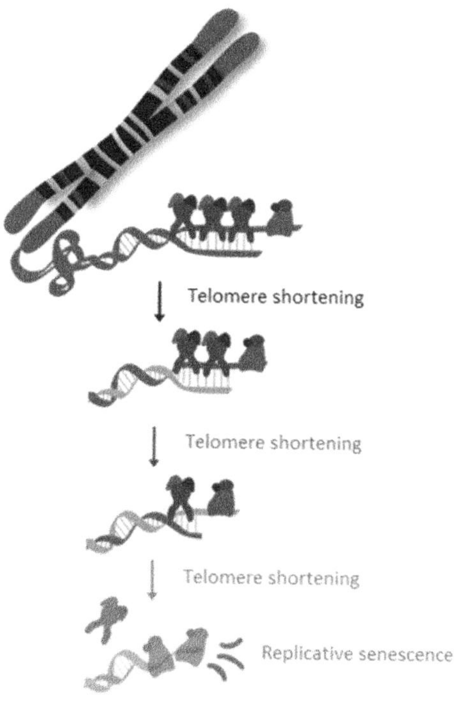

Figure 5.3. *Replicative senescence: the example of the model organism Saccharomyces cerevisiae. The DNA end replication causes the progressive shortening of telomeres. This shortening is accompanied by the loss of telomere-associated proteins such as Rap1, Rif1, Rif2 and the inhibitory effect that these proteins have on telomerase action at the telomere, but also on the DNA damage response. In the absence of telomerase, the presence of a single very short telomere is sufficient to cause checkpoint activation and replicative senescence*

Knowing that telomere size is heterogeneous within a cell, it is legitimate to ask whether the induction of senescence is generated by the shortest telomeres among the short telomeres. In *S. cerevisiae*, the senescence signal in telomerase-negative cells comes from the first telomere to reach a critical reduced size (Abdallah et al. 2009; Khadaroo et al. 2009). These results were consolidated by mathematical simulations and formal genetics (Xu et al. 2013). Consistent with this idea of senescence being driven by the shortest telomeres, five dysfunctional telomeres are sufficient to observe senescence in human cells in culture (Kaul et al. 2011). Telomeres are thus central to the control of senescence, and it is important to understand how they are maintained and recognized during this process. Indeed, as telomeres shorten, their identity evolves toward a state that allows them to be recognized by the double-strand break response machinery (d'Adda di Fagagna et al. 2003; Abdallah et al. 2009). The activation of this DNA damage signaling leads to the initiation of cell division arrest, which is responsible for senescence (Teixeira 2013).

Independent of telomere size, studies suggest that other factors are able to influence the intraclonal setting of senescence.

5.6.2. *Heterogeneity of senescence at the single cell level*

For the majority of organisms where telomerase has been inactivated, the senescence phenotype is very heterogeneous at both the interclonal and intraclonal level. Indeed, if telomerase is inactivated in independent isogenic yeast clones, their proliferative capacity can be very different until the culture stops growing. This is also the case for subclones of the same clone. Due to this heterogeneity, access to molecular data of replicative senescence remains complex and difficult to interpret, which has led to finding ways to circumvent this heterogeneity. This was possible in *S. cerevisiae* and led to the discovery that there was an intrinsic link between the molecular basis of senescence and the generation of this heterogeneity. In other words, the heterogeneity of senescence is not due to a technical problem of manipulation or the ability to measure cell proliferation, but is largely related to the biology of telomeres, their replication and maintenance.

Thus, in order to characterize the key factors involved in the heterogeneity of senescence, time-resolved micromanipulation or microfluidic experiments at the single-cell level have been developed. These data obtained in single cell allow us to unravel the dynamics of cell division from telomerase inactivation to permanent cell cycle arrest and cell death. These studies have shown that there is a large variability in the number of generations from telomerase inactivation to cell cycle arrest (ranging from 10 to 70). In addition, although a majority of cell lineages abruptly change from active cell division to two or three extended cell cycles and finally cell death, other cells show early extended cell cycles followed by a return to normal cell cycles. These prolonged cell cycles that occur before senescence and are followed by cycles of normal duration, called non-terminal prolonged cycles, as well as variations in the number of generations until senescence or cell death, can alone account for a large portion of the heterogeneity of cell growth in the absence of telomerase.

Prolonged non-terminal cell cycles observed in the absence of telomerase at the single cell level are caused by activation of checkpoints induced during DNA damage and are more frequent as telomeres shorten (Xu et al. 2015; Coutelier et al. 2018; Martin et al. 2021). However, the molecular causes of this early activation of checkpoints are not yet known. One possibility is that telomere replication, which is known to be difficult, causes replication fork blockages in telomeric regions, or even double-strand breaks. Since DNA repair mechanisms are repressed at telomeres, these replication problems would be all the more significant and would represent one of the factors contributing to the heterogeneity of cell growth in the absence of telomerase.

One of the intriguing observations of telomerase-negative yeast cells in microfluidics is that prolonged cycles can occur, which raised the question of how cells with activated DNA damage checkpoints manage to divide. The answer is that a phenomenon, called adaptation to DNA damage, comes into play. In fact, in the absence of the removal of checkpoint activation, cells with unrepairable DNA damage can occasionally resume mitosis. It is thought that this process gives the cell the opportunity to try out repair mechanisms present in other phases of the cell cycle (Sandell and Zakian 1993; Toczyski et al. 1997; Lee et al. 1998). However, this mechanism results in increased genomic instability, indicating that survival of the damaged cell is at the expense of genomic integrity. As a result, when adaptation occurs during prolonged non-terminal cell cycles followed by normal cycles, genomic instability accumulates in cells with significant proliferative capacity. Thus, in the absence of telomerase, adaptation to damage appears as an additional mechanism for increasing the heterogeneity of senescence through a mutational effect and is able to generate genomic variants, even though DNA damage checkpoints are still intact (Coutelier et al. 2018).

Thus, single-cell studies have allowed us not only to overcome intercellular variability, but also to identify and better understand molecular mechanisms that may play a role in the heterogeneity of senescence.

5.7. Senescence escape phenomena and genomic instability

In populations, some yeasts manage, despite the absence of telomerase, to resume their cell viability after entering senescence. This recovery is due to the establishment of escape mechanisms that correspond to alternative mechanisms of telomere maintenance by a small number of cells. These cells that have recovered a proliferative capacity and a size similar to that of wild-type cells are called survivors (Lundblad and Blackburn 1993).

In *S. cerevisiae*, two types of survivors can appear and absolutely require the intervention of Rad52 and Pol32 proteins (a subunit of DNA polymerase δ), indicating a mode of telomere maintenance via homologous recombination in the form of an alternative mechanism for telomere maintenance (ALT) (Lundblad and Blackburn 1993; McEachern and Haber 2006; Lydeard et al. 2007).

Type I survivors are obtained by adding the subtelomeric Y' elements on all chromosomes. They retain only a short region of terminal telomeric repeats. In addition to the involvement of Rad52 and pol32, the recombination proteins Rad51, Rad54, Rad55 and Rad57 are also required (Le et al. 1999). In contrast, type II survivors are characterized by a sudden elongation in telomere length. Their telomeric $TG_{1\text{-}3}$ repeats are very long (several kilobases) and very heterogeneous in size. They progressively shorten with cell divisions and undergo constant recombination events to maintain cell proliferation. Their formation seems to be favored when telomeres are longer at the time of senescence (Chen et al. 2001; Larrivee and Wellinger 2006).

Type I survivors, which account for 90% of survivors, therefore appear more frequently than type II survivors. However, they have a slower growth rate, which explains why type II survivors eventually invade liquid cultures. In both types of survivors, reintroduction of telomerase restores normal homeostasis at telomeres (Makovets et al. 2008). Circular molecules of Y' DNA were detected in type I survivors, and circular molecules containing only telomeric repeats, sometimes single stranded, were detected in type II survivors (Larrivee and Wellinger 2006). It was therefore proposed that the telomeres of survivors are maintained by a rolling circle replication mechanism, less favored than the action of telomerase, a mechanism formally demonstrated in the yeast *Kluyveromyces lactis* (Topcu et al. 2005). A "break induced replication" (BIR) mechanism has also been proposed, as telomere emergence is also totally dependent on Pol32, a factor especially required during BIR (Lydeard et al. 2007). Despite the seemingly different genetic dependencies for the establishment of the two types of survivors in *S. cerevisiae*, a unified model has recently been proposed, where a succession of Rad51-dependent events at first, and Rad59-dependent events thereafter, allows for long-term telomere maintenance in the absence of telomerase (Kockler et al. 2021). In this study, the authors also estimate the frequency of emergence of survivors at ~10^{-5}–10^{-4}.

Many other species have senescence escape mechanisms, the nature of which may vary. The most frequent survivors of the yeast *S. pombe* lacking telomerase activity show that all three chromosomes were able to circularize and thus get rid of the problem of shortening of the DNA ends (Nakamura et al. 1998). This ability to fuse the two shortened telomeres of a single chromosome could simply be explained by the small number of telomeres present in each cell (six in this organism) and the non-negligible possibility of selecting the events that lead to the fusions between the two ends of the same chromosome. Remarkably, these cells are viable, in the same way as synthetic *S. cerevisiae* yeast with a single circular chromosome, but are unable to undergo meiosis, demonstrating the importance of chromosome and telomere linearity in sexual reproduction in eukaryotes (Nakamura et al. 1998; Shao et al. 2019). Extensive studies in this organism have revealed other alternative mechanisms of telomere maintenance based on the ability to form heterochromatin (Jain et al. 2010; Tihan et al. 2017).

In humans, telomerase is reactivated in most cancers (Kim et al. 1994; Wright et al. 1996). It is proposed that this reactivation is not necessary for the initiation of tumorigenesis, but that it is necessary for the establishment of immortality in these cells, and thus promotes the progression of already established tumors. In tumors where telomerase is not reactivated, the very long and heterogeneous telomeres are maintained by a mechanism known as "alternative telomere elongation" (Bryan et al. 1995), dependent on recombination events that rely on basic mechanisms discovered in yeast, such as BIR (Dilley et al. 2016; Apte and Cooper 2017).

In *S. cerevisiae*, after senescence due to telomere shortening, survivors that have recovered their proliferative capacity do not show an increase in gross chromosomal rearrangement (GCR) because telomere maintenance in the absence of telomerase is ensured by homologous recombination (Myung et al. 2001). However, the deletion of certain genes causing a problem in telomere maintenance also induces an increase in GCR (Chen and Kolodner 1999). Early studies, using fluctuation assays based on the loss of a reporter gene, estimated the mutation rate during cell divisions in the absence of telomerase (Hackett et al. 2001; Hackett and Greider 2003). These studies showed that the mutation rate at the ends was 10 times higher than at the centromeric regions, reinforcing the idea that a shortened telomere causes major chromosomal rearrangements at the ends. A recent study, suggested that in the absence of telomerase, cells may become genetically unstable and accumulate mutations early in senescence (Coutelier et al. 2018). Thus, thanks to the succession of telomere damage, the activation of checkpoints and the implementation of adaptation mechanisms to DNA damage, telomerase-negative cells are able to maintain a proliferative capacity over the long term. Then, the combined activation of DNA damage signaling, damage repair and cell cycle pathways result in translocations, gene loss and aneuploidy. Although many questions remain unanswered, these data together provide a better understanding of how non-functional telomeres can induce genomic instability.

5.8. Conclusion

A better understanding of the mechanisms regulating telomere biology is essential for the development of effective treatments in the field of aging and oncology. The yeast *S. cerevisiae*, because of its ease of manipulation and its speed of proliferation, has allowed the development of numerous studies of molecular biology and genetics that have largely contributed to the knowledge of the fundamental mechanisms of telomere biology. The development of new high resolution methods (HiC, molecular combing) and the accessibility of high-throughput sequencing platforms (NGS, nanopore technology, PacBio) will certainly allow future studies to better understand the role of telomeres under physiological and pathological conditions and discover new pathways involved in telomere biology.

5.9. References

Abdallah, P., Luciano, P., Runge, K.W., Lisby, M., Géli, V., Gilson, E., Teixeira, M.T. (2009). A two-step model for senescence triggered by a single critically short telomere. *Nat. Cell Biol.*, 11(8), 988–993.

d'Adda di Fagagna, F., Reaper, P.M., Clay-Farrace, L., Fiegler, H., Carr, P., Von Zglinicki, T., Saretzki, G., Carter, N.P., Jackson, S.P. (2003). A DNA damage checkpoint response in telomere-initiated senescence. *Nature*, 426(6963), 194–198.

Apte, M.S. and Cooper, J.P. (2017). Life and cancer without telomerase: ALT and other strategies for making sure ends (don't) meet. *Crit. Rev. Biochem. Mol. Biol.*, 52(1), 57–73.

Arora, R., Lee, Y., Wischnewski, H., Brun, C.M., Schwarz, T., Azzalin, C.M. (2014). RNaseH1 regulates TERRA-telomeric DNA hybrids and telomere maintenance in ALT tumour cells. *Nat. Commun.*, 5, 5220.

Azzalin, C.M., Reichenbach, P., Khoriauli, L., Giulotto, E., Lingner, J. (2007). Telomeric repeat containing RNA and RNA surveillance factors at mammalian chromosome ends. *Science*, 318(5851), 798–801.

Bah, A. and Azzalin, C.M. (2012). The telomeric transcriptome: From fission yeast to mammals. *Int. J. Biochem. Cell Biol.*, 44(7), 1055–1059.

Bah, A., Gilson, E., Wellinger, R.J. (2011). Telomerase is required to protect chromosomes with vertebrate-type T2AG3 3' ends in *Saccharomyces cerevisiae*. *J. Biol. Chem.*, 286(31), 27132–27138.

Balk, B., Dees, M., Bender, K., Luke, B. (2014). The differential processing of telomeres in response to increased telomeric transcription and RNA-DNA hybrid accumulation. *RNA Biol.*, 11(2), 95–100.

Bergstrom, A., Simpson, J.T., Salinas, F., Barre, B., Parts, L., Zia, A., Nguyen Ba, A.N., Moses, A.M., Louis, E.J., Mustonen, V. et al. (2014). A high-definition view of functional genetic variation from natural yeast genomes. *Mol. Biol. Evol.*, 31(4), 872–888.

Bianchi, A. and Shore, D. (2007). Early replication of short telomeres in budding yeast. *Cell*, 128(6), 1051–1062.

Blackburn, E.H. and Gall, J.G. (1978). A tandemly repeated sequence at the termini of the extrachromosomal ribosomal RNA genes in Tetrahymena. *J. Mol. Biol.*, 120(1), 33–53.

Blackburn, E.H., Greider, C.W., Szostak, J.W. (2006). Telomeres and telomerase: The path from maize, Tetrahymena and yeast to human cancer and aging. *Nat. Med.*, 12(10), 1133–1138.

Bodnar, A.G., Ouellette, M., Frolkis, M., Holt, S.E., Chiu, C.P., Morin, G.B., Harley, C.B., Shay, J.W., Lichtsteiner, S., Wright, W.E. (1998). Extension of life-span by introduction of telomerase into normal human cells. *Science*, 279(5349), 349–352.

Bonetti, D., Martina, M., Clerici, M., Lucchini, G., Longhese, M.P. (2009). Multiple pathways regulate 3' overhang generation at *S. cerevisiae* telomeres. *Mol. Cell*, 35(1), 70–81.

Bonetti, D., Clerici, M., Anbalagan, S., Martina, M., Lucchini, G., Longhese, M.P. (2010). Shelterin-like proteins and Yku inhibit nucleolytic processing of *Saccharomyces cerevisiae* telomeres. *PLoS Genet*, 6(5), e1000966.

Brown, C.A., Murray, A.W., Verstrepen, K.J. (2010). Rapid expansion and functional divergence of subtelomeric gene families in yeasts. *Curr. Biol.*, 20(10), 895–903.

Bryan, T.M., Englezou, A., Gupta, J., Bacchetti, S., Reddel, R.R. (1995). Telomere elongation in immortal human cells without detectable telomerase activity. *EMBO J.*, 14(17), 4240–4248.

Chan, C.S. and Tye, B.K. (1983a). Organization of DNA sequences and replication origins at yeast telomeres. *Cell*, 33(2), 563–573.

Chan, C.S. and Tye, B.K. (1983b). A family of *Saccharomyces cerevisiae* repetitive autonomously replicating sequences that have very similar genomic environments. *J. Mol. Biol.*, 168(3), 505–523.

Chen, C. and Kolodner, R.D. (1999). Gross chromosomal rearrangements in *Saccharomyces cerevisiae* replication and recombination defective mutants. *Nat. Genet.*, 23(1), 81–85.

Chen, Q., Ijpma, A., Greider, C.W. (2001). Two survivor pathways that allow growth in the absence of telomerase are generated by distinct telomere recombination events. *Mol. Cell. Biol.*, 21(5), 1819–1827.

Chen, L.Y., Redon, S., Lingner, J. (2012). The human CST complex is a terminator of telomerase activity. *Nature*, 488(7412), 540–544.

Chen, H., Xue, J., Churikov, D., Hass, E.P., Shi, S., Lemon, L.D., Luciano, P., Bertuch, A.A., Zappulla, D.C., Géli, V. et al. (2018). Structural insights into yeast telomerase recruitment to telomeres. *Cell*, 172(1–2), 331–343e313.

Cosgrove, A.J., Nieduszynski, C.A., Donaldson, A.D. (2002). Ku complex controls the replication time of DNA in telomere regions. *Genes Dev.*, 16(19), 2485–2490.

Coutelier, H., Xu, Z., Morisse, M.C., Lhuillier-Akakpo, M., Pelet, S., Charvin, G., Dubrana, K., Teixeira, M.T. (2018). Adaptation to DNA damage checkpoint in senescent telomerase-negative cells promotes genome instability. *Genes Dev.*, 32(23–24), 1499–1513.

Cusanelli, E., Romero, C.A., Chartrand, P. (2013). Telomeric noncoding RNA TERRA is induced by telomere shortening to nucleate telomerase molecules at short telomeres. *Molecular Cell*, 51(6), 780–791.

van Deursen, J.M. (2014). The role of senescent cells in ageing. *Nature*, 509(7501), 439–446.

Diede, S.J. and Gottschling, D.E. (1999). Telomerase-mediated telomere addition *in vivo* requires DNA primase and DNA polymerases alpha and delta. *Cell*, 99(7), 723–733.

Dilley, R.L., Verma, P., Cho, N.W., Winters, H.D., Wondisford, A.R., Greenberg, R.A. (2016). Break-induced telomere synthesis underlies alternative telomere maintenance. *Nature*, 539(7627), 54–58.

Dionne, I. and Wellinger, R.J. (1998). Processing of telomeric DNA ends requires the passage of a replication fork. *Nucleic Acids Res.*, 26(23), 5365–5371.

Donaldson, A.D. (2005). Shaping time: Chromatin structure and the DNA replication programme. *Trends Genet.*, 21(8), 444–449.

Emerson, C.H. and Bertuch, A.A. (2016). Consider the workhorse: Nonhomologous end-joining in budding yeast. *Biochem. Cell Biol.*, 94(5), 396–406.

Evans, S.K. and Lundblad, V. (1999). Est1 and Cdc13 as comediators of telomerase access. *Science*, 286(5437), 117–120.

Fallet, E., Jolivet, P., Soudet, J., Lisby, M., Gilson, E., Teixeira, M.T. (2014). Length-dependent processing of telomeres in the absence of telomerase. *Nucleic Acids Res.*, 42(6), 3648–3665.

Faure, V., Coulon, S., Hardy, J., Géli, V. (2010). Cdc13 and telomerase bind through different mechanisms at the lagging- and leading-strand telomeres. *Molecular Cell*, 38(6), 842–852.

Ferguson, B.M. and Fangman, W.L. (1992). A position effect on the time of replication origin activation in yeast. *Cell*, 68(2), 333–339.

Feuerhahn, S., Iglesias, N., Panza, A., Porro, A., Lingner, J. (2010). TERRA biogenesis, turnover and implications for function. *FEBS Letters*, 584(17), 3812–3818.

Gallardo, F. and Chartrand, P. (2008). Telomerase biogenesis: The long road before getting to the end. *RNA Biol.*, 5(4), 212–215.

Gallardo, F., Laterreur, N., Cusanelli, E., Ouenzar, F., Querido, E., Wellinger, R.J., Chartrand, P. (2011). Live cell imaging of telomerase RNA dynamics reveals cell cycle-dependent clustering of telomerase at elongating telomeres. *Mol. Cell*, 44(5), 819–827.

Gao, H., Cervantes, R.B., Mandell, E.K., Otero, J.H., Lundblad, V. (2007). RPA-like proteins mediate yeast telomere function. *Nat. Struct. Mol. Biol.*, 14(3), 208.

Gilson, E. and Gasser, S.M. (1995). Repressor activator protein 1 and its ligands: Organising chromatin domains. *Nucleic Acids Mol. Biol.*, 9, 308–327.

Gilson, E. and Géli, V. (2007). How telomeres are replicated. *Nat. Rev. Mol. Cell Biol.*, 8(10), 825–838.

Gilson, E., Laroche, T., Gasser, S.M. (1993a). Telomeres and the functional architecture of the nucleus. *Trends Cell Biol.*, 3(4), 128–134.

Gilson, E., Roberge, M., Giraldo, R., Rhodes, D., Gasser, S.M. (1993b). Distortion of the DNA double helix by RAP1 at silencers and multiple telomeric binding sites. *J. Mol. Biol.*, 231(2), 293–310.

Giraud-Panis, M.J., Teixeira, M.T., Géli, V., Gilson, E. (2010). CST meets shelterin to keep telomeres in check. *Mol. Cell*, 39(5), 665–676.

Goto, G.H., Zencir, S., Hirano, Y., Ogi, H., Ivessa, A., Sugimoto, K. (2015). Binding of multiple rap1 proteins stimulates chromosome breakage induction during DNA replication. *PLoS Genet.*, 11(8), e1005283.

Gottschling, D.E., Aparicio, O.M., Billington, B.L., Zakian, V.A. (1990). Position effect at *S. cerevisiae* telomeres: Reversible repression of Pol II transcription. *Cell*, 63(4), 751–762.

Goudsouzian, L.K., Tuzon, C.T., Zakian, V.A. (2006). *S. cerevisiae* Tel1p and Mre11p are required for normal levels of Est1p and Est2p telomere association. *Mol. Cell*, 24(4), 603–610.

Graf, M., Bonetti, D., Lockhart, A., Serhal, K., Kellner, V., Maicher, A., Jolivet, P., Teixeira, M.T., Luke, B. (2017). Telomere length determines TERRA and R-Loop regulation through the cell cycle. *Cell*, 170(1), 72–85.e14.

Grandin, N., Reed, S.I., Charbonneau, M. (1997). Stn1, a new *Saccharomyces cerevisiae* protein, is implicated in telomere size regulation in association with Cdc13. *Genes Dev.*, 11(4), 512–527.

Grandin, N., Damon, C., Charbonneau, M. (2001). Ten1 functions in telomere end protection and length regulation in association with Stn1 and Cdc13. *EMBO J.*, 20(5), 1173–1183.

Gravel, S., Larrivee, M., Labrecque, P., Wellinger, R.J. (1998). Yeast Ku as a regulator of chromosomal DNA end structure. *Science*, 280(5364), 741–744.

Greider, C.W. and Blackburn, E.H. (1985). Identification of a specific telomere terminal transferase activity in Tetrahymena extracts. *Cell*, 43(2 Pt 1), 405–413.

Greider, C.W. and Blackburn, E.H. (1987). The telomere terminal transferase of Tetrahymena is a ribonucleoprotein enzyme with two kinds of primer specificity. *Cell*, 51(6), 887–898.

Greider, C.W. and Blackburn, E.H. (1989). A telomeric sequence in the RNA of Tetrahymena telomerase required for telomere repeat synthesis. *Nature*, 337(6205), 331–337.

Griffith, J.D., Comeau, L., Rosenfield, S., Stansel, R.M., Bianchi, A., Moss, H., de Lange, T. (1999). Mammalian telomeres end in a large duplex loop. *Cell*, 97(4), 503–514.

Grunstein, M. (1997). Molecular model for telomeric heterochromatin in yeast. *Curr. Opin. Cell Biol.*, 9(3), 383–387.

Haber, D.A. (1995). Telomeres, cancer, and immortality. *N. Engl. J. Med.*, 332(14), 955–956.

Hackett, J.A. and Greider, C.W. (2003). End resection initiates genomic instability in the absence of telomerase. *Mol. Cell Biol.*, 23(23), 8450–8461.

Hackett, J.A., Feldser, D.M., Greider, C.W. (2001). Telomere dysfunction increases mutation rate and genomic instability. *Cell*, 106(3), 275–286.

Hafner, L., Lezaja, A., Zhang, X., Lemmens, L., Shyian, M., Albert, B., Follonier, C., Nunes, J.M., Lopes, M., Shore, D. et al. (2018). Rif1 binding and control of chromosome-internal DNA replication origins is limited by telomere sequestration. *Cell Rep.*, 23(4), 983–992.

Hafner, L., Shore, D., Mattarocci, S. (2019). ChECing out Rif1 action in freely cycling cells. *Curr. Genet.*, 65(2), 429–434.

Harley, C.B., Futcher, A.B., Greider, C.W. (1990). Telomeres shorten during ageing of human fibroblasts. *Nature*, 345(6274), 458–460.

Hass, E.P. and Zappulla, D.C. (2015). The Ku subunit of telomerase binds Sir4 to recruit telomerase to lengthen telomeres in *S. cerevisiae*. *Elife*, 4, e07750, 1–19.

Hastie, N.D., Dempster, M., Dunlop, M.G., Thompson, A.M., Green, D.K., Allshire, R.C. (1990). Telomere reduction in human colorectal carcinoma and with ageing. *Nature*, 346(6287), 866–868.

Hayflick, L. and Moorhead, P.S. (1961). The serial cultivation of human diploid cell strains. *Exp. Cell Res.*, 25, 585–621.

Henderson, E.R. and Blackburn, E.H. (1989). An overhanging 3' terminus is a conserved feature of telomeres. *Mol. Cell Biol.*, 9(1), 345–348.

Henderson, E.R., Moore, M., Malcolm, B.A. (1990). Telomere G-strand structure and function analyzed by chemical protection, base analogue substitution, and utilization by telomerase *in vitro*. *Biochemistry*, 29(3), 732–737.

Hiraga, S.I., Monerawela, C., Katou, Y., Shaw, S., Clark, K.R., Shirahige, K., Donaldson, A.D. (2018). Budding yeast Rif1 binds to replication origins and protects DNA at blocked replication forks. *EMBO Rep.*, 19(9), e46222, 1–14.

Hughes, T.R., Weilbaecher, R.G., Walterscheid, M., Lundblad, V. (2000). Identification of the single-strand telomeric DNA binding domain of the *Saccharomyces cerevisiae* Cdc13 protein. *Proc. Natl. Acad. Sci. USA*, 97(12), 6457–6462.

Iglesias, N., Redon, S., Pfeiffer, V., Dees, M., Lingner, J., Luke, B. (2011). Subtelomeric repetitive elements determine TERRA regulation by Rap1/Rif and Rap1/Sir complexes in yeast. *EMBO Rep.*, 12(6), 587–593.

Ivessa, A.S., Zhou, J.Q., Schulz, V.P., Monson, E.K., Zakian, V.A. (2002). Saccharomyces Rrm3p, a 5' to 3' DNA helicase that promotes replication fork progression through telomeric and subtelomeric DNA. *Genes Dev.*, 16(11), 1383–1396.

Jain, D., Hebden, A.K., Nakamura, T.M., Miller, K.M., Cooper, J.P. (2010). HAATI survivors replace canonical telomeres with blocks of generic heterochromatin. *Nature*, 467(7312), 223–227.

Jolivet, P., Serhal, K., Graf, M., Eberhard, S., Xu, Z., Luke, B., Teixeira, M.T. (2019). A subtelomeric region affects telomerase-negative replicative senescence in *Saccharomyces cerevisiae*. *Sci. Rep.*, 9(1), 1845.

Kaizer, H., Connelly, C.J., Bettridge, K., Viggiani, C., Greider, C.W. (2015). Regulation of telomere length requires a conserved N-terminal domain of Rif2 in *Saccharomyces cerevisiae*. *Genetics*, 201(2), 573–586.

Kaul, Z., Cesare, A.J., Huschtscha, L.I., Neumann, A.A., Reddel, R.R. (2011). Five dysfunctional telomeres predict onset of senescence in human cells. *EMBO Reports*, 13(1), 52–59.

Keener, R., Connelly, C.J., Greider, C.W. (2019). Tel1 activation by the MRX complex is sufficient for telomere length regulation but not for the DNA damage response in *Saccharomyces cerevisiae*. *Genetics*, 213(4), 1271–1288.

Khadaroo, B., Teixeira, M.T., Luciano, P., Eckert-Boulet, N., Germann, S.M., Simon, M.N., Gallina, I., Abdallah, P., Gilson, E., Géli, V., Lisby, M. (2009). The DNA damage response at eroded telomeres and tethering to the nuclear pore complex. *Nat. Cell Biol.*, 11(8), 980–987.

Kim, N.W., Piatyszek, M.A., Prowse, K.R., Harley, C.B., West, M.D., Ho, P.L., Coviello, G.M., Wright, W.E., Weinrich, S.L., Shay, J.W. (1994). Specific association of human telomerase activity with immortal cells and cancer. *Science*, 266(5193), 2011–2015.

Kockler, Z.W., Comeron, J.M., Malkova, A. (2021). A unified alternative telomere-lengthening pathway in yeast survivor cells. *Molecular Cell*, 8, 1816–1829.

Konig, P. and Rhodes, D. (1997). Recognition of telomeric DNA. *Trends Biochem. Sci.*, 22(2), 43–47.

Kwapisz, M. and Morillon, A. (2020). Subtelomeric transcription and its regulation. *J. Mol. Biol.*, 432(15), 4199–4219.

Kyrion, G., Liu, K., Liu, C., Lustig, A.J. (1993). RAP1 and telomere structure regulate telomere position effects in *Saccharomyces cerevisiae*. *Genes Dev.*, 7(7A), 1146–1159.

Lalonde, M. and Chartrand, P. (2020). TERRA, a multifaceted regulator of telomerase activity at telomeres. *J. Mol. Biol.*, 432(15), 4232–4243.

Larrivee, M. and Wellinger, R.J. (2006). Telomerase- and capping-independent yeast survivors with alternate telomere states. *Nat. Cell Biol.*, 8(7), 741–747.

Larrivee, M., LeBel, C., Wellinger, R.J. (2004). The generation of proper constitutive G-tails on yeast telomeres is dependent on the MRX complex. *Genes Dev.*, 18(12), 1391–1396.

Le, S., Moore, J.K., Haber, J.E., Greider, C.W. (1999). RAD50 and RAD51 define two pathways that collaborate to maintain telomeres in the absence of telomerase. *Genetics*, 152(1), 143–152.

Lee, S.E., Moore, J.K., Holmes, A., Umezu, K., Kolodner, R.D., Haber, J.E. (1998). Saccharomyces Ku70, mre11/rad50 and RPA proteins regulate adaptation to G2/M arrest after DNA damage. *Cell*, 94(3), 399–409.

Lemon, L.D., Morris, D.K., Bertuch, A.A. (2019). Loss of Ku's DNA end binding activity affects telomere length via destabilizing telomere-bound Est1 rather than altering TLC1 homeostasis. *Sci. Rep.*, 9(1), 10607.

Lewis, K.A. and Wuttke, D.S. (2012). Telomerase and telomere-associated proteins: Structural insights into mechanism and evolution. *Structure*, 20(1), 28–39.

Lin, J.J. and Zakian, V.A. (1996). The Saccharomyces CDC13 protein is a single-strand TG1–3 telomeric DNA-binding protein *in vitro* that affects telomere behavior *in vivo*. *Proc. Natl. Acad. Sci. USA*, 93(24), 13760–13765.

Lingner, J., Cooper, J.P., Cech, T.R. (1995). Telomerase and DNA end replication: No longer a lagging strand problem? *Science*, 269(5230), 1533–1534.

Lingner, J., Hughes, T.R., Shevchenko, A., Mann, M., Lundblad, V., Cech, T.R. (1997). Reverse transcriptase motifs in the catalytic subunit of telomerase. *Science*, 276(5312), 561–567.

Lopes, J., Piazza, A., Bermejo, R., Kriegsman, B., Colosio, A., Teulade-Fichou, M.P., Foiani, M., Nicolas, A. (2011). G-quadruplex-induced instability during leading-strand replication. *EMBO J.*, 30(19), 4033–4046.

Lopez, C.R., Ribes-Zamora, A., Indiviglio, S.M., Williams, C.L., Haricharan, S., Bertuch, A.A. (2011). Ku must load directly onto the chromosome end in order to mediate its telomeric functions. *PLoS Genet.*, 7(8), e1002233.

Louis, E.J. (1995). The chromosome ends of *Saccharomyces cerevisiae*. *Yeast*, 11(16), 1553–1573.

Lue, N.F., Chan, J., Wright, W.E., Hurwitz, J. (2014). The CDC13–STN1–TEN1 complex stimulates Pol alpha activity by promoting RNA priming and primase-to-polymerase switch. *Nat. Commun.*, 5, 5762.

Luke, B., Panza, A., Redon, S., Iglesias, N., Li, Z., Lingner, J. (2008). The Rat1p 5' to 3' exonuclease degrades telomeric repeat-containing RNA and promotes telomere elongation in *Saccharomyces cerevisiae*. *Mol. Cell*, 32(4), 465–477.

Lundblad, V. and Blackburn, E.H. (1993). An alternative pathway for yeast telomere maintenance rescues est1-senescence. *Cell*, 73(2), 347–360.

Lundblad, V. and Szostak, J.W. (1989). A mutant with a defect in telomere elongation leads to senescence in yeast. *Cell*, 57(4), 633–643.

Lydeard, J.R., Jain, S., Yamaguchi, M., Haber, J.E. (2007). Break-induced replication and telomerase-independent telomere maintenance require Pol32. *Nature*, 448(7155), 820–823.

Mak, H.C., Pillus, L., Ideker, T. (2009). Dynamic reprogramming of transcription factors to and from the subtelomere. *Genome Res.*, 19(6), 1014–1025.

Makovets, S., Williams, T.L., Blackburn, E.H. (2008). The telotype defines the telomere state in *Saccharomyces cerevisiae* and is inherited as a dominant non-Mendelian characteristic in cells lacking telomerase. *Genetics*, 178(1), 245–257.

Marcand, S., Wotton, D., Gilson, E., Shore, D. (1997). Rap1p and telomere length regulation in yeast. *Ciba Found. Symp.*, 211, 76–93, discussion 93–103.

Marcand, S., Brevet, V., Gilson, E. (1999). Progressive cis-inhibition of telomerase upon telomere elongation. *EMBO J.*, 18(12), 3509–3519.

Marcand, S., Brevet, V., Mann, C., Gilson, E. (2000). Cell cycle restriction of telomere elongation. *Curr. Biol.*, 10(8), 487–490.

Martin, V., Du, L.-L., Rozenzhak, S., Russell, P. (2007). Protection of telomeres by a conserved Stn1–Ten1 complex. *Proc. Natl. Acad. Sci. USA*, 104(35), 14038–14043.

Martin, H., Doumic, M., Teixeira, M.T., Xu, Z. (2021). Telomere shortening causes distinct cell division regimes during replicative senescence in *Saccharomyces cerevisiae*. *bioRxiv*, 448683.

Mattarocci, S., Shyian, M., Lemmens, L., Damay, P., Altintas, D.M., Shi, T., Bartholomew, C.R., Thomä, N.H., Hardy, C.F., Shore, D. (2014). Rif1 controls DNA replication timing in yeast through the PP1 phosphatase Glc7. *Cell Rep.*, 7(1), 62–69.

Mattarocci, S., Reinert, J.K., Bunker, R.D., Fontana, G.A., Shi, T., Klein, D., Cavadini, S., Faty, M., Shyian, M., Hafner, L. et al. (2017). Rif1 maintains telomeres and mediates DNA repair by encasing DNA ends. *Nat. Struct. Mol. Biol.*, 24(7), 588–595.

McClintock, B. (1939). The behavior in successive nuclear divisions of a chromosome broken at meiosis. *Proc. Natl. Acad. Sci. USA*, 25(8), 405–416.

McClintock, B. (1940). The stability of broken ends of chromosomes in zea mays. *Genetics*, 26, 234–282.

McEachern, M.J. and Blackburn, E.H. (1994). A conserved sequence motif within the exceptionally diverse telomeric sequences of budding yeasts. *Proc. Natl. Acad. Sci. USA*, 91(8), 3453–3457.

McEachern, M.J. and Haber, J.E. (2006). Break-induced replication and recombinational telomere elongation in yeast. *Annu. Rev. Biochem.*, 75, 111–135.

McGee, J.S., Phillips, J.A., Chan, A., Sabourin, M., Paeschke, K., Zakian, V.A. (2010). Reduced Rif2 and lack of Mec1 target short telomeres for elongation rather than double-strand break repair. *Nat. Struct. Mol. Biol.*, 17(12), 1438–1445.

Michelson, R.J., Rosenstein, S., Weinert, T. (2005). A telomeric repeat sequence adjacent to a DNA double-stranded break produces an anticheckpoint. *Genes Dev*, 19(21), 2546–2559.

Miller, K.M., Rog, O., Cooper, J.P. (2006). Semi-conservative DNA replication through telomeres requires Taz1. *Nature*, 440(7085), 824–828.

Mimitou, E.P. and Symington, L.S. (2010). Ku prevents Exo1 and Sgs1–dependent resection of DNA ends in the absence of a functional MRX complex or Sae2. *EMBO Journal*, 29(19), 3358–3369.

Mirman, Z., Lottersberger, F., Takai, H., Kibe, T., Gong, Y., Takai, K., Bianchi, A., Zimmermann, M., Durocher, D., de Lange, T. (2018). 53BP1–RIF1–shieldin counteracts DSB resection through CST- and Polalpha-dependent fill-in. *Nature*, 560(7716), 112–116.

Miyake, Y., Nakamura, M., Nabetani, A., Shimamura, S., Tamura, M., Yonehara, S., Saito, M., Ishikawa, F. (2009). RPA-like mammalian Ctc1–Stn1–Ten1 complex binds to single-stranded DNA and protects telomeres independently of the Pot1 pathway. *Mol. Cell*, 36(2), 193–206.

Moretti, P. and Shore, D. (2001). Multiple interactions in Sir protein recruitment by Rap1p at silencers and telomeres in yeast. *Mol. Cell Biol.*, 21(23), 8082–8094.

Moretti, P., Freeman, K., Coodly, L., Shore, D. (1994). Evidence that a complex of SIR proteins interacts with the silencer and telomere-binding protein RAP1. *Genes Dev.*, 8(19), 2257–2269.

Morse, R.H. (2000). RAP, RAP, open up! New wrinkles for RAP1 in yeast. *Trends Genet.*, 16(2), 51–53.

Moser, B.A. and Nakamura, T.M. (2009). Protection and replication of telomeres in fission yeast. *Biochem. Cell Biol.*, 87(5), 747–758.

Moyzis, R.K., Buckingham, J.M., Cram, L.S., Dani, M., Deaven, L.L., Jones, M.D., Meyne, J., Ratliff, R.L., Wu, J.R. (1988). A highly conserved repetitive DNA sequence, (TTAGGG)n, present at the telomeres of human chromosomes. *Proc. Natl. Acad. Sci. USA*, 85(18), 6622–6626.

Mozdy, A.D., Podell, E.R., Cech, T.R. (2008). Multiple yeast genes, including Paf1 complex genes, affect telomere length via telomerase RNA abundance. *Mol. Cell Biol.*, 28(12), 4152–4161.

Müller, H.J. (1938). The remaking of chromosomes. *The Collecting Net*, 13(8), 181–198.

Myung, K., Chen, C., Kolodner, R.D. (2001). Multiple pathways cooperate in the suppression of genome instability in *Saccharomyces cerevisiae*. *Nature*, 411(6841), 1073–1076.

Nakamura, T.M., Cooper, J.P., Cech, T.R. (1998). Two modes of survival of fission yeast without telomerase. *Science*, 282(5388), 493–496.

Nandakumar, J. and Cech, T.R. (2013). Finding the end: Recruitment of telomerase to telomeres. *Nat. Rev. Mol. Cell Biol.*, 14(2), 69–82.

Ng, L.J., Cropley, J.E., Pickett, H.A., Reddel, R.R., Suter, C.M. (2009). Telomerase activity is associated with an increase in DNA methylation at the proximal subtelomere and a reduction in telomeric transcription. *Nucleic Acids Res.*, 37(4), 1152–1159.

Oganesian, L., Moon, I.K., Bryan, T.M., Jarstfer, M.B. (2006). Extension of G-quadruplex DNA by ciliate telomerase. *EMBO J.*, 25(5), 1148–1159.

Olovnikov, A.M. (1973). A theory of marginotomy. The incomplete copying of template margin in enzymic synthesis of polynucleotides and biological significance of the phenomenon. *J. Theor. Biol.*, 41(1), 181–190.

Osterhage, J.L., Talley, J.M., Friedman, K.L. (2006). Proteasome-dependent degradation of Est1p regulates the cell cycle-restricted assembly of telomerase in *Saccharomyces cerevisiae*. *Nat. Struct. Mol. Biol.*, 13(8), 720–728.

Ottaviani, D., Lever, E., Mitter, R., Jones, T., Forshew, T., Christova, R., Tomazou, E.M., Rakyan, V.K., Krawetz, S.A., Platts, A.E. et al. (2008). Reconfiguration of genomic anchors upon transcriptional activation of the human major histocompatibility complex. *Genome Res.*, 18(11), 1778–1786.

Pardo, B. and Marcand, S. (2005). Rap1 prevents telomere fusions by nonhomologous end joining. *EMBO J.*, 24(17), 3117–3127.

Parkinson, G.N., Lee, M.P., Neidle, S. (2002). Crystal structure of parallel quadruplexes from human telomeric DNA. *Nature*, 417(6891), 876–880.

Paschini, M., Toro, T.B., Lubin, J.W., Braunstein-Ballew, B., Morris, D.K., Lundblad, V. (2012). A naturally thermolabile activity compromises genetic analysis of telomere function in *Saccharomyces cerevisiae*. *Genetics*, 191(1), 79–93.

Pasquier, E. and Wellinger, R.J. (2020). *In vivo* chromatin organization on native yeast telomeric regions is independent of a cis-telomere loopback conformation. *Epigenetics Chromatin*, 13(1), 23.

Porter, S.E., Greenwell, P.W., Ritchie, K.B., Petes, T.D. (1996). The DNA-binding protein Hdf1p (a putative Ku homologue) is required for maintaining normal telomere length in *Saccharomyces cerevisiae*. *Nucleic Acids Res.*, 24(4), 582–585.

Qi, H. and Zakian, V.A. (2000). The saccharomyces telomere-binding protein Cdc13p interacts with both the catalytic subunit of DNA polymerase alpha and the telomerase-associated est1 protein. *Genes & Development*, 14(14), 1777–1788.

Raices, M., Verdun, R.E., Compton, S.A., Haggblom, C.I., Griffith, J.D., Dillin, A., Karlseder, J. (2008). *C. elegans* telomeres contain G-strand and C-strand overhangs that are bound by distinct proteins. *Cell*, 132(5), 745–757.

Ribeyre, C. and Shore, D. (2012). Anticheckpoint pathways at telomeres in yeast. *Nat. Struct. Mol. Biol.*, 19(3), 307–313.

Romano, G.H., Harari, Y., Yehuda, T., Podhorzer, A., Rubinstein, L., Shamir, R., Gottlieb, A., Silberberg, Y., Pe'er, D., Ruppin, E. et al. (2013). Environmental stresses disrupt telomere length homeostasis. *PLoS Genet.*, 9(9), e1003721.

Ruis, P. and Boulton, S.J. (2021). The end protection problem-an unexpected twist in the tail. *Genes Dev.*, 35(1–2), 1–21.

Sabourin, M., Tuzon, C.T., Zakian, V.A. (2007). Telomerase and Tel1p preferentially associate with short telomeres in *S. cerevisiae*. *Mol. Cell*, 27(4), 550–561.

Sandell, L.L. and Zakian, V.A. (1993). Loss of a yeast telomere: Arrest, recovery, and chromosome loss. *Cell*, 75(4), 729–739.

Schoeftner, S. and Blasco, M.A. (2010). Chromatin regulation and non-coding RNAs at mammalian telomeres. *Semin. Cell Dev. Biol.*, 21(2), 186–193.

Sen, D. and Gilbert, W. (1988). Formation of parallel four-stranded complexes by guanine-rich motifs in DNA and its implications for meiosis. *Nature*, 334(6180), 364–366.

Shampay, J. and Blackburn, E.H. (1988). Generation of telomere-length heterogeneity in *Saccharomyces cerevisiae*. *Proc. Natl. Acad. Sci. USA*, 85(2), 534–538.

Shao, Y., Lu, N., Cai, C., Zhou, F., Wang, S., Zhao, Z., Zhao, G., Zhou, J.Q., Xue, X., Qin, Z. (2019). A single circular chromosome yeast. *Cell Res.*, 29(1), 87–89.

Shi, T., Bunker, R.D., Mattarocci, S., Ribeyre, C., Faty, M., Gut, H., Scrima, A., Rass, U., Rubin, S.M., Shore, D., Thoma, N.H. (2013). Rif1 and Rif2 shape telomere function and architecture through multivalent Rap1 interactions. *Cell*, 153(6), 1340–1353.

Shore, D. (1994). RAP1: A protean regulator in yeast. *Trends Genet.*, 10(11), 408–412.

Singer, M.S. and Gottschling, D.E. (1994). TLC1: Template RNA component of *Saccharomyces cerevisiae* telomerase. *Science*, 266(5184), 404–409.

Soudet, J., Jolivet, P., Teixeira, M.T. (2014). Elucidation of the DNA end-replication problem in *Saccharomyces cerevisiae*. *Mol. Cell*, 53(6), 954–964.

Sridhar, A., Kedziora, S., Donaldson, A.D. (2014). At short telomeres Tel1 directs early replication and phosphorylates Rif1. *PLoS Genet.*, 10(10), e1004691.

Stevenson, J.B. and Gottschling, D.E. (1999). Telomeric chromatin modulates replication timing near chromosome ends. *Genes Dev.*, 13(2), 146–151.

Surovtseva, Y.V., Churikov, D., Boltz, K.A., Song, X., Lamb, J.C., Warrington, R., Leehy, K., Heacock, M., Price, C.M., Shippen, D.E. (2009). Conserved telomere maintenance component 1 interacts with STN1 and maintains chromosome ends in higher eukaryotes. *Mol. Cell*, 36(2), 207–218.

Szostak, J.W. and Blackburn, E.H. (1982). Cloning yeast telomeres on linear plasmid vectors. *Cell*, 29(1), 245–255.

Teixeira, M.T. (2013). *Saccharomyces cerevisiae* as a model to study replicative senescence triggered by telomere shortening. *Frontiers in Oncology*, 3, 101.

Teixeira, M.T. and Gilson, E. (2005). Telomere maintenance, function and evolution: The yeast paradigm. *Chromosome Res.*, 13(5), 535–548.

Teixeira, M.T., Arneric, M., Sperisen, P., Lingner, J. (2004). Telomere length homeostasis is achieved via a switch between telomerase- extendible and -nonextendible states. *Cell*, 117(3), 323–335.

Tihan, T., Apte, M.S., Cooper, J.P. (2017). Life and cancer without telomerase: ALT and other strategies for making sure ends (don't) meet. *Journal of Neuro-oncology*, 52(1), 57–73.

Toczyski, D.P., Galgoczy, D.J., Hartwell, L.H. (1997). CDC5 and CKII control adaptation to the yeast DNA damage checkpoint. *Cell*, 90(6), 1097–1106.

Topcu, Z., Nickles, K., Davis, C., McEachern, M.J. (2005). Abrupt disruption of capping and a single source for recombinationally elongated telomeres in Kluyveromyces lactis. *Proc. Natl. Acad. Sci. USA*, 102(9), 3348–3353.

Tsukamoto, Y., Taggart, A.K., Zakian, V.A. (2001). The role of the Mre11–Rad50–Xrs2 complex in telomerase- mediated lengthening of *Saccharomyces cerevisiae* telomeres. *Curr. Biol.*, 11(17), 1328–1335.

Tucey, T.M. and Lundblad, V. (2014). Regulated assembly and disassembly of the yeast telomerase quaternary complex. *Genes Dev.*, 28(19), 2077–2089.

Tuzon, C.T., Wu, Y., Chan, A., Zakian, V.A. (2011). The *Saccharomyces cerevisiae* telomerase subunit Est3 binds telomeres in a cell cycle- and Est1-dependent manner and interacts directly with Est1 *in vitro*. *PLoS Genet.*, 7(5), e1002060.

Vodenicharov, M.D. and Wellinger, R.J. (2006). DNA degradation at unprotected telomeres in yeast is regulated by the CDK1 (Cdc28/Clb) cell-cycle kinase. *Mol. Cell*, 24(1), 127–137.

Vodenicharov, M.D. and Wellinger, R.J. (2007). The cell division cycle puts up with unprotected telomeres: Cell cycle regulated telomere uncapping as a means to achieve telomere homeostasis. *Cell Cycle*, 6(10), 1161–1167.

Vodenicharov, M.D., Laterreur, N., Wellinger, R.J. (2010). Telomere capping in non-dividing yeast cells requires Yku and Rap1. *EMBO J.*, 29(17), 3007–3019.

Wang, S.S. and Zakian, V.A. (1990). Sequencing of saccharomyces telomeres cloned using T4 DNA polymerase reveals two domains. *Mol. Cell Biol.*, 10(8), 4415–4419.

Watson, J.D. (1972). Origin of concatemeric T7 DNA. *Nat. New Biol.*, 239(94), 197–201.

Weinert, T.A. and Hartwell, L.H. (1988). The RAD9 gene controls the cell cycle response to DNA damage in *Saccharomyces cerevisiae*. *Science*, 241(4863), 317–322.

Wellinger, R.J. and Zakian, V.A. (2012). Everything you ever wanted to know about Saccharomyces cerevisiae telomeres: Beginning to end. *Genetics*, 191(4), 1073–1105.

Wellinger, R.J., Wolf, A.J., Zakian, V.A. (1993a). Saccharomyces telomeres acquire single-strand TG1–3 tails late in S phase. *Cell*, 72(1), 51–60.

Wellinger, R.J., Wolf, A.J., Zakian, V.A. (1993b). Origin activation and formation of single-strand TG1–3 tails occur sequentially in late S phase on a yeast linear plasmid. *Mol. Cell Biol.*, 13(7), 4057–4065.

Westmoreland, J.W., Mihalevic, M.J., Bernstein, K.A., Resnick, M.A. (2018). The global role for Cdc13 and Yku70 in preventing telomere resection across the genome. *DNA Repair (Amst)*, 62, 8–17.

Williams, J.M., Ouenzar, F., Lemon, L.D., Chartrand, P., Bertuch, A.A. (2014). The principal role of Ku in telomere length maintenance is promotion of Est1 association with telomeres. *Genetics*, 197(4), 1123–1136.

Wright, J.H., Gottschling, D.E., Zakian, V.A. (1992). Saccharomyces telomeres assume a non-nucleosomal chromatin structure. *Genes Dev.*, 6(2), 197–210.

Wright, W.E., Piatyszek, M.A., Rainey, W.E., Byrd, W., Shay, J.W. (1996). Telomerase activity in human germline and embryonic tissues and cells. *Dev. Genet.*, 18(2), 173–179.

Wu, P., Takai, H., de Lange, T. (2012). Telomeric 3' overhangs derive from resection by Exo1 and Apollo and fill-in by POT1b-associated CST. *Cell*, 150(1), 39–52.

Xu, Z., Duc, K.D., Holcman, D., Teixeira, M.T. (2013). The length of the shortest telomere as the major determinant of the onset of replicative senescence. *Genetics*, 194(4), 847–857.

Xu, Z., Fallet, E., Paoletti, C., Fehrmann, S., Charvin, G., Teixeira, M.T. (2015). Two routes to senescence revealed by real-time analysis of telomerase-negative single lineages. *Nat. Commun.*, 6, 7680.

Yamada, M., Hayatsu, N., Matsuura, A., Ishikawa, F. (1998). Y'-Help1, a DNA helicase encoded by the yeast subtelomeric Y' element, is induced in survivors defective for telomerase. *J. Biol. Chem.*, 273(50), 33360–33366.

Yue, J.X., Li, J., Aigrain, L., Hallin, J., Persson, K., Oliver, K., Bergstrom, A., Coupland, P., Warringer, J., Lagomarsino, M.C. et al. (2017). Contrasting evolutionary genome dynamics between domesticated and wild yeasts. *Nat. Genet.*, 49(6), 913–924.

Zahler, A.M., Williamson, J.R., Cech, T.R., Prescott, D.M. (1991). Inhibition of telomerase by G-quartet DNA structures. *Nature*, 350(6320), 718–720.

Zakian, V.A., Blanton, H.M., Wetzel, L. (1986). Distribution of telomere-associated sequences in yeast. *Basic Life Sci.*, 40, 493–498.

Zappulla, D.C., Goodrich, K.J., Arthur, J.R., Gurski, L.A., Denham, E.M., Stellwagen, A.E. Cech, T.R. (2011). Ku can contribute to telomere lengthening in yeast at multiple positions in the telomerase RNP. *RNA*, 17(2), 298–311.

von Zglinicki, T. (2002). Oxidative stress shortens telomeres. *Trends Biochem. Sci.*, 27(7), 339–344.

6

Dealing with Replication Stress at Telomeres

Stéphane COULON
CNRS UMR7258, INSERM UMR1068,
Institut Paoli-Calmettes, CRCM, Université Aix-Marseille, France

The terminal ends of chromosomes contain G-rich repetitive sequences that are difficult to replicate. Indeed, secondary structures of DNA that form at telomeres increase replication stress and then threaten genome stability. The role of shelterin is to protect telomeres by preventing DNA damage response (DDR), but during the last decade shelterin has been also shown to promote efficient replication of telomeres to ensure their homeostasis. In this chapter, we will present the current knowledge and most recent discoveries documenting how shelterin-associated proteins and accessory factors participate in the replication of telomeric repeated tracks to limit replication stress at telomeres.

6.1. Introduction

6.1.1. *Telomere, telomerase and shelterin*

Most extremities of linear chromosomes are composed of double-strand tandem repeats which ends with a 3′ single-stranded DNA (3′-overhang). Specific proteins

---

For a color version of all the figures in this chapter, see www.iste.co.uk/saintome/telomeres.zip.

Telomeres,
coordinated by Carole SAINTOMÉ.

tightly bind to these repeated sequences to form a nucleoprotein structure called a telomere. The main function of this group of protein is to protect telomeric DNA from degradation, hence the name shelterin.

In humans, terminal repeated sequences (5′-TTAGGG-3′) extend over 10–15 kb pairs and the 3′-overhang is about 50–400 nucleotides. This single-stranded protrusion is able to invade the duplex DNA and presumably folds back into a telomere loop structure (T-loop) that participates in the stabilization and the protection of the chromosome ends. The shleterin complex contains six proteins. The telomeric repeat binding factor 1 and 2 (TRF1 and TRF2) bind to double-strand telomeric tracks, while the protection of telomeres 1 protein (POT1) binds to the 3′ overhang. TRF1 and TRF2 recruit the other components of shelterin: TRF2 and TRF1-interacting nuclear protein 2 (TIN2), the human ortholog of the yeast repressor/activator protein 1 (RAP1), and TPP1, also called TINT1, PTOP and PIP1.

One function of the shelterin is to recruit a reverse transcriptase called telomerase that is able to elongate the 3′ overhang by the addition of telomeric repeats. The telomerase is a ribonucleoprotein minimally comprised of the catalytic subunit (telomerase reverse transcriptase [TERT]) and its intrinsic RNA (telomerase RNA component [TERC]). In stem cells and germinal cells in which telomerase activity is present, the TERT subunit is recruited to telomeres through its association with TPP1 during the S phase of the cell cycle (Nandakumar et al. 2012; Zhong et al. 2012) and is inhibited by the CTC1–STN1–TEN1 complex (CST) that acts as a terminator of telomerase activity through its interaction with TPP1-POT1 heterodimer (Chen et al. 2012). The CST complex is a single-stranded DNA (ssDNA) binding protein that shares homology with RPA (replication protein A). CST stimulates DNA polymerase α/primase for the synthesis of the complementary strand, ensuring telomere replication completion (Casteel et al. 2008; Miyake et al. 2009; Surovtseva et al. 2009).

Chromosomes ends are natural DNA double-stranded breaks that need to be distinguished from internal double-strand breaks. So, in addition to telomere length control through telomerase recruitment, the main function of the shelterin is to protect terminal ends from the DDR that includes classical and alternative non-homologous end joining (NHEJ) and the homology-directed repair (HDR), by preventing the activation of the checkpoint kinases ataxia-telangiectasia mutated (ATM) and ataxia-telangiectasia and Rad3-Related (ATR) (Arnoult and Karlseder 2015; Lange 2018). The T-loop formation that is promoted by the telomeric proteins is also a key mechanism involved in the telomere protection (Doksani et al. 2013).

Human telomeres also form a compact chromatin structure *in vivo* thanks to the specific protein–protein and protein–DNA interactions between the shelterin proteins and telomeric DNA (Bandaria et al. 2016; Benarroch-Popivker et al. 2016). This DNA compaction also participates in the protection of the telomeric repeats by preventing DDR signalling at telomeres. Telomeres localize throughout the nuclear volume, but they are not free to roam across the nucleus. They are attached to the nuclear matrix (NM) through shelterin and lamins (Kaminker et al. 2009; Ludérus et al. 1996), and a subset of telomeres is found at the nuclear periphery (Arnoult et al. 2010; Crabbe et al. 2012). This specific localization to the nuclear envelope (NE) has a potential impact on nuclear organization, gene expression and telomere maintenance.

6.1.2. *Aging, cancer and telomere syndromes*

In most somatic cells, the telomerase enzyme is not express and consequently telomeres undergo gradual shortening with ongoing cell division. This natural erosion is caused by the incomplete replication of the linear DNA molecules by the conventional DNA polymerases, often referred to as the *end-replication problem* (Lingner et al. 1995). Telomere shortening provided the first molecular explanation for why many cells cease to divide in culture (Harley et al. 1990). Indeed, when a subset of telomeres reaches a critically shortened length, these dysfunctional telomeres trigger the DDR, which causes cell cycle arrest through the activation of p53 pathway (Shay and Wright 2010), and at least five damaged telomeres within a cell are sufficient to provoke this p53-dependent pathway (Kaul et al. 2011). This response is often termed telomere-initiated cellular senescence, which was formally described five decades ago when Hayflick and colleagues (1965) showed that human fibroblasts had a limited ability to proliferate in culture. Thus, the telomeres not only protect chromosome ends from being recognized as DNA damage, but also serve as a gauge for the mitotic age of the cell. As a result, accumulation of senescent cells in an organism contributes to aging which is characterized by rapid decline of the regenerative capacity of tissues.

Importantly, cells that eventually lose critical cell cycle checkpoint functions (e.g. in the absence of p53 and/or Rb) escape replicative senescence and continue to divide until many telomeres are critically shortened (Campisi and d'Adda di Fagagna 2007; Shay and Wright 2010). They enter into crisis (a second growth arrest state) which is characterized by chromosomal end-to-end fusions and cycles of chromosome fusion-bridge-breakage almost universally leading to apoptotic cell death (Artandi et al. 2000; Hayashi et al. 2015). However, in rare cases this genome instability, which is a major hallmark of cancer, promotes immortalization of the cell either by re-activating telomerase (Kim et al. 1994) or by using a homologous recombination telomere-copying mechanism known as ALT for alternative lengthening of telomeres (Bryan et al. 1997). Both mechanisms stabilize telomeres and then allow the cells to grow continuously (i.e. become immortal), and this is generally believed to be a critical step in tumorigenesis.

In humans, telomere dysfunction that may result from telomerase deficiency or alteration of the shelterin components leads to premature telomere shortening. As for replicative senescence, these damaged telomeres trigger the cell cycle arrest through activation of the DDR and cause the loss of regenerative capacity of different tissues, which are related to monogenic inherited diseases. These telomere syndromes display signatures of human premature ageing and give rise to the bone marrow failure syndrome dyskeratosis congenita (DC), aplastic anemia and idiopathic pulmonary fibrosis (IPF) (Armanios and Blackburn 2012; Savage 2018; Shay and Wright 2019). Although clinical symptoms are diverse among these diseases, they are all defined by the short telomere phenotype.

6.1.3. *The telomere replication problem*

It is clear that short and dysfunctional telomeres is the key feature of the complex relationships that exist between telomere biology, aging, human diseases and cancer. As described earlier, the *end-replication problem* accounts for the progressive telomere attrition in dividing cells that do not express the telomerase. Aside from the *end-replication problem*, the replication of telomeric repeated sequences may cause abrupt telomere shortening and it is well established that it is the shortest telomeres, not average telomere length, that are able to activate DDA and subsequently trigger p53-dependent senescence (Hemann et al. 2001). Indeed, out of all the genome, telomeres are one of the most difficult regions to replicate because they encompass multiple obstacles that impede replication fork progression and then alter telomere length homeostasis. This defines the *telomere replication problem*.

Telomere replication mostly proceeds unidirectionally from a centromere-proximal origin. DNA two-dimensional gel electrophoresis analysis (2D-gel) in yeast demonstrated that replication forks naturally slow and eventually stall as they approach telomeric chromatin (Ivessa et al. 2002; Makovets et al. 2004; Miller et al. 2006). If stalling persists, the fork may collapse and the newly generated double-strand break triggers a restart mechanism through homologous recombination that risks the loss of genetic material. Alternatively, a stalled fork may reverse allowing fork stabilization and restart (Margalef et al. 2018; Matmati et al. 2020) (Figure 6.1). In mammals and mice, replication stress leads to telomere aberrations characterized by abnormal structures on metaphase chromosomes that can be visualized by fluorescent in situ hybridization (FISH) (Martínez et al. 2009; Sfeir et al. 2009). The nature of these aberrant structures is not clearly established but they are thought to arise from partially replicated and broken telomeres or may even represent entangled telomeres. Therefore, telomeres are fragile sites that are hard to replicate and can generate a stress when replicated.

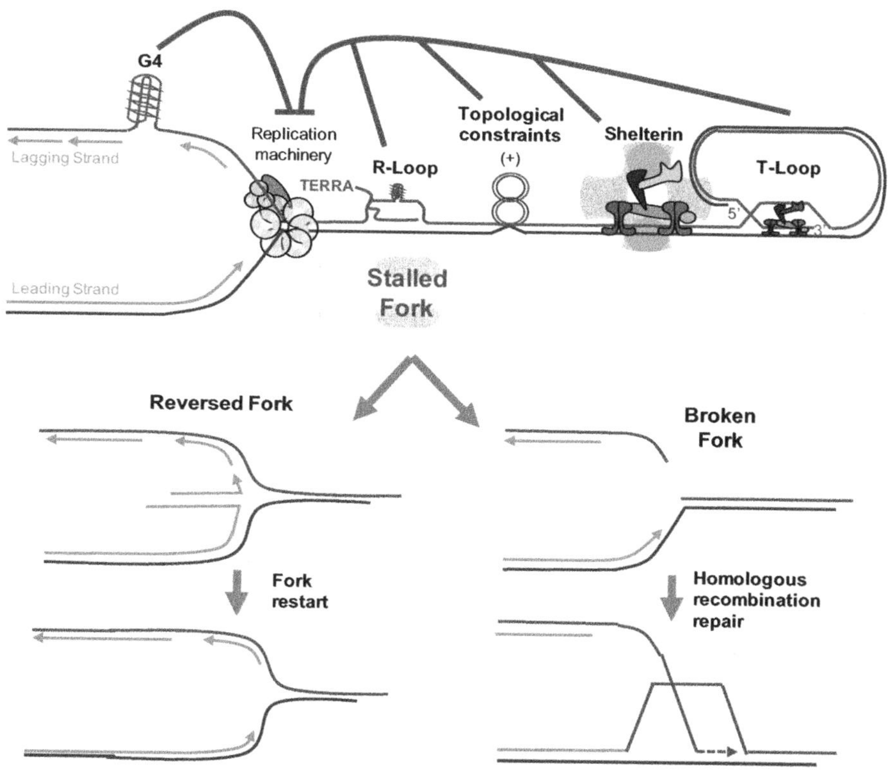

Figure 6.1. *The source of replication stress at terminal ends. Telomere replication is a major challenge for cells because many obstacles to the progression of the replication fork are concentrated at the ends of the chromosomes. T-loop structures, G-quadruplexes, R-loops, topological constraints and telomeric bound proteins are diverse sources of endogenous blocks that may lead to fork stalling. Stalled forks may collapse and either break or reverse. According to the current knowledge, the repair of broken forks relies on homologous recombination pathway while fork reversal allows fork restart*

What causes replication fork stalling at telomeres? Heterochromatin, the highly repetitive nature of G-rich regions that are prone to form secondary structures such as G-quadruplexes (G4), T-loop super-structures, TERRA transcription including R-loop and RNA:DNA hybrids as well as telomere compaction and telomere attachment to the NE are diverse sources of endogenous blocks that impede the progression of the fork through telomeric tracts (Figure 6.1). Although telomere-bound proteins have been thought to impede replication fork progression (Ohki and Ishikawa 2004), it was subsequently established that TRF1, TRF2 and their fission yeast counterpart Taz1 are required to promote efficient replication of telomeres and to prevent fork stalling (Martínez et al. 2009; Sfeir et al. 2009; Ye et al. 2010). This highlights the active role played by shelterin in ensuring efficient replication.

In this chapter, we will present the current knowledge of how shelterin associated proteins and accessory factors participate in the replication of telomeric repeated tracks to limit replication stress at telomeres.

6.2. Removal of the DNA structures that impede RF progression

6.2.1. *G-quadruplex dissolution by helicases*

Four guanines can associate through Hoogsteen hydrogen bonding to form a planar G-quartet, stabilized by a monovalent cation. Three or more G-quartets can stack on top of each other to form a G-quadruplex structure or G4. G-rich sequences such as telomeric DNA that contain four runs of at least three guanine bases can form stable intramolecular G4. G4 formation is promoted in single-stranded DNA during DNA replication and transcription. Thus, G4 structures can constitute prominent barriers to replication fork progression and are intrinsically recombinogenic and mutagenic leading to the idea that G4 may promote chromosome instability (Tarsounas and Tijsterman 2013; Rhodes and Lipps 2015). At telomeres, G4 can primarily form on the G-rich lagging strand template but can also arise ahead of the fork on the G-rich strand of R-loops and cause fork arrest (Figure 6.1). If not properly resolved, G4 can eventually cause fork breakage and telomere loss. On the other hand, G4 might also participate in telomere protection when formed in the G-rich telomeric overhang (Gilson and Géli 2007; Nandakumar and Cech 2013).

To avoid genetic instability and telomeric alterations, G4 must be unwound. Many helicases have been implicated in this process (Figure 6.2). These include the 3′-5′ directed RecQ helicase family, WRN and BLM (Sun et al. 1998; Mohaghegh et al. 2001; Huber et al. 2002). WRN and BLM defects are responsible for the high predisposition to cancer in people with Werner's and Bloom's syndromes, respectively. Cells lacking WRN exhibit loss of the telomeric lagging strand which contains the G-rich sequence capable of forming G4 structures (Crabbe et al. 2004; Arnoult et al. 2009; Barefield and Karlseder 2012). WRN likely plays a major role in unwinding G4 during telomere replication, however the molecular bases of WRN involvement in telomere replication are unknown. Recruitment of WRN at G4 sites could be mediated through interactions with several partners as RPA complex, PCNA (proliferating cell nuclear antigen) and Pol δ (Shen and Loeb 2001), as well as TRF2 (Opresko et al. 2002). Unlike WRN, a BLM-deficient cell line shows a fragile telomere phenotype and this effect is epistatic with TRF1 depletion, suggesting that BLM facilitates telomere replication in a TRF1-dependent manner, presumably by removing G4 (Sfeir et al. 2009; Barefield and Karlseder 2012; Drosopoulos et al. 2015). Indeed, BLM contains the FxLxP TRF1 binding motif and binds TRF1 *in vitro* (Lillard-Wetherell 2004). Along the same lines, it has been proposed more recently that TRF1 recruits BLM *in vivo* to remove secondary structures such as G4 to prevent lagging-telomere fragility (Zimmermann et al. 2014). Thus, the shelterin component TRF1 seems to play an active role in removing G4 formed on the lagging strand during replication of telomeric repeats by recruiting the specific G4 helicase BLM.

Similarly to WRN and BLM helicases, RTEL1 (regulator of telomere length 1) has also been implicated in G4 processing in humans and mice (Ding et al. 2004; Sfeir et al. 2009; Vannier et al. 2012). RTEL1 is an essential helicase involved in DNA replication and recombination, and it is required for the maintenance of telomere integrity (reviewed in Vannier et al. (2014)). Importantly, in humans, mutations in RTEL1 have also been implicated in Hoyeraal–Hreidarsson syndrome, a severe form DC with strong telomere dysfunction (Ballew et al. 2013; Deng et al. 2013; Guen et al. 2013). *In vitro*, RTEL1 is capable of unwinding G4 with a 5′-3′ polarity (Vannier et al. 2013). Remarkably, RTEL1 exhibits a PIP box domain that mediates the interaction with PCNA, suggesting that RTEL1 is recruited at telomeres by the replisome to remove G4 (Vannier et al. 2013). In mice, Rtel1 knock-out animals are embryonic lethal and cells derived from these mice exhibit various chromosomal abnormalities and telomere loss (Ding et al. 2004). Additionally, telomere fragility in RTEL1-deficient mouse cell line is exacerbated by loss of BLM indicating that BLM and RTEL1 function in distinct pathways (Vannier et al. 2012). Overall, these results suggest that both the replisome through PCNA and shelterin through TRF1 recruit specific helicases such as BLM, WRN and RTEL1 to suppress G4 structures to allow the proper replication of telomeres (Figure 6.2).

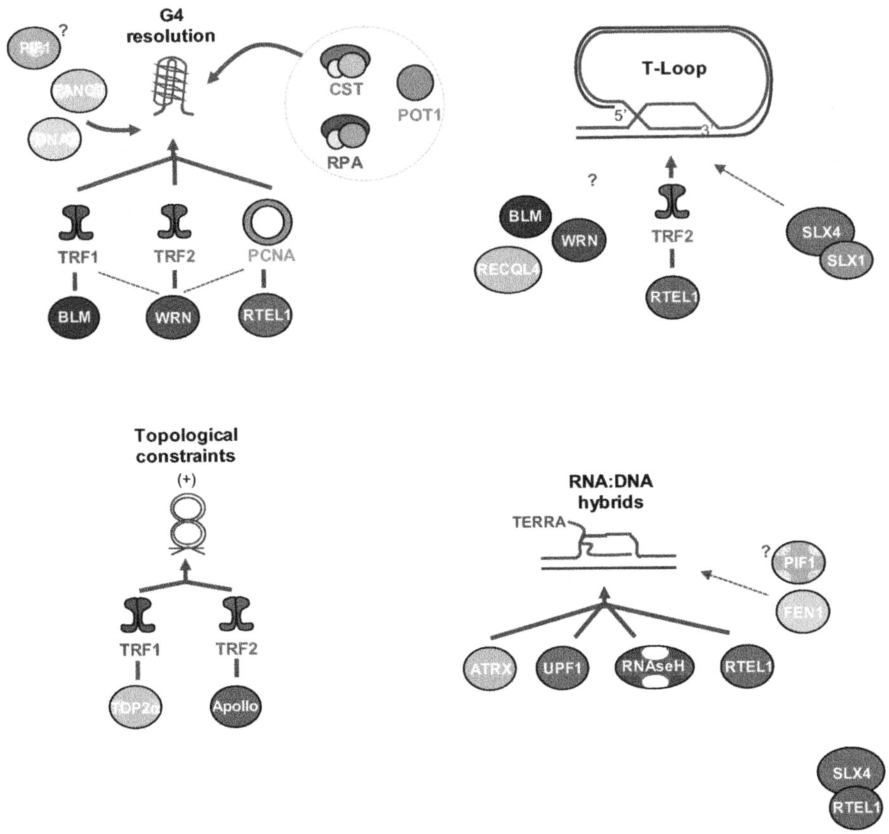

Figure 6.2. *Replication stress control at telomeres. Telomeric and non-telomeric factors are required to remove physical constraints that are formed at terminal sequences. TRF1 and TRF2 serve as docking platforms to recruit accessory factors such as helicases and nucleases. Other elements are recruited independently of the shelterin proteins*

The Pifl helicase family is a group of 5′-3′ directed helicases found in nearly all eukaryotes. The *Saccharomyces cerevisiae* Pifl (ScPifl) is the best-characterized member of this family. Among the multiple nuclear functions that ScPifl fulfills in a cell (reviewed in Bochman et al. (2010)), this helicase is a potent G4 unwinder and promotes replication through G4 motifs in the genome (Ribeyre et al. 2009; Paeschke et al. 2011). At telomeres, ScPifl could probably unwind G4 (Paeschke et al. 2011), but presumably the main function of ScPifl is to inhibit telomerase action by releasing the telomerase RNA from telomeric ends (Boulé et al. 2005; Boulé and Zakian 2007). However, although the role of ScPif1 in G4 unwinding is clear, there is no evidence that ScPif1 is associated with the replisome or binds to telomeric proteins to process telomeric G4 at telomeres. Instead ScPif1 may preferentially patrol DNA and anchor itself to a 3′-tailed DNA junction to unfold G4 (Zhang et al. 2006). Budding yeast is unique because it encodes yet another helicase, Rrm3, that belongs to the Pif1 family (Bochman et al. 2010). Unlike Pif1, Rrm3 appears to travel with the replication fork and contributes to the efficient replication of telomeric repeats (Azvolinsky et al. 2006). Although Rrm3 is able to suppress G4-induced genome instability when the level of Pif1 is low (Paeschke et al. 2013), it is not established whether it is required to remove G4 at telomeres.

The human PIF1 (hPIF1) is also able to unwind G4 structures *in vitro* (Sanders 2010) and localizes to sites thought to be G4 structures (Rodriguez et al. 2012). Nevertheless, a direct role of hPIF1 in G4 processing at mammalian telomeres has not been established yet, although hPIF1 and mouse PIF1 seem to associate with TERT *in vivo* (Mateyak and Zakian 2006; Snow et al. 2007). The unique member of the Pif1 family in the fission yeast is the helicase Pfh1 (Pinter et al. 2008). Pfh1 is essential for the replication of regions that are difficult to replicate and promotes fork movement past G4 (Sabouri et al. 2012, 2014). At telomeres, Pfh1 facilitates telomeric DNA replication, likely by unwinding the G4 structures that are formed (McDonald et al. 2014; Wallgren et al. 2016) . In contrast to its budding yeast counterpart Pif1, but like Rrm3, one could speculate that Pfh1 may travel with the fork to facilitate telomere replication.

Despite the large number of helicases involved in G4 resolution at chromosome ends, it appears that under certain conditions telomeric G4 can be cleaved by a nuclease. This is possibly one function of DNA2, a 5′-3′ directed helicase with a 3′ exo-/endonuclease activity involved in the maintenance of genome stability (Lin et al. 2013). Indeed, mammalian DNA2 is able to cleave G4 *in vitro*, and DNA2 deficiency leads to telomere replication defects. DNA2 could associate with telomeres through interaction with TRF1-2, as these proteins co-immunoprecipitate (Lin et al. 2013), unless its recruitment depends on RPA (Masuda-Sasa et al. 2008; Zhou et al. 2015). These findings are supported by the observations that in budding yeast Dna2 may process G4 at telomeres (Markiewicz-Potoczny et al. 2018). Cleavage of G4 structures causes DNA breakage and would be expected to be deleterious for telomere replication. Thus, one might imagine that G4 cleavage is not the favored option for limiting replication stress during the replication of telomeric sequences.

6.2.2. *G-quadruplex dissolution by single-strand binding proteins*

Aside from helicases, several single-strand DNA binding proteins have been described as preventing the formation of G4 structures *in vitro* (Figure 6.2). These include the telomeric protein POT1 (Zaug et al. 2005; Torigoe and Furukawa 2006; Wang et al. 2011) and the replication protein A complex RPA (Salas et al. 2006). POT1, rather than being an active G4 unwinder like a helicase, would act as a steric driver that binds to telomeric tails and then destabilizes or prevents G4 formation (Wang et al. 2011). For its part, RPA is not a core component of shelterin but would rather be brought to telomeres during replication by the incoming fork. *In vitro*, RPA is able to bind and unfold G4 structures according to a 5′-3′ directionality (Salas et al. 2006; Safa et al. 2016). Indeed, RPA binds preferentially to 5′ single-stranded tail of the G4 structure. A mutation in the DNA binding domain A of human RPA1 (D228Y) alters the DNA binding activity of the RPA complex and therefore its G4 unwinding function (Audry et al. 2015). In fission yeast, the corresponding mutation (D223Y) provokes the shortening of telomeres (Ono et al. 2003) and this phenotype is rescued by overexpression of the Pif1 helicase members (Audry et al. 2015). This suggests that *in vivo* the RPA complex prevents the formation of G-rich structures at lagging strand telomeres to facilitate telomerase activity. Thus, it is likely that RPA unwinds G4 *in vivo*. Another possibility that is not mutually exclusive with the previous one is that RPA recruits or stimulates helicases to resolve G4 structures (Wang et al. 2021). Indeed, the direct physical interaction between RPA and WRN suggests that both proteins may function together *in vivo* (Brosh et al. 1999). In yeast, RPA and Pif1 could cooperate to remove G4 (Maestroni et al. 2020). Along the same lines, huRPA and ScRPA stimulate the G4 cleavage activity of huDNA2 and ScDna2, respectively (Masuda-Sasa et al. 2008). RPA also interacts with FANCJ (Fanconia anemia [FA] Complementation Group J), a 5′-3′ helicase involved in interstrand crosslink repair (ICL) (Gupta et al. 2007). Interestingly, FANCJ is also capable of unwinding G4 structures *in vitro* (London et al. 2008; Wu et al. 2008) and this activity is stimulated by RPA (Wu et al. 2008). *In vivo*, FANCJ promotes DNA synthesis through G4 structures independently of its function in the FA pathway (Schwab et al. 2013; Bosch et al. 2014). Despite these findings, a direct role of FANCJ at telomeres has not yet been established, although FANCJ has been identified at telomeres in ALT cells (Déjardin and Kingston 2009). Thus, RPA might play a key role in preventing or unwinding G4 structures either by acting alone or by recruiting a complementary factor such as a helicase.

As presented above, the CST complex is an ssDNA binding protein that shares homology with RPA. During telomere replication, CST participates in many of the steps required for telomere maintenance (see below and reviewed in Stewart et al.

(2018)). Recently, *in vitro* biochemical analysis revealed that CST binds and resolves G4s (Bhattacharjee et al. 2017; Zhang et al. 2019) and *in vivo* CST prevents accumulation of G4 (Zhang et al. 2019). Thus, in addition to RPA and POT1, CST has proven to be a major actor in G4 processing during telomere replication.

According to these observations, cells have evolved multiple pathways to resolve G4 structures that threaten the integrity of telomere during replication. This complex network may act sequentially or work together to prevent the formation of secondary structures. This network would likely rely on the central component of the replisome PCNA that may act as a tool belt for the recruitment of different replication factors, as well as the shelterin proteins TRF1-2 that may be used as a scaffold, but also POT1, the RPA complex, the CST complex and probably other factors that remain to be discovered.

6.2.3. *T-loop dissolution*

The T-loop, where the 3′ overhang of telomeres invades the double-stranded part of the telomeric repeats through strand displacement to form a D-loop, participates in telomere protection but also represents an obstacle to the progression of the replication fork. Mechanisms that dismantle the T-loop to allow the telomerase access to the 3′ overhang and to avoid collision with the replisome during S phase are therefore necessary (Figure 6.2). In addition to its ability to unwind G4, RTEL1 is involved in T-loop disassembly during DNA replication (Uringa et al. 2012; Vannier et al. 2012). For this function, RTEL1 recruitment required the direct interaction with TRF2 through a C4C4 motif (Sarek et al. 2015). While it is established that RTEL1 associates with the replisome through PCNA binding to promote telomere and genome wide replication (Vannier et al. 2013), the TRF2-RTEL1 interaction is likely to be tightly controlled to anticipate T-loop dissolution prior to the arrival of the replication fork. Indeed, dephosphorylation of TRF2 in S-phase provides a transient access to RTEL1 likely to unwind T-loop to facilitate telomere replication (Sarek et al. 2019). There is no doubt that additional modifications on TRF2, PCNA and RTEL1 might further coordinate the action of RTEL1 throughout the cell cycle to discriminate between different replication barriers such as G4, T-loops or other barriers.

In the absence of RTEL1, T-loops are inappropriately resolved by the SLX1-SLX4 nucleases leading to catastrophic telomere events such as t-circle formation (extrachromosomal circular DNAs that contain telomeric repeat sequences) and rapid telomere shortening (Vannier et al. 2012). However, RTEL1 might not be the sole helicase involved in T-loop disassembly. WRN and BLM are also capable of dissociating telomeric D-loop *in vitro* (Lillard-Wetherell 2004; Opresko et al. 2004) and these two helicases are known to interact directly with TRF2 (Opresko et al. 2002; Lillard-Wetherell 2004) and POT1 (Opresko et al. 2005). T-loop resolution might also involve an additional member of the RECQ helicase family, RECQL4, which is mutated in the Rothmund-Thomson syndrome (Ghosh et al. 2011).

The resolution of complex structures at chromosome ends relies on several helicases while the SLX1-SLX4 nucleases seem to be used as a last resort (see below). Indeed, the shelterin protein TRF2 seems to play a key role in recruiting and/or stimulating WRN, BLM and RTEL1 helicases to disassemble the T-loop (Opresko et al. 2002; Lillard-Wetherell 2004; Sarek et al. 2015). It is likely that many other actors involved in this process remain to be identified. We now need to discover how the cell orchestrates T-loop resolution when the replication fork approaches, and the nature of the cellular signal that triggers T-loop disassembly. Notably, TRF2 also functions with the Apollo 5′-exonuclease to protect telomeres (Lenain et al. 2006; van Overbeek and de Lange 2006). It is currently not known whether Apollo is involved in T-loop resolution, nevertheless it prevents the formation of topological constraints by the removal of superhelical stress caused by the nucleoprotein complex that stabilizes the base of the T-loop (Ye et al. 2010). Mechanisms that orchestrate T-loop resolution are likely to rely on a complex network of post-translational modifications presumably involving the shelterin proteins. This may represent an active area of investigation in the future.

6.2.4. *RNA:DNA hybrid removal*

In most eukaryotes, telomeres are actively transcribed into TElomere Repeat containing RNA (TERRA) (Azzalin et al. 2007; Schoeftner and Blasco 2007). TERRA transcription initiates from the subtelomeric regions towards the TTAGGG tract. Under certain circumstances, RNA molecules can anneal to their genomic template co- or post-transcriptionally to generate RNA:DNA hybrids. Strand displacement by the RNA:DNA hybrids forms a typical structure known as an R-loop (Costantino and Koshland 2015). TERRA R-loops are natural structures that are formed at telomeres in human and yeast cells and may contribute to replication stress and chromosomal instability as they represent natural barriers to the progression of the replication fork (for review, see (Rippe and Luke 2015)). More dramatically, if G4 structures are formed at the displaced G-rich strand, the R-loop may be a major determinant of replication fork progression impairment, double-strand break and telomeric loss. TERRA expression is cell-cycle regulated, it peaks at the G1-S transition and declines from S to G2 in telomerase positive mammalian cells (Flynn et al. 2015) and also in budding yeast (Graf et al. 2017), probably to avoid collision between RNA Pol II-mediated transcription and the replication fork. This also implies that the TERRA R-loop must be dissolved prior to the passage of the fork (Figure 6.2). A major enzyme involved in RNA:DNA hybrid resolution is the RNA endonuclease H (RNase H) that degrades the RNA moiety of the duplex (Arudchandran et al. 2000). TERRA transcription is enhanced at telomeres in the absence of telomerase, and the R-loop is mainly dissolved by RNAse H in budding yeast (Balk et al. 2013) and by RNAse H1 in ALT cells (Arora et al. 2014). Moreover, ATRX mutations are frequently found in ALT cancer cells (Lovejoy et al. 2012). ATRX is a chromatin remodeler involved in the establishment of silent heterochromatin by deposition of the histone H3.3 variant in repetitive regions such as the pericentric chromatin and telomeres (Goldberg et al. 2010). Knockdown of ATRX results in persistent TERRA levels in G2/M and R-loop accumulation, suggesting that ATRX promotes R-loop removal through TERRA displacement (Flynn et al. 2015; Nguyen et al. 2017). The mechanism by which ATRX promotes TERRA displacement is currently not clear. One possibility would be that the recruitment of ATRX is associated with the presence of R-loops (Nguyen et al. 2017) or that ATRX recognizes and/or modifies G4 structures by itself, since ATRX has been found to bind G4 *in vitro* (Law et al. 2010). Another hypothesis would be that ATRX influences gene expression by recognizing unusual DNA structures and then converting them to regular forms by facilitating the incorporation of histone variant H3.3 (Clynes et al. 2015).

UPF1 (Up-frameshift 1) is a conserved eukaryotic phosphoprotein with nucleic acid-dependent ATPase and 5′-3′ helicase activity. UPF1 has a dual function, first in cytoplasmic RNA quality control and second in S-phase progression and genome stability (Azzalin and Lingner 2006). At chromosome ends, UPF1 is recruited through its physical interaction with TPP1 and its depletion induces severe telomeric aberrations and TERRA accumulation (Azzalin et al. 2007; Chawla et al. 2011). Interestingly, the telomeric defects observed upon UPF1 depletion arise mainly from incomplete leading-strand telomere replication (Chawla et al. 2011). Taken together, these studies suggest that UPF1 participates in telomere replication, raising the possibility that UPF1 may displace TERRA molecules from telomeric chromatin to avoid R-loop accumulation which might induce replication stress on the leading telomere.

Helicases are also potential enzymes that could control RNA:DNA hybrid levels. For example, a connection between RTEL1 and the abundance of TERRA has been depicted (Ghisays et al. 2021). In that process, RTEL1 would recognize G4 which form in TERRA and stabilize TERRA-containing telomeric R-loops. This stabilization would further limit transcription and reduce the TERRA level. In addition to RTEL1, the ability of budding yeast Pif1 and hPIF1 to unwind

RNA:DNA hybrids (Zhang et al. 2006; Boulé and Zakian 2007) makes Pif1 helicase a potential candidate for resolution of the R-loop (Chib et al. 2016).

Endonuclease could also process RNA:DNA hybrids. This could be the case of the FEN1 flap endonuclease that can limit RNA:DNA hybrid accumulation at the leading strand (Teasley et al. 2015). These findings suggest that the replisome can bypass the RNA:DNA hybrid to avoid co-directional collision between the replisome and RNA Pol II. The role of FEN1 would be to remove RNA:DNA hybrid flaps to avoid their accumulation, which could lead to telomere fragility.

Although TERRA transcription and R-loops represent real threats to the replication of telomeric repeats, it becomes clear that TERRA fulfills numerous and important functions in the biology of telomeres including telomere length regulation, telomere replication, telomere end protection, telomeric chromatin composition changes and telomere mobility (Feuerhahn et al. 2010; Azzalin and Lingner 2014; Maicher et al. 2014). Thus, telomere transcription should be tightly controlled, particularly when replication is engaged so that collisions are avoided. Here, we have listed several proteins that promote degradation or displacement of TERRA, but how these mechanisms are orchestrated and regulated when the replication fork approaches will require more investigation.

6.2.5. *Topological constraints*

Telomeres form a compact chromatin structure *in vivo* through specific protein–protein and protein–DNA interactions between the shelterin proteins and telomeric DNA (Bandaria et al. 2016). In line with these findings, TRF2 confers a topological state to telomeric DNA that participates in telomere protection (Benarroch-Popivker et al. 2016). This DNA compaction participates in the protection of the telomeric repeats by preventing DDR signaling at telomeres. However, the condensation and topological constraints of the telomeric tracts also represent a barrier to the progression of the replication fork. This implies that during S-phase the telomeric DNA must be decondensed to allow passage of the replication fork and then recompacted post-replication (Figure 6.2). This raises the question of how telomere decompaction is performed. As mentioned above, Apollo 5′-exonuclease functions with TRF2 to protect telomeres (Lenain et al. 2006; van Overbeek and de Lange 2006) and cooperate together to remove superhelical constraints (Ye et al. 2010). Moreover, they seem to act in synergy with the DNA topoisomerase 2α (TOP2α) by removing the topological barrier generated by fork progression through telomeric chromatin (Ye et al. 2010). The role of TOP2α in telomere protection has also been observed in mice, where its recruitment depends on TRF1 (d'Alcontres et al. 2014). Thus, it seems clear that TRF2 and likely TRF1 plays an essential function in controlling the topological state of telomeric DNA. Additional work will be necessary to elucidate mechanisms that are used by the cell to regulate telomere condensation as the incoming fork approaches.

Telomere anchoring highlights another type of superhelical constraint that is introduced by tethering. In yeasts, telomeres adopt the "Rabl" conformation. Telomeres localize to the NE on one side of the nucleus while centromeres occupy the other side (Funabiki et al. 1993; Nagai et al. 2010). In budding yeast, telomere anchoring depends on the redundant Esc1-Sir4-Rap1 and yKu-Mps3 pathways. In fission yeast, NE tethering is mediated by the interaction between the inner nuclear membrane component Bqt4 and Rap1 (Chikashige et al. 2009). In addition, the Fun30 chromatin remodeler Fft3 also participates in telomere anchoring independently of Bqt4 (Steglich et al. 2015). In contrast, human telomeres localize throughout the nuclear volume. However, human telomeres are not free to roam across the nucleus but are attached to the NM through shelterin and lamins (Ludérus et al. 1996; Kaminker et al. 2009) and only a subset of telomeres are found at the nuclear periphery (Arnoult et al. 2010). Thus, to allow efficient replication of telomeres, these topological constraints need to be released by promoting telomere detachment from NE or NM. This may represent another active area of research for the next decade.

6.3. Accessory factors that contribute to telomere replication I

6.3.1. *The CST complex*

As described in the introduction, CST promotes the synthesis of the complementary strand through the recruitment of the DNA polymerase α/primase complex (Pol α). Independently of its C-strand fill in reaction, CST has been shown to be essential for the replication of the duplex telomeric DNA (Gu et al. 2012; Stewart et al. 2012; Wang et al. 2012) and also helps to resolve replication problems throughout the genome likely in GC-rich repetitive sequences (Stewart et al. 2012; Wang et al. 2012). In fission yeast, in which CTC1 homolog is missing, Stn1-Ten1 is essential for the replication of telomeres and is also required for the replication of subtelomeric regions (Takikawa et al. 2016; Matmati et al. 2018). It is unclear how CST participates in telomere replication, however CST likely contributes to fork restart by facilitating Okazaki fragment synthesis by resolving or preventing G4 on the single-stranded G-rich template DNA or avoids fork stalling by preventing G4 arising ahead of the fork at transcription sites (Zhang et al. 2019). It has also been proposed that CST could also promote replication recovery partly by controlling dormant origins throughout the genome and at telomeres (Stewart et al. 2012; Wang et al. 2014). Indeed, a recent study has reported first that CST can suppress origin licensing by preventing origin replication complex (ORC) association and second that CST can enhance replisome assembly by promoting Pol α chromatin association (Wang et al. 2019). CST could also promote efficient replication restart when replication forks stall by recruitment of recombination factors such as Rad51 (Chastain et al. 2016), indicating that CST could initiate recombination-based replication restart to prevent genome instability. In addition, CST is thought to protect stalled replication forks by directly blocking MRE11 degradation of nascent-strand DNA (Lyu et al. 2020) (Figure 6.3).

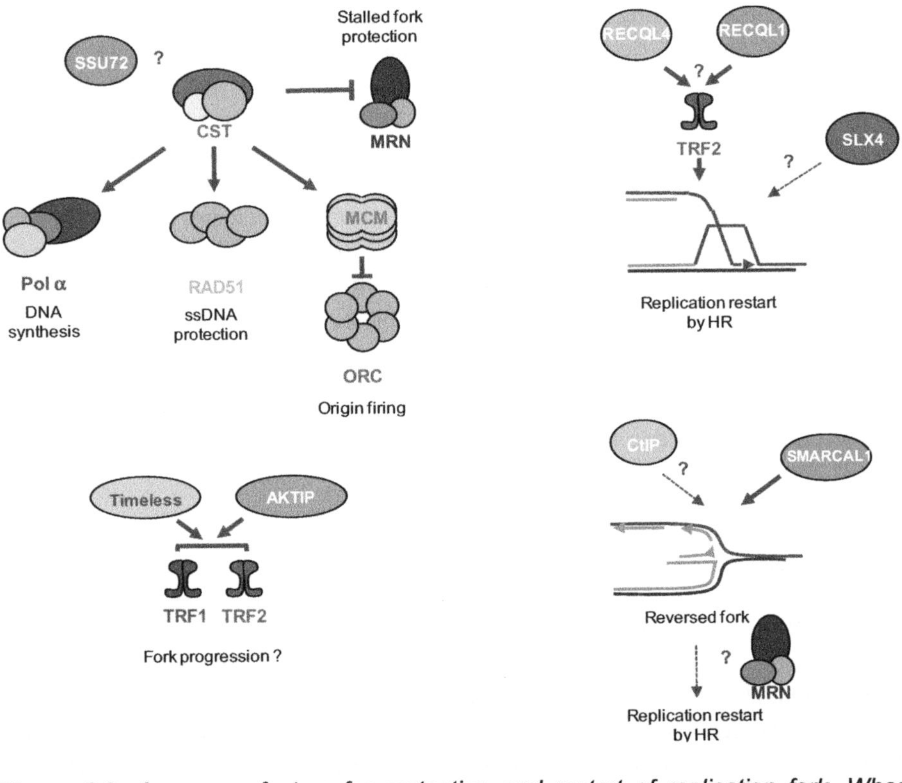

Figure 6.3. *Accessory factors for protection and restart of replication fork. When replication forks stall, they may collapse and break. Accessory factors are required to protect stalled forks or restart broken forks at telomeres. CST complex and other accessory factors certainly play a key role in that process. Although these mechanisms are not clearly identified, we presume that fork reversal and homologous recombination-dependent fork restart are possible pathways to ensure telomere stability*

Over the last decade, CST has proven to be a key factor required for telomere replication and telomere length maintenance. Importantly, mutations in CTC1 and STN1 cause the Coats plus syndrome, a lethal disease in which some symptoms mirror those of DC (Anderson et al. 2012; Simon et al. 2016). Although insights were gained into CST involvement in telomere replication, further characterization is required to determine how CST coordinates with the other replication factors such as RPA and the several helicases and how CST recruitment is achieved. In budding yeast, Cdk1-dependent phosphorylation of $Cdc13^{CTC1}$ and Stn1 subunits have been shown to regulate telomerase activity (Li et al. 2009; Liu et al. 2013). In fission yeast, Stn1 is phosphorylated and its dephosphorylation, by the Ssu72 phosphatase, regulates telomere replication by controlling recruitment of Stn1 (Escandell et al. 2019). This mechanism is evolutionarily conserved since SSU72 controls STN1 recruitment to telomeres in human cells, although no evidence of phosphorylation of STN1 has been established yet (Escandell et al. 2019). Further analysis will be necessary to confirm if CST subunits are modified and if these post-translational modifications control human telomere replication.

6.3.2. *The MRN complex and CtIP*

The MRN complex (MRE11-RAD50-NBS1) regulates signaling and DDR at DSB and stalled forks (Syed and Tainer 2018). It is well established that, at dysfunctional telomeres, MRN is recruited by TRF2 and promotes NHEJ-mediated repair through ATM activation (Deng et al. 2009; Dimitrova and Lange 2009; Rai et al. 2017); however, its role in telomere replication is more uncertain.

In fission yeast, MRN was shown to process terminally arrested forks to promote restart through HR (Teixeira-Silva et al. 2017). In this attractive model, MRN in cooperation with Ctp1 (homolog of human CtIP, see below) is required to evict Ku dimer from reversed forks and then to initiate resection to promote fork restart. So far, there is no evidence that MRN is needed at telomeres to process stalled forks in a normal context in yeast. Nevertheless, MRN is required to ensure telomere maintenance in the absence of telomerase (Matmati et al. 2020). Thus, it is possible that MRN is recruited at a subset of stalled/reversed forks to ensure replication restart in telomerase positive cells and/or that MRN is necessary to promote fork restart in cells lacking telomerase activity. This hypothesis needs to be further investigated in mammalian cells.

Another key factor that functions in HR pathway is CtIP, known to control resection at DNA double-stranded breaks by regulating MRN activities (Andres and Williams 2017). Recently, CtIP was shown to have a MRN-independent role in protecting stalled forks from DNA2-dependent degradation (Przetocka et al. 2018).

In agreement with this study, CtIP was also shown to be required for faithful replication through telomeres (Stroik et al. 2019). Although the action of CtIP at telomeric replication forks is unclear, CtIP facilitates the repair, the restart or the rescue of stalled forks and avoids telomeric loss and t-circle formation.

Although the detailed molecular mechanisms in which MRN and CtIP are involved are not clearly identified, it is likely possible that these factors are required to process replication stress at telomeres to maintain telomere integrity (Figure 6.3). However, the described inhibition of MRE11 nuclease activity by CST indicates that MRN action should be tightly controlled to avoid inappropriate resection at stalled forks.

6.3.3. *SLX4*

SLX4 is known to act as a scaffold to assemble a DNA repair complex consisting of the structure-specific endonucleases SLX1, XPF and MUS81 (for review, see Guervilly and Gaillard (2018)). In human cells, SLX4 primarily associates with relatively long telomeres, and telomeric localization of this complex is dependent on direct interaction between SLX4 and TRF2 (Wan et al. 2013; Wilson et al. 2013). Knowing the ability of SLX4 and its partners to cleave DNA structures including Holliday junctions and D-loops (Fekairi et al. 2009; Muñoz et al. 2009; Svendsen et al. 2009; Wan et al. 2013), the T-loop structures may represent a substrate for SLX4-endonuclease complexes. Nevertheless, TRF2 seems to restrain endonucleolytic activities of SLX4-associated nucleases to prevent abrupt telomere shortening *in vivo* (Poulet et al. 2009; Saint-Léger et al. 2014; Sarkar et al. 2015), likely by the excision of the t-circles through the processing of homologous recombination intermediates. Contradictorily, SLX4 associates with telomeres throughout the cell cycle, peaking in late S phase (Sarkar et al. 2015), and dysregulation of SLX4 leads to broken telomeres and telomeric dysfunction induced foci (TIFs) (Wilson et al. 2013; Sarkar et al. 2015), suggesting that SLX4 is necessary to process joint molecule intermediates during telomere replication. Thus, it might possible that the role of SLX4 and its partners in replication of terminal ends is restricted to long telomeres, such as ALT or murine telomeres (Wan et al. 2013; Wilson et al. 2013; Sarkar et al. 2015), that may exhibit high propensity to form recombination intermediates that necessitate endonucleolytic cleavage. Recently, SLX4 has also been described to recruit RTEL1 to prevent replication-transcription conflicts at hard-to-replicate loci, including telomeres, independently of its nuclease scaffold functions (Takedachi et al. 2020; Wu et al. 2020).

Overall, current evidence converges toward a multiple role of SLX4: first acting as a negative regulator of telomere length through the cleavage of the T-loop, second as a replication accessory factor by preventing fragility of long telomeres and finally as a recruiter of RTEL1 to resolve replication-transcription conflicts.

6.3.4. *Timeless and AKTIP*

Interestingly, TRF1 and TRF2 also associate with the Timeless protein, a component of the fork protection complex (FPC) (Leman et al. 2012) . The FPC is a part of the replisome and ensures proper replication fork pausing and the smooth passage of the replication forks at hard-to-replicate regions, including telomeres (reviewed in Leman and Noguchi (2013)). Remarkably, Timeless downregulation or deletion of the Timeless ortholog *swi1* leads to shortened telomeres in human cells and in fission yeast, respectively (Leman and Noguchi 2012). In fission yeast, Swi1 ensures the correct replication of telomeres, presumably by maintaining the fork in a natural conformation to prevent fork collapse and telomeric repeat instability (Gadaleta et al. 2016). These observations show, first, that replisome integrity is important for telomere maintenance, and second, the replisome and the shelterin act together to guarantee telomeric tract stability by preventing the stalling and collapse of replication forks (Figure 6.3).

In line with this, AKTIP interacts with telomeric DNA and binds the shelterin components TRF1 and TRF2 (Burla et al. 2015). AKTIP is a mammalian shelterin-bound factor that shares homology with the E2 variant ubiquitin-conjugating (UEV) enzymes. AKTIP depletion results in the formation of TIFs and in a partial loss of telomeric-bound TRF1. In MEFs, Ft1-depletion (the mouse AKTIP ortholog) leads to telomere dysfunction. Thus, AKTIP seems to be a telomeric accessory factor that works in concert with TRF1 to facilitate telomeric DNA replication, although the precise AKTIP function remains unknown (Figure 6.3).

6.3.5. *RECQL helicases and SMARCAL1 translocase*

Two members of the RECQL helicase family (RECQL1 and RECQL4) seem to also play a role in telomere replication (Ghosh et al. 2011; Popuri et al. 2014). Both are recruited to telomere in S-phase in TRF2-dependant manner and their downregulation causes telomere dysfunction. In contrast to WRN and BLM, RECQL1 and RECQL4 are unable to unwind G4 structures *in vitro* (Rossi et al. 2010; Popuri et al. 2014). Yet, the telomeric functions of these two helicases are not known, but both would rather be required for telomeric D-loop resolution, and for the processing of Holliday junctions in the case of RECQL1 (Ghosh et al. 2011; Popuri et al. 2014) (Figure 6.3). Besides, the role of RECQL1 is predominant in

ALT cells rather than in telomerase-positive cells. Although the precise role of these RECQL helicases needs to be determined, both are likely involved in the processing of recombination intermediates during replication or eventually the T-loop disassembling.

SMARCAL1 belongs to the SNF2 family translocases that are involved in the genome-wide replication stress response (Bansbach et al. 2009; Ciccia et al. 2009; Yusufzai et al. 2009; Bétous et al. 2012). *In vitro*, SMARCAL1 can catalyze fork regression, which involves the reannealing of cDNA strands and can also catalyze the restoration of a regressed fork back to a normal replication fork structure (Yusufzai and Kadonaga 2008; Ciccia et al. 2009; Bétous et al. 2012, 2013). Importantly, SMARCAL1-deficient cells display telomere-associated DNA damage and accumulate T-circles, and these telomeric functions are not shared by the other translocases of the SNF2 family (Poole et al. 2015). Here again, the exact mechanism of SMARCAL1 action at telomeres remains unknown, but this translocase could catalyze fork reversal and/or fork restoration during replication of terminal ends (Figure 6.3).

6.4. Concluding remarks

The passage of the replication fork through the telomeric tract is probably one of the most difficult processes that occurs during chromosome duplication. In this chapter, we have listed the different obstacles that a replication fork encounters when approaching the telomeric chromatin. Efficient replication of telomeres requires a number of interactions between shelterin and non-telomeric proteins to support fork progression. In the absence of these factors, replication forks may stall and break. A broken fork necessitates recombination-driven repair and may give rise to aberrant recombination, leading to telomere fragility. Alternatively, a stalled fork may reverse, providing time to resolve blocking structures (Figure 6.1). Reversed replication forks adopt a four-way branched structure that can also be recognized by telomerase (Margalef et al. 2018; Matmati et al. 2020). In fission yeast, it has also been proposed that telomerase could extend the exposed G-rich single-strand DNA of this structure to promote *de novo* telomere synthesis (Matmati et al. 2020). Then, it is tempting to speculate that telomerase contributes to telomere replication in mammals too, where it could either stabilize reversed forks or promote *de novo* telomere synthesis to complete replication. The interplay between shelterin and telomerase in promoting telomere replication fork restart may represent an active area of investigation for the future.

6.5. References

d'Alcontres, M.S., Palacios, J.A., Mejias, D., Blasco, M.A. (2014). TopoIIα prevents telomere fragility and formation of ultra thin DNA bridges during mitosis through TRF1-dependent binding to telomeres. *Cell Cycle*, 13(9), 1463–1481.

Anderson, B.H., Kasher, P.R., Mayer, J., Szynkiewicz, M., Jenkinson, E.M., Bhaskar, S.S., Urquhart, J.E., Daly, S.B., Dickerson, J.E., O'Sullivan, J. et al. (2012). Mutations in CTC1, encoding conserved telomere maintenance component 1, cause Coats plus. *Nature Genetics*, 44(3), 338–342.

Andres, S.N. and Williams, R.S. (2017). CtIP/Ctp1/Sae2, molecular form fit for function. *DNA Repair*, 56, 109–117.

Armanios, M. and Blackburn, E.H. (2012). The telomere syndromes. *Nature Reviews Genetics*, 13(10), 693–704.

Arnoult, N. and Karlseder, J. (2015). Complex interactions between the DNA-damage response and mammalian telomeres. *Nature Structural & Molecular Biology*, 22(11), 859–866.

Arnoult, N., Saintomé, C., Ourliac-Garnier, I., Riou, J.-F., Londoño-Vallejo, A. (2009). Human POT1 is required for efficient telomere C-rich strand replication in the absence of WRN. *Genes & Development*, 23(24), 2915–2924.

Arnoult, N., Schluth-Bolard, C., Letessier, A., Drascovic, I., Bouarich-Bourimi, R., Campisi, J., Kim, S.-H., Boussouar, A., Ottaviani, A., Magdinier, F. et al. (2010). Replication timing of human telomeres is chromosome arm-specific, influenced by subtelomeric structures and connected to nuclear localization. *PLoS Genetics*, 6(4), e1000920.

Arora, R., Lee, Y., Wischnewski, H., Brun, C.M., Schwarz, T., Azzalin, C.M. (2014). RNaseH1 regulates TERRA-telomeric DNA hybrids and telomere maintenance in ALT tumour cells. *Nature Communications*, 5, 5220.

Artandi, S.E., Chang, S., Lee, S.L., Alson, S., Gottlieb, G.J., Chin, L., DePinho, R.A. (2000). Telomere dysfunction promotes non-reciprocal translocations and epithelial cancers in mice. *Nature*, 406(6796), 641–645.

Arudchandran, A., Cerritelli, S., Narimatsu, S., Itaya, M., Shin, D.Y., Shimada, Y., Crouch, R.J. (2000). The absence of ribonuclease H1 or H2 alters the sensitivity of Saccharomyces cerevisiae to hydroxyurea, caffeine and ethyl methanesulphonate: Implications for roles of RNases H in DNA replication and repair. *Genes to Cells: Devoted to Molecular & Cellular Mechanisms*, 5(10), 789–802.

Audry, J., Maestroni, L., Delagoutte, E., Gauthier, T., Nakamura, T.M., Gachet, Y., Saintomé, C., Géli, V., Coulon, S. (2015). RPA prevents G-rich structure formation at lagging-strand telomeres to allow maintenance of chromosome ends. *The EMBO Journal*, 34(14), 1942–1958.

Azvolinsky, A., Dunaway, S., Torres, J.Z., Bessler, J.B., Zakian, V.A. (2006). The *S. cerevisiae* Rrm3p DNA helicase moves with the replication fork and affects replication of all yeast chromosomes. *Genes & Development*, 20(22), 3104–3116.

Azzalin, C.M. and Lingner, J. (2006). The human RNA surveillance factor UPF1 is required for S phase progression and genome stability. *Current Biology: CB*, 16(4), 433–439.

Azzalin, C.M. and Lingner, J. (2014). Telomere functions grounding on TERRA firma. *Trends in Cell Biology*, 25(1), 29–36.

Azzalin, C.M., Reichenbach, P., Khoriauli, L., Giulotto, E., Lingner, J. (2007). Telomeric repeat containing RNA and RNA surveillance factors at mammalian chromosome ends. *Science*, 318(5851), 798–801.

Balk, B., Maicher, A., Dees, M., Klermund, J., Luke-Glaser, S., Bender, K., Luke, B. (2013). Telomeric RNA-DNA hybrids affect telomere-length dynamics and senescence. *Nature Structural & Molecular Biology*, 20(10), 1199–1205.

Ballew, B.J., Yeager, M., Jacobs, K., Giri, N., Boland, J., Burdett, L., Alter, B.P., Savage, S.A. (2013). Germline mutations of regulator of telomere elongation helicase 1, RTEL1, in *Dyskeratosis congenita*. *Human Genetics*, 132(4), 473–480.

Bandaria, J.N., Qin, P., Berk, V., Chu, S., Yildiz, A. (2016). Shelterin protects chromosome ends by compacting telomeric chromatin. *Cell*, 164(4), 735–746.

Bansbach, C.E., Bétous, R., Lovejoy, C.A., Glick, G.G., Cortez, D. (2009). The annealing helicase SMARCAL1 maintains genome integrity at stalled replication forks. *Genes & Development*, 23(20), 2405–2414.

Barefield, C. and Karlseder, J. (2012). The BLM helicase contributes to telomere maintenance through processing of late-replicating intermediate structures. *Nucleic Acids Research*, 40(15), 7358–7367.

Benarroch-Popivker, D., Pisano, S., Mendez-Bermudez, A., Lototska, L., Kaur, P., Bauwens, S., Djerbi, N., Latrick, C.M., Fraisier, V., Pei, B. et al. (2016). TRF2-mediated control of telomere DNA topology as a mechanism for chromosome-end protection. *Molecular Cell*, 61(2), 274–286.

Bétous, R., Mason, A.C., Rambo, R.P., Bansbach, C.E., Badu-Nkansah, A., Sirbu, B.M., Eichman, B.F., Cortez, D. (2012). SMARCAL1 catalyzes fork regression and Holliday junction migration to maintain genome stability during DNA replication. *Genes & Development*, 26(2), 151–162.

Bétous, R., Couch, F.B., Mason, A.C., Eichman, B.F., Manosas, M., Cortez, D. (2013). Substrate-selective repair and restart of replication forks by DNA translocases. *Cell Reports*, 3(6), 1958–1969.

Bhattacharjee, A., Wang, Y., Diao, J., Price, C.M. (2017). Dynamic DNA binding, junction recognition and G4 melting activity underlie the telomeric and genome-wide roles of human CST. *Nucleic Acids Research*, 45(21), 12311–12324.

Bochman, M.L., Sabouri, N., Zakian, V.A. (2010). Unwinding the functions of the Pif1 family helicases. *DNA Repair*, 9(3), 237–249.

Bosch, P.C., Segura-Bayona, S., Koole, W., van Heteren, J.T., Dewar, J.M., Tijsterman, M., Knipscheer, P. (2014). FANCJ promotes DNA synthesis through G-quadruplex structures. *The EMBO Journal*, 33(21), 2521–2533.

Boulé, J.-B. and Zakian, V.A. (2007). The yeast Pif1p DNA helicase preferentially unwinds RNA DNA substrates. *Nucleic Acids Research*, 35(17), 5809–5818.

Boulé, J.-B., Vega, L.R., Zakian, V.A. (2005). The yeast Pif1p helicase removes telomerase from telomeric DNA. *Nature*, 438(7064), 57–61.

Brosh, R.M., Orren, D.K., Nehlin, J.O., Ravn, P.H., Kenny, M.K., Machwe, A., Bohr, V.A. (1999). Functional and physical interaction between WRN helicase and human replication protein A. *Journal of Biological Chemistry*, 274(26), 18341–18350.

Bryan, T.M., Englezou, A., Dalla-Pozza, L., Dunham, M.A., Reddel, R.R. (1997). Evidence for an alternative mechanism for maintaining telomere length in human tumors and tumor-derived cell lines. *Nature Medicine*, 3(11), 1271–1274.

Burla, R., Carcuro, M., Raffa, G.D., Galati, A., Raimondo, D., Rizzo, A., Torre, M.L., Micheli, E., Ciapponi, L., Cenci, G. et al. (2015). AKTIP/Ft1, a new shelterin-interacting factor required for telomere maintenance. *PLoS Genetics*, 11(6), e1005167.

Campisi, J. and d'Adda di Fagagna, F. (2007). Cellular senescence: When bad things happen to good cells. *Nature Reviews. Molecular Cell Biology*, 8(9), 729–740.

Casteel, D.E., Zhuang, S., Zeng, Y., Perrino, F.W., Boss, G.R., Goulian, M., Pilz, R.B. (2008). A DNA polymerase-α·primase cofactor with homology to replication protein A-32 regulates DNA replication in mammalian cells. *Journal of Biological Chemistry*, 284(9), 5807–5818.

Chastain, M., Zhou, Q., Shiva, O., Whitmore, L., Jia, P., Dai, X., Huang, C., Fadri-Moskwik, M., Ye, P., Chai, W. (2016). Human CST facilitates genome-wide RAD51 recruitment to GC-rich repetitive sequences in response to replication stress. *Cell Reports*, 16(5), 1300–1314.

Chawla, R., Redon, S., Raftopoulou, C., Wischnewski, H., Gagos, S., Azzalin, C.M. (2011). Human UPF1 interacts with TPP1 and telomerase and sustains telomere leading-strand replication. *The EMBO Journal*, 30(19), 4047–4058.

Chen, L.-Y., Redon, S., Lingner, J. (2012). The human CST complex is a terminator of telomerase activity. *Nature*, 1–7.

Chib, S., Byrd, A.K., Raney, K.D. (2016). Yeast helicase Pif1 unwinds RNA:DNA hybrids with higher processivity than DNA:DNA duplexes. *Journal of Biological Chemistry*, 291(11), 5889–5901.

Chikashige, Y., Yamane, M., Okamasa, K., Tsutsumi, C., Kojidani, T., Sato, M., Haraguchi, T., Hiraoka, Y. (2009). Membrane proteins Bqt3 and -4 anchor telomeres to the nuclear envelope to ensure chromosomal bouquet formation. *The Journal of Cell Biology*, 187(3), 413–427.

Ciccia, A., Bredemeyer, A.L., Sowa, M.E., Terret, M.-E., Jallepalli, P.V., Harper, J.W., Elledge, S.J. (2009). The SIOD disorder protein SMARCAL1 is an RPA-interacting protein involved in replication fork restart. *Genes & Development*, 23(20), 2415–2425.

Clynes, D., Jelinska, C., Xella, B., Ayyub, H., Scott, C., Mitson, M., Taylor, S., Higgs, D.R., Gibbons, R.J. (2015). Suppression of the alternative lengthening of telomere pathway by the chromatin remodelling factor ATRX. *Nature Communications*, 6, 7538.

Costantino, L. and Koshland, D. (2015). The Yin and Yang of R-loop biology. *Current Opinion in Cell Biology*, 34, 39–45.

Crabbe, L., Verdun, R.E., Haggblom, C.I., Karlseder, J. (2004). Defective telomere lagging strand synthesis in cells lacking WRN helicase activity. *Science*, 306(5703), 1951–1953.

Crabbe, L., Cesare, A.J., Kasuboski, J.M., Fitzpatrick, J.A.J., Karlseder, J. (2012). Human telomeres are tethered to the nuclear envelope during postmitotic nuclear assembly. *Cell Reports*, 2(6), 1521–1529.

Déjardin, J. and Kingston, R.E. (2009). Purification of proteins associated with specific genomic loci. *Cell*, 136(1), 175–186.

Deng, Y., Guo, X., Ferguson, D.O., Chang, S. (2009). Multiple roles for MRE11 at uncapped telomeres. *Nature*, 460(7257), 914–918.

Deng, Z., Glousker, G., Molczan, A., Fox, A.J., Lamm, N., Dheekollu, J., Weizman, O.-E., Schertzer, M., Wang, Z., Vladimirova, O. et al. (2013). Inherited mutations in the helicase RTEL1 cause telomere dysfunction and Hoyeraal-Hreidarsson syndrome. *Proceedings of the National Academy of Sciences*, 110(36), E3408–E3416.

Dimitrova, N. and de Lange, T. (2009). Cell cycle-dependent role of MRN at dysfunctional telomeres: ATM signaling-dependent induction of nonhomologous end joining (NHEJ) in G1 and resection-mediated inhibition of NHEJ in G2. *Molecular and Cellular Biology*, 29(20), 5552–5563.

Ding, H., Schertzer, M., Wu, X., Gertsenstein, M., Selig, S., Kammori, M., Pourvali, R., Poon, S., Vulto, I., Chavez, E. et al. (2004). Regulation of murine telomere length by Rtel. *Cell*, 117(7), 873–886.

Doksani, Y., Wu, J.Y., de Lange, T., Zhuang, X. (2013). Super-resolution fluorescence imaging of telomeres reveals TRF2-dependent T-loop formation. *Cell*, 155(2), 345–356.

Drosopoulos, W.C., Kosiyatrakul, S.T., Schildkraut, C.L. (2015). BLM helicase facilitates telomere replication during leading strand synthesis of telomeres. *The Journal of Cell Biology*, 210(2), 191–208.

Escandell, J.M., Carvalho, E.S., Gallo-Fernández, M., Reis, C.C., Matmati, S., Luís, I.M., Abreu, I.A., Coulon, S., Ferreira, M.G. (2019). Ssu72 phosphatase is a conserved telomere replication terminator. *The EMBO Journal*, 38(7), e100476.

Fekairi, S., Scaglione, S., Chahwan, C., Taylor, E.R., Tissier, A., Coulon, S., Dong, M.-Q., Ruse, C., Yates, J.R., Russell, P. et al. (2009). Human SLX4 is a Holliday junction resolvase subunit that binds multiple DNA repair/recombination endonucleases. *Cell*, 138(1), 78–89.

Feuerhahn, S., Iglesias, N., Panza, A., Porro, A., Lingner, J. (2010). TERRA biogenesis, turnover and implications for function. *FEBS Letters*, 584(17), 3812–3818.

Flynn, R.L., Cox, K.E., Jeitany, M., Wakimoto, H., Bryll, A.R., Ganem, N.J., Bersani, F., Pineda, J.R., Suva, M.L., Benes, C.H. et al. (2015). Alternative lengthening of telomeres renders cancer cells hypersensitive to ATR inhibitors. *Science*, 347(6219), 273–277.

Funabiki, H., Hagan, I., Uzawa, S., Yanagida, M. (1993). Cell cycle-dependent specific positioning and clustering of centromeres and telomeres in fission yeast. *The Journal of Cell Biology*, 121(5), 961–976.

Gadaleta, M.C., Das, M.M., Tanizawa, H., Chang, Y.-T., Noma, K., Nakamura, T.M., Noguchi, E. (2016). Swi1Timeless prevents repeat instability at fission yeast telomeres. *PLoS Genetics*, 12(3), e1005943.

Ghisays, F., Garzia, A., Wang, H., Canasto-Chibuque, C., Hohl, M., Savage, S.A., Tuschl, T., Petrini, J.H.J. (2021). RTEL1 influences the abundance and localization of TERRA RNA. *Nature Communications*, 12(1), 3016.

Ghosh, A.K., Rossi, M.L., Singh, D.K., Dunn, C., Ramamoorthy, M., Croteau, D.L., Liu, Y., Bohr, V.A. (2011). RECQL4, the protein mutated in Rothmund-Thomson syndrome, functions in telomere maintenance. *Journal of Biological Chemistry*, 287(1), 196–209.

Gilson, E. and Géli, V. (2007). How telomeres are replicated. *Nature Reviews. Molecular Cell Biology*, 8(10), 825–838.

Goldberg, A.D., Banaszynski, L.A., Noh, K.-M., Lewis, P.W., Elsaesser, S.J., Stadler, S., Dewell, S., Law, M.J., Guo, X., Li, X. et al. (2010). Distinct factors control histone variant H3.3 localization at specific genomic regions. *Cell*, 140(5), 678–691.

Graf, M., Bonetti, D., Lockhart, A., Serhal, K., Kellner, V., Maicher, A., Jolivet, P., Teixeira, M.T., Luke, B. (2017). Telomere length determines TERRA and R-loop regulation through the cell cycle. *Cell*, 170(1), 72-85.e14.

Gu, P., Min, J.-N., Wang, Y., Huang, C., Peng, T., Chai, W., Chang, S. (2012). CTC1 deletion results in defective telomere replication, leading to catastrophic telomere loss and stem cell exhaustion. *The EMBO Journal*, 31(10), 2309–2321.

Guen, T.L., Jullien, L., Touzot, F., Schertzer, M., Gaillard, L., Perderiset, M., Carpentier, W., Nitschke, P., Picard, C., Couillault, G. et al. (2013). Human RTEL1 deficiency causes Hoyeraal-Hreidarsson syndrome with short telomeres and genome instability. *Human Molecular Genetics*, 22(16), 3239–3249.

Guervilly, J.-H. and Gaillard, P.-H.L. (2018). SLX4: Multitasking to maintain genome stability. *Critical Reviews in Biochemistry and Molecular Biology*, 53(5), 475–514.

Gupta, R., Sharma, S., Sommers, J.A., Kenny, M.K., Cantor, S.B., Brosh, R.M. (2007). FANCJ (BACH1) helicase forms DNA damage inducible foci with replication protein A and interacts physically and functionally with the single-stranded DNA-binding protein. *Blood*, 110(7), 2390–2398.

Harley, C.B., Futcher, A.B., Greider, C.W. (1990). Telomeres shorten during ageing of human fibroblasts. *Nature*, 345(6274), 458–460.

Hayashi, M.T., Cesare, A.J., Rivera, T., Karlseder, J. (2015). Cell death during crisis is mediated by mitotic telomere deprotection. *Nature*, 522(7557), 492–496.

Hayflick, L. (1965). The limited *in vitro* lifetime of human diploid cell strains. *Experimental Cell Research*, 37(3), 614–636.

Hemann, M.T., Strong, M.A., Hao, L.Y., Greider, C.W. (2001). The shortest telomere, not average telomere length, is critical for cell viability and chromosome stability. *Cell*, 107(1), 67–77.

Huber, M.D., Lee, D.C., Maizels, N. (2002). G4 DNA unwinding by BLM and Sgs1p: Substrate specificity and substrate-specific inhibition. *Nucleic Acids Research*, 30(18), 3954–3961.

Ivessa, A.S., Zhou, J.-Q., Schulz, V.P., Monson, E.K., Zakian, V.A. (2002). Saccharomyces Rrm3p, a 5' to 3' DNA helicase that promotes replication fork progression through telomeric and subtelomeric DNA. *Genes & Development*, 16(11), 1383–1396.

Kaminker, P.G., Kim, S.-H., Desprez, P.-Y., Campisi, J. (2009). A novel form of the telomere-associated protein TIN2 localizes to the nuclear matrix. *Cell Cycle*, 8(6), 931–939.

Kaul, Z., Cesare, A.J., Huschtscha, L.I., Neumann, A.A., Reddel, R.R. (2011). Five dysfunctional telomeres predict onset of senescence in human cells. *EMBO Reports*, 13(1), 52–59.

Kim, N.W., Piatyszek, M.A., Prowse, K.R., Harley, C.B., West, M.D., Ho, P.L., Coviello, G.M., Wright, W.E., Weinrich, S.L., Shay, J.W. (1994). Specific association of human telomerase activity with immortal cells and cancer. *Science*, 266(5193), 2011–2015.

de Lange, T. (2018). Shelterin-mediated telomere protection. *Annual Review of Genetics*, 52, 223–247.

Law, M.J., Lower, K.M., Voon, H.P.J., Hughes, J.R., Garrick, D., Viprakasit, V., Mitson, M., Gobbi, M.D., Marra, M., Morris, A. et al. (2010). ATR-X syndrome protein targets tandem repeats and influences allele-specific expression in a size-dependent manner. *Cell*, 143(3), 367–378.

Leman, A.R. and Noguchi, E. (2012). Local and global functions of Timeless and Tipin in replication fork protection. *Cell Cycle*, 11(21), 3945–3955.

Leman, A.R. and Noguchi, E. (2013). The replication fork: Understanding the eukaryotic replication machinery and the challenges to genome duplication. *Genes*, 4(1), 1–32.

Leman, A.R., Dheekollu, J., Deng, Z., Lee, S.W., Das, M.M., Lieberman, P.M., Noguchi, E. (2012). Timeless preserves telomere length by promoting efficient DNA replication through human telomeres. *Cell Cycle*, 11(12), 2337–2347.

Lenain, C., Bauwens, S., Amiard, S., Brunori, M., Giraud-Panis, M.-J., Gilson, E. (2006). The Apollo 5' exonuclease functions together with TRF2 to protect telomeres from DNA repair. *Current Biology: CB*, 16(13), 1303–1310.

Li, S., Makovets, S., Matsuguchi, T., Blethrow, J.D., Shokat, K.M., Blackburn, E.H. (2009). Cdk1-dependent phosphorylation of cdc13 coordinates telomere elongation during cell-cycle progression. *Cell*, 136(1), 50–61.

Lillard-Wetherell, K. (2004). Association and regulation of the BLM helicase by the telomere proteins TRF1 and TRF2. *Human Molecular Genetics*, 13(17), 1919–1932.

Lin, W., Sampathi, S., Dai, H., Liu, C., Zhou, M., Hu, J., Huang, Q., Campbell, J., Shin-Ya, K., Zheng, L. et al. (2013). Mammalian DNA2 helicase/nuclease cleaves G-quadruplex DNA and is required for telomere integrity. *The EMBO Journal*, 32(10), 1425–1439.

Lingner, J., Cooper, J., Cech, T. (1995). Telomerase and DNA end replication: No longer a lagging strand problem? *Science*, 269(5230), 1533–1534.

Liu, C.-C., Gopalakrishnan, V., Poon, L.-F., Yan, T., Li, S. (2013). Cdk1 regulates the temporal recruitment of telomerase and CST complex for telomere replication. *Molecular and Cellular Biology*. https://doi.org/10.1128/mcb.01235-13.

London, T.B.C., Barber, L.J., Mosedale, G., Kelly, G.P., Balasubramanian, S., Hickson, I.D., Boulton, S.J., Hiom, K. (2008). FANCJ is a structure-specific DNA helicase associated with the maintenance of genomic G/C tracts. *Journal of Biological Chemistry*, 283(52), 36132–36139.

Lovejoy, C.A., Li, W., Reisenweber, S., Thongthip, S., Bruno, J., de Lange, T., De, S., Petrini, J.H.J., Sung, P.A., Jasin, M. et al. (2012). Loss of ATRX, genome instability, and an altered DNA damage response are hallmarks of the alternative lengthening of telomeres pathway. *PLoS Genetics*, 8(7), e1002772.

Ludérus, M.E., van Steensel, B., Chong, L., Sibon, O.C., Cremers, F.F., de Lange, T. (1996). Structure, subnuclear distribution, and nuclear matrix association of the mammalian telomeric complex. *Journal of Cell Biology*, 135(4), 867–881.

Lyu, X., Lei, K.H., Sang, P.B., Shiva, O., Chastain, M., Chi, P., Chai, W. (2020). Human CST complex protects stalled replication forks by directly blocking MRE11 degradation of nascent-strand DNA. *The EMBO Journal*, e103654.

Maestroni, L., Audry, J., Luciano, P., Coulon, S., Géli, V., Corda, Y. (2020). RPA and Pif1 cooperate to remove G-rich structures at both leading and lagging strand. *Cell Stress*, 4(3), 48–63.

Maicher, A., Kastner, L., Luke, B. (2014). Telomeres and disease: Enter TERRA. *RNA Biology*, 9(6), 843–849.

Makovets, S., Herskowitz, I., Blackburn, E.H. (2004). Anatomy and dynamics of DNA replication fork movement in yeast telomeric regions. *Molecular and Cellular Biology*, 24(9), 4019–4031.

Margalef, P., Kotsantis, P., Borel, V., Bellelli, R., Panier, S., Boulton, S.J. (2018). Stabilization of reversed replication forks by telomerase drives telomere catastrophe. *Cell*, 172(3), 439–453.e14.

Markiewicz-Potoczny, M., Lisby, M., Lydall, D. (2018). A critical role for Dna2 at unwound telomeres. *Genetics*, 209(1), 129–141.

Martínez, P., Thanasoula, M., Muñoz, P., Liao, C., Tejera, A., McNees, C., Flores, J.M., Fernández-Capetillo, O., Tarsounas, M., Blasco, M.A. (2009). Increased telomere fragility and fusions resulting from TRF1 deficiency lead to degenerative pathologies and increased cancer in mice. *Genes & Development*, 23(17), 2060–2075.

Masuda-Sasa, T., Polaczek, P., Peng, X.P., Chen, L., Campbell, J.L. (2008). Processing of G4 DNA by Dna2 helicase/nuclease and replication protein A (RPA) provides insights into the mechanism of Dna2/RPA substrate recognition. *Journal of Biological Chemistry*, 283(36), 24359–24373.

Mateyak, M.K. and Zakian, V.A. (2006). Human PIF helicase is cell cycle regulated and associates with telomerase. *Cell Cycle*, 5(23), 2796–2804.

Matmati, S., Vaurs, M., Escandell, J.M., Maestroni, L., Nakamura, T.M., Ferreira, M.G., Géli, V., Coulon, S. (2018). The fission yeast Stn1-Ten1 complex limits telomerase activity via its SUMO-interacting motif and promotes telomeres replication. *Sci Adv.*, (4), 1–14.

Matmati, S., Lambert, S.A.E., Géli, V., Coulon, S. (2020). Telomerase repairs collapsed replication forks at telomeres. *Cell Reports*, 30(10), 3312-3322.e3.

McDonald, K.R., Sabouri, N., Webb, C.J., Zakian, V.A. (2014). The Pif1 family helicase Pfh1 facilitates telomere replication and has an RPA-dependent role during telomere lengthening. *DNA Repair*, 24, 80–86.

Miller, K.M., Rog, O., Cooper, J.P. (2006). Semi-conservative DNA replication through telomeres requires Taz1. *Nature*, 440(7085), 824–828.

Miyake, Y., Nakamura, M., Nabetani, A., Shimamura, S., Tamura, M., Yonehara, S., Saito, M., Ishikawa, F. (2009). RPA-like mammalian Ctc1-Stn1-Ten1 complex binds to single-stranded DNA and protects telomeres independently of the Pot1 pathway. *Molecular Cell*, 36(2), 193–206.

Mohaghegh, P., Karow, J.K., Brosh, R.M., Bohr, V.A., Hickson, I.D. (2001). The Bloom's and Werner's syndrome proteins are DNA structure-specific helicases. *Nucleic Acids Research*, 29(13), 2843–2849.

Muñoz, I.M., Hain, K., Déclais, A.-C., Gardiner, M., Toh, G.W., Sanchez-Pulido, L., Heuckmann, J.M., Toth, R., Macartney, T., Eppink, B. et al. (2009). Coordination of structure-specific nucleases by human SLX4/BTBD12 is required for DNA repair. *Molecular Cell*, 35(1), 116–127.

Nagai, S., Heun, P., Gasser, S.M. (2010). Roles for nuclear organization in the maintenance of genome stability. *Epigenomics*, 2(2), 289–305.

Nandakumar, J. and Cech, T.R. (2013). Finding the end: Recruitment of telomerase to telomeres. *Nature Reviews. Molecular Cell Biology*, 14(2), 69–82.

Nandakumar, J., Bell, C.F., Weidenfeld, I., Zaug, A.J., Leinwand, L.A., Cech, T.R. (2012). The TEL patch of telomere protein TPP1 mediates telomerase recruitment and processivity. *Nature*, 492(7428), 285–289.

Nguyen, D.T., Voon, H.P.J., Xella, B., Scott, C., Clynes, D., Babbs, C., Ayyub, H., Kerry, J., Sharpe, J.A., Sloane-Stanley, J.A. et al. (2017). The chromatin remodelling factor ATRX suppresses R-loops in transcribed telomeric repeats. *EMBO Reports*, 18(6), 914–928.

Ohki, R. and Ishikawa, F. (2004). Telomere-bound TRF1 and TRF2 stall the replication fork at telomeric repeats. *Nucleic Acids Research*, 32(5), 1627–1637.

Ono, Y., Tomita, K., Matsuura, A., Nakagawa, T., Masukata, H., Uritani, M., Ushimaru, T., Ueno, M. (2003). A novel allele of fission yeast rad11 that causes defects in DNA repair and telomere length regulation. *Nucleic Acids Research*, 31(24), 7141.

Opresko, P.L., von Kobbe, C., Laine, J.-P., Harrigan, J., Hickson, I.D., Bohr, V.A. (2002). Telomere-binding protein TRF2 binds to and stimulates the Werner and Bloom syndrome helicases. *The Journal of Biological Chemistry*, 277(43), 41110–41119.

Opresko, P.L., Otterlei, M., Graakjaer, J., Bruheim, P., Dawut, L., Kølvraa, S., May, A., Seidman, M.M., Bohr, V.A. (2004). The Werner syndrome helicase and exonuclease cooperate to resolve telomeric D loops in a manner regulated by TRF1 and TRF2. *Molecular Cell*, 14(6), 763–774.

Opresko, P.L., Mason, P.A., Podell, E.R., Lei, M., Hickson, I.D., Cech, T.R., Bohr, V.A. (2005). POT1 stimulates RecQ helicases WRN and BLM to unwind telomeric DNA substrates. *Journal of Biological Chemistry*, 280(37), 32069–32080.

van Overbeek, M. and de Lange, T. (2006). Apollo, an artemis-related nuclease, interacts with TRF2 and protects human telomeres in S phase. *Current Biology*, 16(13), 1295–1302.

Paeschke, K., Capra, J.A., Zakian, V.A. (2011). DNA replication through G-quadruplex motifs is promoted by the *Saccharomyces cerevisiae* Pif1 DNA helicase. *Cell*, 145(5), 678–691.

Paeschke, K., Bochman, M.L., Garcia, P.D., Cejka, P., Friedman, K.L., Kowalczykowski, S.C., Zakian, V.A. (2013). Pif1 family helicases suppress genome instability at G-quadruplex motifs. *Nature*, 497(7450), 458–462. https://doi.org/10.1038/nature12149.

Pinter, S.F., Aubert, S.D., Zakian, V.A. (2008). The *Schizosaccharomyces pombe* Pfh1p DNA helicase is essential for the maintenance of nuclear and mitochondrial DNA. *Molecular and Cellular Biology*, 28(21), 6594–6608.

Poole, L.A., Zhao, R., Glick, G.G., Lovejoy, C.A., Eischen, C.M., Cortez, D. (2015). SMARCAL1 maintains telomere integrity during DNA replication. *Proceedings of the National Academy of Sciences*, 112(48), 14864–14869.

Popuri, V., Hsu, J., Khadka, P., Horvath, K., Liu, Y., Croteau, D.L., Bohr, V.A. (2014). Human RECQL1 participates in telomere maintenance. *Nucleic Acids Research*, 42(9), 5671–5688.

Poulet, A., Buisson, R., Faivre-Moskalenko, C., Koelblen, M., Amiard, S., Montel, F., Cuesta-Lopez, S., Bornet, O., Guerlesquin, F., Godet, T. et al. (2009). TRF2 promotes, remodels and protects telomeric Holliday junctions. *The EMBO Journal*, 28(6), 641–651.

Przetocka, S., Porro, A., Bolck, H.A., Walker, C., Lezaja, A., Trenner, A., von Aesch, C., Himmels, S.-F., D'Andrea, A.D., Ceccaldi, R. et al. (2018). CtIP-mediated fork protection Synergizes with BRCA1 to suppress genomic instability upon DNA replication stress. *Molecular Cell*, 72(3), 568–582.e6.

Rai, R., Hu, C., Broton, C., Chen, Y., Lei, M., Chang, S. (2017). NBS1 phosphorylation status dictates repair choice of dysfunctional telomeres. *Molecular Cell*, 65(5), 801–817.e4.

Rhodes, D. and Lipps, H.J. (2015). G-quadruplexes and their regulatory roles in biology. *Nucleic Acids Research*, gkv862.

Ribeyre, C., Lopes, J., Boulé, J.-B., Piazza, A., Guédin, A., Zakian, V.A., Mergny, J.-L., Nicolas, A. (2009). The yeast Pif1 helicase prevents genomic instability caused by G-quadruplex-forming CEB1 sequences *in vivo*. *PLoS Genetics*, 5(5), e1000475.

Rippe, K. and Luke, B. (2015). TERRA and the state of the telomere. *Nature Structural & Molecular Biology*, 22(11), 853–858.

Rodriguez, R., Miller, K.M., Forment, J.V., Bradshaw, C.R., Nikan, M., Britton, S., Oelschlaegel, T., Xhemalce, B., Balasubramanian, S., Jackson, S.P. (2012). Small-molecule-induced DNA damage identifies alternative DNA structures in human genes. *Nature Chemical Biology*, 8(3), 301–310.

Rossi, M.L., Ghosh, A.K., Kulikowicz, T., Croteau, D.L., Bohr, V.A. (2010). Conserved helicase domain of human RecQ4 is required for strand annealing-independent DNA unwinding. *DNA Repair*, 9(7), 796–804.

Sabouri, N., McDonald, K.R., Webb, C.J., Cristea, I.M., Zakian, V.A. (2012). DNA replication through hard-to-replicate sites, including both highly transcribed RNA Pol II and Pol III genes, requires the *S. pombe* Pfh1 helicase. *Genes & Development*, 26(6), 581–593.

Sabouri, N., Capra, J.A., Zakian, V.A. (2014). The essential *Schizosaccharomyces pombe* Pfh1 DNA helicase promotes fork movement past G-quadruplex motifs to prevent DNA damage. *BMC Biology*, 12(1), 101.

Safa, L., Gueddouda, N.M., Thiebaut, F., Delagoutte, E., Petruseva, I., Lavrik, O., Mendoza, O., Bourdoncle, A., Alberti, P., Riou, J.-F., Saintomé, C. (2016). 5' to 3' unfolding directionality of DNA secondary structures by replication protein A: G-quadruplexes and duplexes. *Journal of Biological Chemistry*. https://doi.org/10.1074/jbc.m115.709667.

Saint-Léger, A., Koelblen, M., Civitelli, L., Bah, A., Djerbi, N., Giraud-Panis, M.-J., Londoño-Vallejo, A., Ascenzioni, F., Gilson, E. (2014). The basic N-terminal domain of TRF2 limits recombination endonuclease action at human telomeres. *Cell Cycle*, 13(15), 2469–2474.

Salas, T.R., Petruseva, I., Lavrik, O., Bourdoncle, A., Mergny, J.-L., Favre, A., Saintomé, C. (2006). Human replication protein A unfolds telomeric G-quadruplexes. *Nucleic Acids Research*, 34(17), 4857–4865.

Sanders, C.M. (2010). Human Pif1 helicase is a G-quadruplex DNA-binding protein with G-quadruplex DNA-unwinding activity. *The Biochemical Journal*, 430(1), 119–128.

Sarek, G., Vannier, J.-B., Panier, S., Petrini, J.H.J., Boulton, S.J. (2015). TRF2 recruits RTEL1 to telomeres in S phase to promote T-loop unwinding. *Molecular Cell*, 57(4), 622–635.

Sarek, G., Kotsantis, P., Ruis, P., Ly, D.V., Margalef, P., Borel, V., Zheng, X.-F., Flynn, H.R., Snijders, A.P., Chowdhury, D. et al. (2019). CDK phosphorylation of TRF2 controls t-loop dynamics during the cell cycle. *Nature*, 155(7438), 345–355. https://doi.org/10.1038/s41586-019-1744-8.

Sarkar, J., Wan, B., Yin, J., Vallabhaneni, H., Horvath, K., Kulikowicz, T., Bohr, V.A., Zhang, Y., Lei, M., Liu, Y. (2015). SLX4 contributes to telomere preservation and regulated processing of telomeric joint molecule intermediates. *Nucleic Acids Research*, 43(12), 5912–5923.

Savage, S.A. (2018). Beginning at the ends: Telomeres and human disease. *F1000Research*, 7. https://doi.org/10.12688/f1000research.14068.1.

Schoeftner, S. and Blasco, M.A. (2007). Developmentally regulated transcription of mammalian telomeres by DNA-dependent RNA polymerase II. *Nature*, 10(2), 228–236.

Schwab, R.A., Nieminuszczy, J., Shin-Ya, K., Niedzwiedz, W. (2013). FANCJ couples replication past natural fork barriers with maintenance of chromatin structure. *The Journal of Cell Biology*, 201(1), 33–48. https://doi.org/10.1083/jcb.201208009.

Sfeir, A., Kosiyatrakul, S.T., Hockemeyer, D., MacRae, S.L., Karlseder, J., Schildkraut, C.L., de Lange, T. (2009). Mammalian telomeres resemble fragile sites and require TRF1 for efficient replication. *Cell*, 138(1), 90–103.

Shay, J.W. and Wright, W.E. (2010). Telomeres and telomerase in normal and cancer stem cells. *FEBS Letters*, 584(17), 3819–3825.

Shay, J.W. and Wright, W.E. (2019). Telomeres and telomerase: Three decades of progress. *Nature Reviews Genetics*, 20(5), 299–309.

Shen, J. and Loeb, L.A. (2001). Unwinding the molecular basis of the Werner syndrome. *Mechanisms of Ageing and Development*, 122(9), 921–944.

Simon, A.J., Lev, A., Zhang, Y., Weiss, B., Rylova, A., Eyal, E., Kol, N., Barel, O., Cesarkas, K., Soudack, M. et al. (2016). Mutations in STN1 cause Coats plus syndrome and are associated with genomic and telomere defects. *The Journal of Experimental Medicine*, 213(8), 1429–1440.

Snow, B.E., Mateyak, M.K., Paderova, J., Wakeham, A., Iorio, C., Zakian, V., Squire, J., Harrington, L. (2007). Murine Pif1 interacts with telomerase and is dispensable for telomere function *in vivo*. *Molecular and Cellular Biology*, 27(3), 1017–1026.

Steglich, B., Strålfors, A., Khorosjutina, O., Persson, J., Smialowska, A., Javerzat, J.-P., Ekwall, K. (2015). The Fun30 chromatin remodeler Fft3 controls nuclear organization and chromatin structure of insulators and subtelomeres in fission yeast. *PLoS Genetics*, 11(3), e1005101.

Stewart, J.A., Wang, F., Chaiken, M.F., Kasbek, C., Chastain, P.D., Wright, W.E., Price, C.M. (2012). Human CST promotes telomere duplex replication and general replication restart after fork stalling. *The EMBO Journal*, 31(17), 3537–3549.

Stewart, J.A., Wang, Y., Ackerson, S.M., Schuck, P.L. (2018). Emerging roles of CST in maintaining genome stability and human disease. *Frontiers in Bioscience (Landmark Edition)*, 23, 1564–1586.

Stroik, S., Kurtz, K., Hendrickson, E.A. (2019). CtIP is essential for telomere replication. *Nucleic Acids Research*, 47(17), 8927–8940. https://doi.org/10.1093/nar/gkz652.

Sun, H., Karow, J.K., Hickson, I.D., Maizels, N. (1998). The Bloom's syndrome helicase unwinds G4 DNA. *Journal of Biological Chemistry*, 273(42), 27587–27592.

Surovtseva, Y.V., Churikov, D., Boltz, K.A., Song, X., Lamb, J.C., Warrington, R., Leehy, K., Heacock, M., Price, C.M., Shippen, D.E. (2009). Conserved telomere maintenance component 1 interacts with STN1 and maintains chromosome ends in higher eukaryotes. *Molecular Cell*, 36(2), 207–218.

Svendsen, J.M., Smogorzewska, A., Sowa, M.E., O'Connell, B.C., Gygi, S.P., Elledge, S.J., Harper, J.W. (2009). Mammalian BTBD12/SLX4 assembles a Holliday junction resolvase and is required for DNA repair. *Cell*, 138(1), 63–77.

Syed, A. and Tainer, J.A. (2018). The MRE11-RAD50-NBS1 complex conducts the orchestration of damage signaling and outcomes to stress in DNA replication and repair. *Annual Review of Biochemistry*, 87, 263–294.

Takedachi, A., Despras, E., Scaglione, S., Guérois, R., Guervilly, J.H., Blin, M., Audebert, S., Camoin, L., Hasanova, Z., Schertzer, M. et al. (2020). SLX4 interacts with RTEL1 to prevent transcription-mediated DNA replication perturbations. *Nature Structural & Molecular Biology*, 27(5), 438–449.

Takikawa, M., Tarumoto, Y., Ishikawa, F. (2016). Fission yeast Stn1 is crucial for semi-conservative replication at telomeres and subtelomeres. *Nucleic Acids Research*. https://doi.org/10.1093/nar/gkw1176.

Tarsounas, M. and Tijsterman, M. (2013). Genomes and G-quadruplexes: For better or for worse. *Journal of Molecular Biology*, 425(23), 4782–4789.

Teasley, D.C., Parajuli, S., Nguyen, M., Moore, H.R., Alspach, E., Lock, Y.J., Honaker, Y., Saharia, A., Piwnica-Worms, H., Stewart, S. A. (2015). Flap endonuclease 1 limits telomere fragility on the leading strand. *Journal of Biological Chemistry*, 290(24), 15133–15145.

Teixeira-Silva, A., Saada, A.A., Hardy, J., Iraqui, I., Nocente, M.C., Fréon, K., Lambert, S.A.E. (2017). The end-joining factor Ku acts in the end-resection of double strand break-free arrested replication forks. *Nature Communications*, 8(1), 1982.

Torigoe, H. and Furukawa, A. (2006). Tetraplex structure of fission yeast telomeric DNA and unfolding of the tetraplex on the interaction with telomeric DNA binding protein Pot1. *Journal of Biochemistry*, 141(1), 57–68.

Uringa, E.-J., Lisaingo, K., Pickett, H.A., Brind'Amour, J., Rohde, J.-H., Zelensky, A., Essers, J., Lansdorp, P.M. (2012). RTEL1 contributes to DNA replication and repair and telomere maintenance. *Molecular Biology of the Cell*, 23(14), 2782–2792.

Vannier, J.-B., Pavicic-Kaltenbrunner, V., Petalcorin, M.I.R., Ding, H., Boulton, S.J. (2012). RTEL1 dismantles T loops and counteracts telomeric G4-DNA to maintain telomere integrity. *Cell*, 149(4), 795–806.

Vannier, J.-B., Sandhu, S., Petalcorin, M.I., Wu, X., Nabi, Z., Ding, H., Boulton, S.J. (2013). RTEL1 is a replisome-associated helicase that promotes telomere and genome-wide replication. *Science*, 342(6155), 239–242.

Vannier, J.-B., Sarek, G., Boulton, S.J. (2014). RTEL1: Functions of a disease-associated helicase. *Trends in Cell Biology*, 24(7), 416–425.

Wallgren, M., Mohammad, J.B., Yan, K.-P., Pourbozorgi-Langroudi, P., Ebrahimi, M., Sabouri, N. (2016). G-rich telomeric and ribosomal DNA sequences from the fission yeast genome form stable G-quadruplex DNA structures *in vitro* and are unwound by the Pfh1 DNA helicase. *Nucleic Acids Research*, 44(13), 6213–6231.

Wan, B., Yin, J., Horvath, K., Sarkar, J., Chen, Y., Wu, J., Wan, K., Lu, J., Gu, P., Yu, E.Y. et al. (2013). SLX4 assembles a telomere maintenance toolkit by bridging multiple endonucleases with telomeres. *Cell Reports*, 4(5), 861–869.

Wang, H., Nora, G.J., Ghodke, H., Opresko, P.L. (2011). Single molecule studies of physiologically relevant telomeric tails reveal POT1 mechanism for promoting G-quadruplex unfolding. *Journal of Biological Chemistry*, 286(9), 7479–7489.

Wang, F., Stewart, J.A., Kasbek, C., Zhao, Y., Wright, W.E., Price, C.M. (2012). Human CST has independent functions during telomere duplex replication and C-strand fill-in. *Cell Reports*. https://doi.org/10.1016/j.celrep.2012.10.007.

Wang, F., Stewart, J.A., Price, C.M. (2014). Human CST abundance determines recovery from diverse forms of DNA damage and replication stress. *Cell Cycle*, 13(22), 3488–3498.

Wang, Y., Brady, K.S., Caiello, B.P., Ackerson, S.M., Stewart, J.A. (2019). Human CST suppresses origin licensing and promotes AND-1/Ctf4 chromatin association. *Life Science Alliance*, 2(2). https://doi.org/10.26508/lsa.201800270.

Wang, Y.-R., Guo, T.-T., Zheng, Y.-T., Lai, C.-W., Sun, B., Xi, X.-G., Hou, X.-M. (2021). Replication protein A plays multifaceted roles complementary to specialized helicases in processing G-quadruplex DNA. *IScience*, 24(5), 102493.

Wilson, J.S.J., Tejera, A.M., Castor, D., Toth, R., Blasco, M.A., Rouse, J. (2013). Localization-dependent and -independent roles of SLX4 in regulating telomeres. *Cell Reports*, 4(5), 853–860.

Wu, Y., Shin-ya, K., Brosh, R.M. (2008). FANCJ helicase defective in fanconia anemia and breast cancer unwinds G-quadruplex DNA to defend genomic stability. *Molecular and Cellular Biology*, 28(12), 4116–4128.

Wu, W., Bhowmick, R., Vogel, I., Özer, Ö., Ghisays, F., Thakur, R.S., de Leon, E.S., Richter, P.H., Ren, L., Petrini, J.H. et al. (2020). RTEL1 suppresses G-quadruplex-associated R-loops at difficult-to-replicate loci in the human genome. *Nature Structural & Molecular Biology*, 27(5), 424–437.

Ye, J., Lenain, C., Bauwens, S., Rizzo, A., Saint-Léger, A., Poulet, A., Benarroch, D., Magdinier, F., Morere, J., Amiard, S. et al. (2010). TRF2 and apollo cooperate with topoisomerase 2alpha to protect human telomeres from replicative damage. *Cell*, 142(2), 230–242.

Yusufzai, T. and Kadonaga, J.T. (2008). HARP is an ATP-driven annealing helicase. *Science*, 322(5902), 748–750.

Yusufzai, T., Kong, X., Yokomori, K., Kadonaga, J.T. (2009). The annealing helicase HARP is recruited to DNA repair sites via an interaction with RPA. *Genes & Development*, 23(20), 2400–2404.

Zaug, A.J., Podell, E.R., Cech, T.R. (2005). Human POT1 disrupts telomeric G-quadruplexes allowing telomerase extension *in vitro*. *Proceedings of the National Academy of Sciences*, 102(31), 10864–10869.

Zhang, D.-H., Zhou, B., Huang, Y., Xu, L.-X., Zhou, J.-Q. (2006). The human Pif1 helicase, a potential *Escherichia coli* RecD homologue, inhibits telomerase activity. *Nucleic Acids Research*, 34(5), 1393–1404.

Zhang, M., Wang, B., Li, T., Liu, R., Xiao, Y., Geng, X., Li, G., Liu, Q., Price, C.M., Liu, Y., Wang, F. (2019). Mammalian CST averts replication failure by preventing G-quadruplex accumulation. *Nucleic Acids Research*, 47(10), 5243–5259.

Zhong, F.L., Batista, L.F.Z., Freund, A., Pech, M.F., Venteicher, A.S., Artandi, S. E. (2012). TPP1 OB-fold domain controls telomere maintenance by recruiting telomerase to chromosome ends. *Cell*, 150(3), 481–494.

Zhou, C., Pourmal, S., Pavletich, N.P. (2015). Dna2 nuclease-helicase structure, mechanism and regulation by Rpa. *ELife*, 4. https://doi.org/10.7554/elife.09832.

Zimmermann, M., Kibe, T., Kabir, S., de Lange, T. (2014). TRF1 negotiates TTAGGG repeat-associated replication problems by recruiting the BLM helicase and the TPP1/POT1 repressor of ATR signaling. *Genes & Development*, 28(22), 2477–2491.

7

A Brief History of Telomerase: Exceptional Women in Command

Vincent GÉLI
Institut Paoli-Calmettes,
CNRS, INSERM, Université Aix-Marseille, France

Like the discovery of the potential of the bacterial CRISPR/Cas9 immunity system by Emmanuelle Charpentier and Jennifer Doudna, which has had an immense impact on our ability to modify the genome, the discovery of telomerase and its role in the proliferative potential of eukaryotic cells by Elizabeth Blackburn, Carol Greider and Jack Szostak came from basic research using model organisms. These two major discoveries, with multiple applications in biomedical research, or the recent development, in record time, of RNA vaccines underline the importance of basic research to offer therapeutic solutions to the major health challenges we are facing.

The discovery of telomerase is a scientific adventure whose outcome continues to condition the activity of several hundred laboratories around the world. This kind of laboratory odyssey, which began in the late 1970s in a laboratory interested in the biology of unicellular protozoa, was recounted in a beautiful article by de Lange (2006) celebrating the Albert Lasker Prize obtained by Elizabeth Blackburn, Carol Greider and Jack Szostak in 2006, a prize that preceded their award of the Nobel Prize in 2009.

This story begins in the 1930s with Müller's observation in Drosophila of the ends generated when irradiated chromosomes were broken. They were unstable, unlike the natural ends of chromosomes (Müller 1938). Müller then evoked a protective structure at the end of the chromosomes, which he named the telomere, a word derived from the fusion of the Greek terms "telos" and "meros" meaning, respectively, "end" and "part" or "element". At the same time, McClintock (1941) independently observed in maize that chromosomes broken by different treatments have the ability to fuse together and generate dicentric chromosomes (with two centromeres) that inevitably break when these centromeres are pulled toward opposite poles of the spindle during cell division. These observations initiated by Müller and McClintock led to the concept that a protective structure exists at the end of chromosomes, one of the properties of which is to stabilize the chromosomes and prevent their fusion (Chuaire 2006). In this respect, McClintock, also a Nobel Prize winner and a great humanist, can be considered as the first woman to have worked on *telomeres*. She had a correspondence with Blackburn, in which she summarized her work on the behavior of chromosome ends and mentioned an enzyme produced in embryonic cells that could repair telomeres.

Interest in telomeres was revived 30 years later when Watson (1972), while working on the replication of the T7 phage, raised the question of replication of the end of chromosomes. Watson, who received the Nobel Prize in 1962 with Crick and Wilkins for the discovery of the structure of the DNA double helix (and who would be stripped of his honorary titles in 2007 after he made comments deemed "unsubstantiated and unwise" about race and intelligence), wonders in particular how replication of the delayed DNA strand, which relies on the synthesis of Okazaki fragments, initiated posterior to the region to be replicated, can occur at the end of the chromosome. He deduced that degradation of the last RNA primer at the chromosome end must induce shortening of the chromosome ends and that there are mechanisms at the chromosome end that compensate for this shortening (Watson 1972). At about the same time, Olovnikov (1973) proposed a link between the shortening of telomeres evoked by Watson and the concept developed by Hayflick (1961, 1965) stipulating that human fetal cells perform a limited number of divisions in cell culture (between 40 and 60 times) before entering a senescence phase. The foundations of the problem of telomere length maintenance and a putative link with the proliferative potential of cells had been laid. However, how this complex puzzle could be put together remained to be discovered (Greider 1998).

The first part of the solution would come through the work of Gall and Blackburn who, in the 1970s and 1980s, were interested in the biology of *Tetrahymena thermophila,* a freshwater ciliate. In this organism, the repetitive rDNA sequences, coding for ribosomal RNA, are organized as small linear DNA fragments. Blackburn realized that tandem repeated sequences, of the TTGGG type, were present at the end of these fragments (we must put this in the technological context of the time, before PCR and modern sequencing techniques) (Blackburn and Gall 1978). The question arose as to the origin of these sequences. The existence of the *terminal transferase* (TdT) catalyzing, independently of the presence of a template, the repetitive addition of a deoxy-ribonucleotide to the 3'-OH end of DNA was known (Motea and Berdis 2010), but this could not explain the repetitive addition of the TTGGG motif. It soon became clear that this new activity was carried by an enzyme that is widely distributed in eukaryotes, notably thanks to experiments conducted in baker's yeast (Szostak and Blackburn 1982). Indeed, Szostak and Blackburn had introduced into baker's yeast a linear yeast plasmid containing at its ends TTGGG sequences from Tetrahymena rDNA. They found that this linear plasmid was stable after introduction into yeast and that the specific characteristics of the *Tetrahymena*-derived ends were maintained in yeast. These experiments suggested that the yeast recognized and used DNA ends from *Tetrahymena*. Working with Shampay, Szostak and Blackburn continued these experiments to determine the sequence of telomeres from the recombinant linear plasmid. The sequence indicated that the original *Tetrahymena* telomeres had been extended by a terminal addition of specific sequences added in yeast (Shampay et al. 1984). These sequences were heterogeneous in size and consisted of irregular repeats made up of TG sequences on the 3' strand exiting the end (Shampay et al. 1984). Compiling these different observations, Szostak and Blackburn deduced that the structural features required for replication and telomere resolution in *Tetrahymena* were highly conserved in yeast and that a novel, evolutionarily conserved activity was capable of lengthening chromosome ends (Blackburn 2010).

In her Nobel lecture, Greider (2009) wonderfully explains how when she arrived in Blackburn's laboratory, the two women decided to identify this new activity through its purification by postulating that the enzyme responsible for the addition of the TTGGG motif must be present in large quantities in *Tetrahymena*. Indeed, ciliates have the particularity of possessing a macro- and micro-nucleus involved in, respectively, vegetative life functions and reproductive functions. During the conjugation of ciliates, a new macro-nucleus is formed from the micro-nuclei and that necessitates the making of thousands of telomeres. The idea was that generating and maintaining all these telomeres required a high amount of this putative elongation activity. The challenge was to find an assay that would track the activity of the enzyme. It was a constant year-long development to find a telomere-mimicking substrate for which Greider would be able to track the elongation by the desired activity. The use of an oligonucleotide with the telomere sequence was the solution to the problem. The oligonucleotide elongated, producing products that varied in size by six bases, reflecting the addition of a repeating motif by this activity. Greider and Blackburn (1985) performed the first visualization of telomerase activity. It remained to purify this activity and understand what type of enzyme was capable of elongating a 3' end of DNA. Obviously, it could not be a DNA polymerase that adds nucleotides to the DNA from 5' to 3', nor a terminal transferase, a new enzyme had to do the job. A first crucial piece of information was Greider's discovery that pre-treatment with ribonuclease (RNase) abolished the activity of the enzyme, suggesting that it could use RNA as a template to elongate the telomeric oligonucleotide. Then, Greider used different chromatography columns to purify the enzyme. She then started to look for the RNA that was systematically present in the active fractions of the enzyme. The identification of this RNA with the techniques of the time was not immediate. Blackburn and Greider (1987) wrote a paper describing the purification scheme of telomerase and in which they proposed the existence of an RNA associated with the enzyme. Greider and Blackburn (1989) identified the telomerase RNA of *Tetrahymena* and delineated a sequence within this RNA (5' CAACCCCAA 3'), serving as a template for the addition of TTGGG motifs on the telomeric oligonucleotide. The telomerase template RNA (called TERC) was then identified, cloned and sequenced in ciliates, baker's yeast and mammals (Shippen-Lentz and Blackburn 1990; Romero and Blackburn 1991; Lingner et al. 1994; Singer and Gottschling 1994; Feng et al. 1995; Blasco et al. 1995).

The identification of tandem repeats in *Tetrahymena* telomeres was followed by the identification of similar repeats in the telomeres of other organisms, including *Oxytricha*, and the trypanosome (Boswell et al. 1982; Bernards et al. 1983). David Prescott's work in *Oxytricha* revealed that telomeres have a single-stranded extension of the G-rich (Klobutcher et al. 1981) 3' end. This almost universal characteristic of telomeres was to have a considerable impact because it raised the question of the regeneration of this single-stranded end after replication of the advanced strand of DNA. Wellinger and Zakian revealed in baker's yeast that a 3' single-stranded end of telomeric DNA (of about 30 nucleotides) was transiently observed during the S phase of the cell cycle, suggesting a regulation of this process during the cell cycle. Wright and Shay, Makarov and Langmore, as well as McElligoot and Wellinger showed a few years later that human telomeres also have longer, single-stranded 3'-ends. In fact, the generation of single-stranded 3'-ends can involve a degradation of the C-strand and it was postulated that it occurs differently between the advanced and delayed DNA strands (Makarov et al. 1997; McElligott and Wellinger 1997; Wright et al. 1997). These discoveries would lead to a paradigm shift: end replication became more of a problem related to the regeneration of the single-stranded 3' end on the leading strand by the transient action of nucleases (Lingner et al. 1995; Wellinger et al. 1996), rather than to the incomplete replication of the lagging strand initially mentioned by Watson.

The next step was taken by Lingner and Cech (Nobel Prize in Chemistry in 1989 for the discovery of ribozymes) with the demonstration that telomerase functions as a reverse transcriptase (RT). This was first made possible by the identification in Euplotes (a genus of ciliates) of two proteins p123 and p43 co-purifying with telomerase activity and present in stoichiometric quantities with telomerase RNA (Lingner and Cech 1996). The identification of the catalytic subunit was possible by comparing the sequence of p123 with the sequence of the yeast genes initially called "ESTs" (for ever shorter telomeres) by Lundblad and Szostak, for which their inactivation caused telomere shortening in yeast (Lundblad and Szostak 1989; Lendvay et al. 1996). Sequence comparison of p123 with one of the EST genes (EST2) revealed a 20% identity between their sequences and a high similarity. Most remarkable was the presence of a common motif in both proteins that is a hallmark of RTs and contains the famous catalytic triad specific to the RTactive site (Lendvay et al. 1996). Mutation of the catalytic triad residues in yeast telomerase (Est2) abolished telomerase activity, indicating that telomerase behaves as a specialized RT (Lingner et al. 1997; Counter et al. 1997). In the same year, the telomerase gene was identified in fissiparous yeast and humans, and was given the name TERT (TElomerase Reverse Transcriptase) (Nakamura et al. 1997; Harrington et al. 1997; Kilian et al. 1997).

The existence of the single-stranded 3' end would prove crucial to understanding how telomerase (TERT/TERC) was recruited to the telomere and would engage the end to elongate the 3' strand of the chromosomes (Wellinger et al. 1996). Another challenge was to discover the proteins bound to telomeric duplexes and to the 3' single stranded end. In 1986 and 1987, Gottschling and Zakian, and Price and Cech identified two closely interacting subunits (alpha and beta) in *Oxytricha nova* that are capable of binding to the 3' single stranded end. These were the first proteins specifically attached to the telomere (Gottschling and Zakian 1986; Price and Cech 1987). Rap1 and Cdc13 proteins binding to yeast telomeric duplexes and single-stranded telomeric DNA, respectively, were identified a few years later by (Lustig et al. 1990; Conrad et al. 1990; Garvik et al. 1995; Nugent et al. 1996). Chong et al. (1995) identified TRF1 in humans, Bilaud et al. (1997) identified TRF2 and Cooper et al. (1997) discovered Taz1 in fissiparous yeast. Baumann and Cech (2001) identified the POT1 protein that binds to the single 3' end of telomeric DNA. In mammals, the initial discovery of TRF1 and TRF2 led to the identification of other telomeric proteins (TIN2, RAP1 and TPP1) associated with TRF1 or TRF2. This led to the concept of "shelterin" introduced by de Lange (2005), a telomere-protecting shelter resulting from the association of these telomeric proteins. A myriad of studies followed demonstrating the multiple functions of shelterin proteins, including telomere protection, replication, positioning within the nucleus and telomerase recruitment, to name a few (Gilson and Géli 2007).

As mentioned above, the history of the discovery of telomeres and telomerase cannot be dissociated from the discovery of one of the major mechanisms inducing the onset of cell senescence. Lundblad and Szostak (1989) discovered the baker's yeast EST1 gene whose inactivation produced a gradual shortening of telomere size, affected chromosome stability and above all induced a senescence phenotype correlated with telomere shortening. This was the first demonstration that excessive telomere shortening induced replicative senescence of cells. Other EST genes were later discovered by Lundblad with an identical phenotype, including the EST2 gene coding for the catalytic subunit of telomerase (Lendvay et al. 1996), and the TLC1 gene coding for yeast telomerase RNA, discovered by Singer and Gottschling (1994). At about the same time as the discovery of EST1, Blackburn et al. (1990) showed that alteration of the Tetrahymena RNA template sequence produced a senescence phenotype. In his Nobel lecture, Szostak (2010) recounts how he foresaw that this discovery linking telomere length to the ability of cells to proliferate would revolutionize the fields of cancer and cellular aging. Then, Greider et al. (1990) demonstrated that the telomeres of human fibroblasts shortened during successive passages. Shay and Wright drove the point home a few years later by demonstrating that ectopic expression of the catalytic subunit of telomerase (hTERT) in human fibroblasts re-extended the telomeres of the cells and extended the number of passages they were able to complete (Bodnar et al. 1998). These experiments joined those of Greider and Szostak and thus established a causal relationship between telomere shortening and cellular senescence. The story continued with the first mouse model built by Blasco and Greider, carrying a deletion of the TERC (mTR) gene (Blasco et al. 1997). Fourth-generation $mTR^{-/-}$ mice had virtually undetectable telomeres and had numerous chromosomal abnormalities (Lee et al. 1998). The link between telomere shortening and aging was established a year later by Greider and DePinho (Rudolph et al. 1999) when Hahn and Weinberg created a human tumor line by combining the expression of telomerase with that of the SV 40 large-T antigen and an oncogenic allele of H-ras (Hahn et al. 1999). The first loop in this odyssey of telomerase and telomeres had been completed. This work would lead to countless studies in hundreds of laboratories, some of the highlights of which are admirably described in a very recent review by DePinho et al. (2021).

Remarkably, 20 years later, Collins and colleagues reconstituted TERT/TERC in human cells, allowing the purification of the human holoenzyme associated with the TCAB1 protein and the dyskerin, NOP10, NHP2 and GAR1 proteins. It determined the structure of human telomerase associated with a telomeric fragment by cryo-electron microscopy at a scale of 10 Å (Nguyen et al. 2018). At the same time, Feigon and colleagues determined the structure of Tetrahymena telomerase associated with telomeric DNA at a resolution of 4.8 Å, revealing previously unsuspected mechanistic details about telomerase activity (Jiang et al. 2018). This marked a new decisive step in the understanding of the functioning of this enzyme discovered 40 years earlier.

The world of telomeres has some singularities. The first is the extraordinary scientific lineage of the pioneers of the field. Liz Blackburn, Carol Greider, Ginger Zakian, Titia de Lange, Tom Cech and Ron de Pihno, to name but a few, have generated an impressive number of outstanding scientists who are now leaders in the field. The second is the number of women who have been instrumental in this scientific adventure. Twenty years ago (in the 2000s), in the amphitheater at Cold-Spring Harbor, the temple of telomere and telomerase conferences, there were, sitting in the front row, mostly women, all of them leading scientists. They included Liz Blackburn, Carol Greider, Titia de Lange, Ginger Zakian, Vicky Lundblad, Susan Gasser, Maria Blasco, Dorothy Shippen, Julia Cooper, Carolin Price and many others. At a time when we are striving for parity in science, the field of telomeres is a model in which women have held a prominent place. In addition to the immense scientific advances they have made, we can thank them for their role in promoting the place of women in science.

7.1. Acknowledgments

I thank Marie-Noelle Simon, Eric Gilson and Raymund Wellinger for their critical reading of this chapter.

7.2. References

Baumann, P. and Cech, T.R. (2001). Pot1, the putative telomere end-binding protein in fission yeast and humans. *Science*, 292(5519), 1171–1175.

Bernards, A., Michels, P.A., Lincke, C.R., Borst, P. (1983). Growth of chromosome ends in multiplying trypanosomes. *Nature*, 303(5918), 592–597.

Bilaud, T., Brun, C., Ancelin, K., Koering, C.E., Laroche, T., Gilson, E. (1997). Telomeric localization of TRF2, a novel human telobox protein. *Nat. Genet.*, 17(2), 236–239.

Blackburn, E.H. (2010) Telomeres and telomerase: The means to the end. Nobel Lecture. *Angew. Chem. Int. Ed. Engl.*, 49, 7405–7421.

Blackburn, E.H. and Gall, J.G. (1978). A tandemly repeated sequence at the termini of the extrachromosomal ribosomal RNA genes in Tetrahymena. *J. Mol. Biol.,* 120(1), 33–53.

Blasco, M.A., Funk, W., Villeponteau, B., Greider, C.W. (1995). Functional characterization and developmental regulation of mouse telomerase RNA. *Science*, 269(5228), 1267–1270.

Blasco, M.A., Lee, H.W., Hande, M.P., Samper, E., Lansdorp, P.M., DePinho, R.A., Greider, C.W. (1997). Telomere shortening and tumor formation by mouse cells lacking telomerase RNA. *Cell*, 91, 25–34.

Bodnar, A.G., Ouellette, M., Frolkis, M., Holt, S.E., Chiu, C.P., Morin, G.B., Harley, C.B., Shay, J.W., Lichtsteiner, S., Wright, W.E. (1998). Extension of life-span by introduction of telomerase into normal human cells. *Science*, 279(5349), 349–352.

Boswell, R.E., Klobutcher, L.A., Prescott, D.M. (1982). Inverted terminal repeats are added to genes during macronuclear development in *Oxytricha nova*. *Proc. Natl. Acad. Sci. USA*, 79(10), 3255–3259.

Chakravarti, D., LaBella, K.A., DePinho, R.A. (2021). Telomeres: History, health, and hallmarks of aging. *Cell*, 184(2), 306–322.

Chong, L., van Steensel, B., Broccoli, D., Erdjument-Bromage, H., Hanish, J., Tempst, P., de Lange, T. (1995). A human telomeric protein. *Science*, 270(5242), 1663–1667.

Chuaire, L. (2006). Telomere and telomerase: Brief review of a history initiated by Hermann Müller and Barbara McClintock. *Colomb. Med.*, 37(4), 332–335.

Conrad, M.N., Wright, J.H., Wolf, A.J., Zakian, V.A. (1990). RAP1 protein interacts with yeast telomeres in vivo: Overproduction alters telomere structure and decreases chromosome stability. *Cell*, 63(4), 739–750.

Cooper, J.P., Nimmo, E.R., Allshire, R.C., Cech, T.R. (1997). Regulation of telomere length and function by a Myb-domain protein in fission yeast. *Nature*, 385(6618), 744–747.

Counter, C.M., Meyerson, M., Eaton, E.N., Weinberg, R.A. (1997). The catalytic subunit of yeast telomerase. *Proc. Natl. Acad. Sci. USA*, 94(17), 9202–9207.

Feng, J., Funk, W.D., Wang, S.S., Weinrich, S.L., Avilion, A.A., Chiu, C.P., Adams, R.R., Chang, E., Allsopp, R.C., Yu, J. et al. (1995). The RNA component of human telomerase. *Science*, 269(5228), 1236–1241.

Garvik, B., Carson, M., Hartwell, L. (1995). Single-stranded DNA arising at telomeres in cdc13 mutants may constitute a specific signal for the RAD9 checkpoint. *Mol. Cell. Biol.*, 15(11), 6128–6138.

Gilson, E. and Géli, V. (2007). How telomeres are replicated. *Nat. Rev. Mol. Cell Biol.*, 8(10), 825–838.

Gottschling, D.E. and Zakian, V.A. (1986). Telomere proteins: Specific recognition and protection of the natural termini of Oxytricha macronuclear DNA. *Cell*, 47(2), 195–205.

Greider, C.W. (1998). Telomeres and senescence: The history, the experiment, the future. *Curr. Biol.*, 8(5), R178–R181.

Greider, C.W. (2009). Telomerase discovery: The excitement of putting together pieces of the puzzle. Nobel Lecture. *Angew. Chem. Int. Ed. Engl.*, 49, 7422–7439.

Greider, C.W. and Blackburn, E.H. (1985). Identification of a specific telomere terminal transferase activity in Tetrahymena extracts. *Cell*, 43, 405–413.

Greider, C.W. and Blackburn, E.H. (1987). The telomere terminal transferase of Tetrahymena is a ribonucleoprotein enzyme with two kinds of primer specificity. *Cell*, 51(6), 887–898.

Greider, C.W. and Blackburn, E.H. (1989). A telomeric sequence in the RNA of Tetrahymena telomerase required for telomere repeat synthesis. *Nature*, 337(6205), 331–337.

Hahn, W.C., Counter, C.M., Lundberg, A.S., Beijersbergen, R.L., Brooks, M.W., Weinberg, R.A. (1999). Creation of human tumour cells with defined genetic elements. *Nature*, 400(6743), 464–468.

Harley, C.B., Futcher, A.B., Greider, C.W. (1990). Telomeres shorten during ageing of human fibroblasts. *Nature*, 345(6274), 458–460.

Harrington, L., McPhail, T., Mar, V., Zhou, W., Oulton, R., Bass, M.B., Arruda, I., Robinson, M.O. (1997). A mammalian telomerase-associated protein. *Science*, 275(5302), 973–977.

Hayflick, L. (1965). The limited in vitro lifetime of human diploid cell strains. *Exp. Cell Res.*, 37(3), 614–636.

Hayflick, L. and Moorhead, P.S. (1961). The serial cultivation of human diploid cell strains. *Exp. Cell Res.*, 25, 585–621.

Jiang, J., Wang, Y., Sušac, L., Chan, H., Basu, R., Zhou, Z.H., Feigon, J. (2018). Structure of telomerase with telomeric DNA. *Cell*, 173(5), 1179–1190.

Kilian, A., Bowtell, D.D., Abud, H.E., Hime, G.R., Venter, D.J., Keese, P.K., Duncan, E.L., Reddel, R.R., Jefferson, R.A. (1997). Isolation of a candidate human telomerase catalytic subunit gene, which reveals complex splicing patterns in different cell types. *Hum. Mol. Genet.*, 6(12), 2011–2019.

Klobutcher, L.A., Swanton, M.T., Donini, P., Prescott, D.M. (1981). All gene-sized DNA molecules in four species of hypotrichs have the same terminal sequence and an unusual 3' terminus. *Proc. Natl. Acad. Sci. USA*, 78(5), 3015–3019.

de Lange, T. (2005). Shelterin: The protein complex that shapes and safeguards human telomeres. *Genes Dev.*, 19(18), 2100–2110.

de Lange, T. (2006). Lasker Laurels for telomerase. *Cell*, 126(6), 1017–1020.

Lee, H.W., Blasco, M.A., Gottlieb, G.J., Horner, J.W., Greider, C.W., DePinho, R.A. (1998). Essential role of mouse telomerase in highly proliferative organs. *Nature*, 392(6676), 569–574.

Lendvay, T.S., Morris, D.K., Sah, J., Balasubramanian, B., Lundblad, V. (1996). Senescence mutants of *Saccharomyces cerevisiae* with a defect in telomere replication identify three additional EST genes. *Genetics*, 144(4), 1399–1412.

Lingner, J. and Cech, T.R. (1996). Purification of telomerase from *Euplotes aediculatus*: requirement of a primer 3' overhang. *Proc. Natl. Acad. Sci. USA*, 93(20), 10712–10717.

Lingner, J., Hendrick, L.L., Cech, T.R. (1994). Telomerase RNAs of different ciliates have a common secondary structure and a permuted template. *Genes Dev.*, 8(16), 1984–1998.

Lingner, J., Cooper, J.P., Cech, T.R. (1995). Telomerase and DNA end replication: No longer a lagging strand problem? *Science*, 269(5230), 1533–1534.

Lingner, J., Hughes, T.R, Shevchenko, A., Mann, M., Lundblad, V., Cech, T.R. (1997). Reverse transcriptase motifs in the catalytic subunit of telomerase. *Science*, 276(5312), 561–567.

Lundblad, V. and Szostak, J.W. (1989). A mutant with a defect in telomere elongation leads to senescence in yeast. *Cell*, 57(4), 633–643.

Lustig, A.J., Kurtz, S., Shore, D. (1990). Involvement of the silencer and UAS binding protein RAP1 in regulation of telomere length. *Science*, 250(4980), 549–553.

Makarov, V.L., Hirose, Y., Langmore, J.P. (1997). Long G tails at both ends of human chromosomes suggest a C strand degradation mechanism for telomere shortening. *Cell*, 88(5), 657–666.

McClintock, B. (1941). The stability of broken ends of chromosomes in *Zea mays*. *Genetics*, 26, 234–282.

McElligott, R. and Wellinger, R.J. (1997). The terminal DNA structure of mammalian chromosomes. *EMBO J.*, 16(12), 3705–3714.

Motea, E.A. and Berdis, A.J. (2010). Terminal deoxynucleotidyl transferase: The story of a misguided DNA polymerase. *Biochim. Biophys. Acta*, 1804(5), 1151–1166.

Müller, H. (1938). The remaking of chromosomes. *Collecting Net.*, 13, 181–198.

Nakamura, T.M., Morin, G.B., Chapman, K.B., Weinrich, S.L., Andrews, W.H., Lingner, J., Harley, C.B., Cech, T.R. (1997). Telomerase catalytic subunit homologs from fission yeast and human. *Science*, 277(5328), 955–959.

Nguyen, T.H.D., Tam, J., Wu, R.A., Greber, B.J., Toso, D., Nogales, E., Collins, K. (2018). Cryo-EM structure of substrate-bound human telomerase holoenzyme. *Nature*, 557(7704), 190–195.

Nugent, C.I., Hughes, T.R., Lue, N.F., Lundblad, V. (1996). Cdc13p: A single-strand telomeric DNA-binding protein with a dual role in yeast telomere maintenance. *Science*, 274(5285), 249–252.

Olovnikov, A.M. (1973). A theory of marginotomy. *J. Theor. Biol.*, 41, 181–190.

Price, C.M. and Cech, T.R. (1987). Telomeric DNA-protein interactions of Oxytricha macronuclear DNA. *Genes Dev.*, 1(8), 783–793.

Romero, D.P. and Blackburn, E.H. (1991). A conserved secondary structure for telomerase RNA. *Cell*, 67(2), 343–353.

Rudolph, K.L., Chang, S., Lee, H.W., Blasco, M., Gottlieb, G.J., Greider, C., DePinho, R.A. (1999). Longevity, stress response, and cancer in aging telomerase-deficient mice. *Cell*, 96(5), 701–712.

Shampay, J., Szostak, J.W., Blackburn, E.H. (1984). DNA sequences of telomeres maintained in yeast. *Nature*, 310(5973), 154–157.

Shippen-Lentz, D. and Blackburn, E.H. (1990). Functional evidence for an RNA template in telomerase. *Science*, 247(4942), 546–552.

Singer, M.S. and Gottschling, D.E. (1994). TLC1: Template RNA component of *Saccharomyces cerevisiae* telomerase. *Science*, 266(5184), 404–409.

Szostak, J.W. (2010). DNA ends: Just the begining. Nobel Lecture. *Angew. Chem. Int. Ed. Engl.*, 49, 7386–704.

Szostak, J.W. and Blackburn, E.H. (1982). Cloning yeast telomeres on linear plasmid vectors. *Cell*, 29(1), 245–255.

Watson, J.D. (1972). Origin of concatemeric T7 DNA. *Nat. New Biol.*, 239, 197–201.

Wellinger, R.J., Wolf, A.J., Zakian, V.A. (1993). Saccharomyces telomeres acquire single-strand TG1-3 tails late in S phase. *Cell*, 72(1), 51–60.

Wellinger, R.J., Ethier, K., Labrecque, P., Zakian, V.A. (1996). Evidence for a new step in telomere maintenance. *Cell*, 85(3), 423–433.

Wright, W.E., Tesmer, V.M., Huffman, K.E., Levene, S.D., Shay, J.W. (1997). Normal human chromosomes have long G-rich telomeric overhangs at one end. *Genes Dev.*, 11(21), 2801–2809.

Yu, G.L., Bradley, J.D., Attardi, L.D., Blackburn, E.H. (1990). In vivo alteration of telomere sequences and senescence caused by mutated Tetrahymena telomerase RNAs. *Nature*, 344(6262), 126–132.

Zakian, VA. (2012). Telomeres: The beginnings and ends of eukaryotic chromosomes. *Exp. Cell Res.*, 318(12), 1456–1460.

8

Telomerase: Structure and Mechanism

Carole SAINTOMÉ[1,2]

[1]*Structure and Instability of Genomes, National Museum of Natural History, CNRS, INSERM, Paris, France*

[2]*UFR 927, Sorbonne University, Paris, France*

8.1. Introduction

In the 1970s, it was shown that the replication of chromosome ends was incomplete and led to the shortening of telomeres (see Chapter 5). This terminal replication problem was solved with the discovery of telomerase in 1985, in the ciliate, *Tetrahymenathermophila*, as a result of the work of Elisabeth Blackburn and her collaborators, who were awarded the Nobel Prize in Medicine in 2009 (see Chapter 7). This protozoan is a very good model for the study of telomeres; it is the most studied organism in research on telomeres and telomerase. Indeed, it has two nuclei per cell, including a macronucleus containing hundreds of different chromosomes, in several copies (about 50), but especially the rDNA chromosome, which is present in several thousands of copies (for review, see Kapler (1993)). This abundance of chromosomes, and thus of telomeres, allowed Blackburn's team to study and characterize the first telomeric motif (Budarf and Blackburn 1986; Spangler et al. 1988), the TTGGGG sequence. While continuing their work on this animal model, Greider and Blackburn (1985) observed that the telomere sequence

For a color version of all the figures in this chapter, see www.iste.co.uk/saintome/telomeres.zip.

Telomeres,
coordinated by Carole SAINTOMÉ.

in *Tetrahymena* was lengthened *in vitro* by sequences identical to the telomere repeat pattern in the presence of *Tetrahymena* nuclear extracts.

Telomerase activity was then demonstrated in many other organisms by an *in vitro* elongation assay with cell extracts; telomerase activity was demonstrated in other ciliates such as *Oxytrichia* (Zahler and Prescott 1988) and *Euplote* (Shippen-Lentz and Blackburn 1989), and in yeast (Lundblad and Szostak 1989). Human telomerase was identified in HeLa tumor cells (Morin 1989).

This chapter is mainly dedicated to the structure of telomerase and its mechanism of action. The structures are available on the PDB or on the online tool Telomerase database, which compiles all known information on telomerase, but also on its substrate, the telomeres.

8.2. Roles

The role of telomerase is therefore to maintain the size of telomeres by adding nucleotides to the 3' end of the chromosome (conventional role). In unicellular organisms, telomerase is constitutively expressed, which allows the long-term division of the organism and its viability. In metazoans, and particularly in humans, its expression depends on the cell type. In somatic (differentiated) cells, telomerase is expressed very little and consequently, telomeres shorten at each cell division, leading inevitably to the entry into senescence of the cells, that is, the aging of the cells. However, in more than 80% of human cancers, the expression of telomerase is increased, allowing it to participate in cell immortalization and leading to unlimited proliferation (Blasco 2003; Stewart and Weinberg 2006; Artandi and DePinho 2010) (see Chapter 11). In stem (undifferentiated) and germ cells, telomerase is constitutively expressed to maintain telomere size and thus unlimited division, preventing stem cell aging. Decreased telomerase activity (absence or mutation) in stem cells leads to premature aging and stem cell failure related diseases such as dyskeratosis congenita, aplastic anemia and pulmonary fibrosis (Townsley et al. 2014; Sareket al. 2015; Shay 2016). Chapter 12 is dedicated to this topic.

In summary, telomerase plays a fundamental role in maintaining unlimited cell division. Depending on the cell type, its presence is either beneficial with an anti-aging role (unicellular organism, stem and germ cells), or harmful with a pro-tumor action (somatic cells).

Telomerase has other roles that are independent of its telomere elongation activity. It regulates a large number of biological functions such as signal transduction, gene expression and protects against oxidative damage due to the presence of reactive oxygen species present in the vast majority of mitochondria. Indeed, it has been shown that protein and RNA subunits of telomerase are imported into mitochondria during oxidative stress (Zheng et al. 2019), suggesting, in particular, a protective role in neurons (Ségal and Géli 2019). A recent review by Ségal et al. (2019) lists these different so-called non-canonical or alternative roles that should be taken into account for the anti-cancer strategy targeting telomerase.

8.3. Structures

Telomerase is a multi-protein complex associated with an RNA sequence that is essential for catalytic activity, and is referred to as a ribonucleoprotein (RNP) complex. The catalytic core of telomerase consists of the TERT (TElomerase Reverse Transcriptase) protein and an RNA molecule containing the template sequence for the TERT protein; these two elements are sufficient for telomerase activity *in vitro*. Numerous *in vivo*, genetic and biochemical studies have revealed interactions of this catalytic core with other proteins, which are either directly involved in telomerase enzymatic activity or in the structure, biogenesis, nuclear localization and regulation of telomerase (Nandakumar and Cech 2013; Schmidt and Cech 2015).

8.3.1. *The catalytic core*

8.3.1.1. *The TERT subunit, a highly conserved domain of telomerase*

In the vast majority of organisms, the TERT subunit is structured into four functional subdomains (Figure 8.1(A)): (i) a central RT-like (reverse transcriptase-like) domain corresponding to the reverse transcriptase activity of telomerase, which contains the seven conserved motifs (1, 2, A, B', C, D and E) of HIV reverse transcriptase (Counter et al. 1997; Nakumara et al. 1997; Lingner et al. 1997); (ii) an N-terminal domain (named TEN for telomerase essential N-domain) of variable size and sequence, allowing telomerase to specifically anchor its telomeric DNA; this domain is connected via a long linker to (iii) the TRBD domain, for telomerase RNA binding domain, binding with a very high affinity and specificity the telomerase RNA; (iv) a C-terminal domain (C terminal element (CTE)) that is not highly conserved and is involved in modulating telomerase processivity and regulating its localization. The TRBD, RT and CTE domains form a larger domain called the TERT ring.

Crystallography has been the method of choice to study the structure of the TERT subunit; while the isolated subdomains of various species have been extensively studied (Jacobs et al. 2006; Rouda and Skordalakes 2007; Harkisheimer et al. 2013; Huang et al. 2014; Jansson et al. 2015), only the crystal structure of the whole *Triboliumcasaneum* TERT has been determined to date (Gillis et al. 2008; Mitchell et al. 2010). We will see later that all efforts have been focused on determining the structure of the holoenzyme.

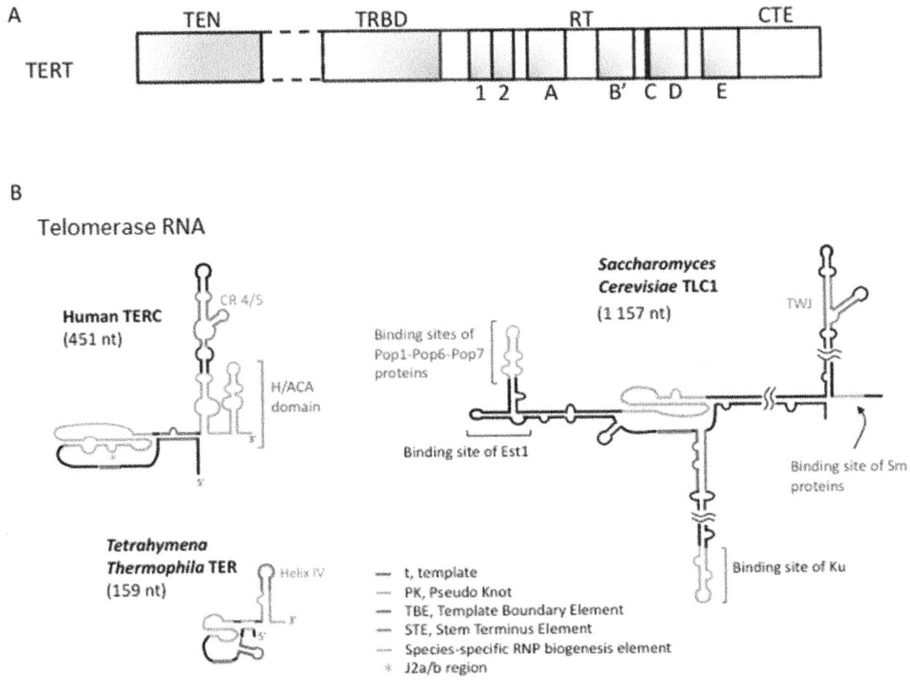

Figure 8.1. *Structural domains of TERT (A) and telomerase RNAs (B)*

8.3.1.2. *The RNA subunit*

Greider and Blackburn (1987) found that there was an RNA sequence that would contribute to the elongation of telomeric repeats in *Tetrahymena*. This hypothesis was based on the fact that RNAse treatment inactivated the synthesis of telomeric repeats *in vitro*. Later, this hypothesis was validated by a cloning experiment of an RNA-encoding gene in *Tetrahymena*. Transcription of this gene produced an RNA with part of the sequence CAACCCAA, a sequence complementary to the telomeric DNA repeat motif in the same organism.

The RNA portion has been cloned in several different organisms and shows variability in both length and sequence, as well as in overall structure (Figure 8.1(B)).

For example, the RNA size can vary from 147 to 205 nt in ciliates, 312 to 559 nt in vertebrates and 928 to over 2,425 nt in yeast. Although their sequence and structure are highly variable, there are, however, conserved subdomains in all species: a template sequence (t, template), complementary to the telomeric sequence; a hairpin motif TBE (template boundary sequence), which prevents DNA synthesis outside the template region (Jansson et al. 2015), and a transactivator domain, named STE (stem terminus element), containing a highly conserved hairpin structure and on which the TERT subunit is anchored (Podlevsky et al. 2016). These domains have been conserved during evolution because they play an essential role for telomerase activity. In the majority of known RNAs, a pseudo-knot structure called PK (*Pseudo Knot*) is present and localized near the template; even if the size and shape of this pseudo-knot vary according to species, its presence is essential for telomerase activity and allows the modulation of telomerase activity in some organisms, such as the ciliate *Tetrahymena*. However, in some organisms, such as the *Trypanosoma* flagellum, it is absent. This lack of pseudo-knot in basal eukaryotic species suggests that the pseudo-knot arose during evolution in more evolved eukaryotes (Podlevsky and Chen 2016). Thus, the common ancestor of the telomerase RNA was not expected to possess a pseudo-knot in the template region.

The PK structure associated with the (t)-template and the TBE motif constitutes the necessary and sufficient t/PK domain for *in vitro* activity of TERT. The STE domain is essential for *in vivo* telomerase activity.

NMR has been widely used to study RNA structure (for review, read Wang et al. (2019)). The first structures were published in the early 2000s and focused on isolated RNA subdomains, as the large size of these RNAs made it impossible to study them whole. It was necessary to ensure that these partial RNA structures, isolated from the protein context, were biologically relevant. The determination of the structure of telomerase by Cryo-Electro-Microscopy (Cryo-EM) allowed for the validation, *a posteriori*, of all the proposed structures. NMR studies in solution also allow for the dynamics of molecules to be studied and, combined with computer calculations, to propose dynamic structural models. Thus, the study of the t/PK motif in humans and *Medaka* fish revealed a highly flexible J2a/b region (Figure 8.1(B)) that imposes an overall bent shape, which appears to be highly conserved as it is found in both species studied (Wang et al. 2016). Furthermore, it has been shown in the ciliate *Tetrahymena* that the pseudo-knot does not form in RNA alone, but only in the presence of the proteins that constitute telomerase (Cash and Feigon 2017). In the absence of proteins, the adopted structure would not allow telomerase assembly. These data suggest that pseudo-knot formation may be a mechanism for regulating telomerase activity. Other regions also show structural changes when

RNA is complexed to the TERT subunit, such as the *Medaka* fish CR4/5 region in the presence of the TERT TRBD domain (Huang et al. 2014),

While telomerase RNAs possess relatively evolutionarily conserved domains involved in the catalytic activity of telomerase, they also have other structural domains, but this time, very diverse, located mainly on the 3' side (Figure 8.1(B)) (Podlevsky and Chen 2016). These domains (species-specific RNP biogenesis element) are involved in telomerase biogenesis and recruitment to telomeres; for this purpose, they constitute binding sites for the various proteins involved in the different biogenesis and telomere recruitment pathways. This aspect will be discussed in detail when telomerase biogenesis and recruitment is presented.

In conclusion, although the RNAs of these various species are highly divergent in size, sequence and structural motifs, all exhibit a highly conserved TERT interaction domain, corresponding to the domain involved in the catalytic activity of telomerase, as well as a variable domain of interaction with numerous and diverse proteins, involved in the biogenesis of the ribonucleoprotein complex. Thus, during evolution, telomerase RNA has maintained the essential motifs for interacting with the highly conserved TERT subunit, while assimilating new and diverse recognition motifs for the various proteins involved in telomerase biogenesis.

8.3.2. *The holoenzyme*

The holoenzyme is the RNP complex that is catalytically active *in vivo*. This holoenzyme contains the two TERT and RNA subunits, as well as many other species-specific proteins, the nature and stoichiometry of which have been difficult to determine (Schmidt and Cech 2015; Wu et al. 2017; Nguyen et al. 2019) (Figure 8.2).

Other proteins may interact with telomerase, either constitutively or transiently, during DNA synthesis. For example, in humans, anchoring of telomerase to the chromosome end and modulation of its processivity involves transient interactions with the POT1-TPP1 complex (Wang et al. 2007) (see shelterin proteins, Chapter 3); recruitment of proteins required for C-strand synthesis and elongation termination occurs via transient interactions with the CST complex, a heterotrimeric complex consisting of the proteins CTC1, Stn1 and Ten1 (Chen et al. 2012, 2013; Schmidt and Cech 2015; Chan et al. 2017; Wu et al. 2017). In contrast, in the ciliate *Tetrahymena*, the p75-p45-p19 complex, homologous to the human CST complex, and the p50 protein, homologous to human TPP1, are integral parts of telomerase (Figure 8.2).

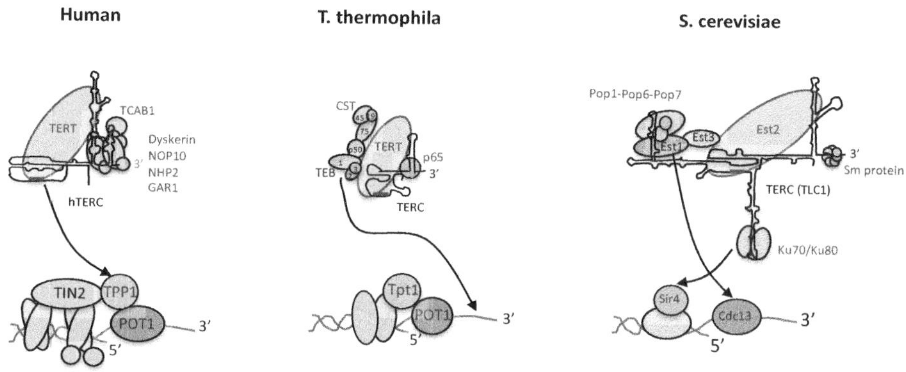

Figure 8.2. *Composition of holoenzymes of various telomerases and mode of recruitment to telomeres. Adapted from Nguyen et al. (2019)*

Determining the structure of telomerase remains a challenge because telomerase is low in abundance in the majority of cells (e.g. approximately 240 in human cells) and tends to aggregate (Xi and Cech 2014). To overcome the low cellular abundance of telomerase, electron microscopy (negative staining), not requiring as large quantities as NMR and crystallography, was employed. The structures of *Tetrahymena* and human telomerase were first obtained in 2013 with a resolution of ~25–30 Å (Jiang et al. 2013; Sauerwaldet al. 2013). But it is cryo-electro-microscopy, a method based on very rapid freezing of protein extracts to observe them in their natural state, that has allowed us to propose telomerase superstructures with a resolution of 9 Å (Jiang et al. 2015), 4.8 Å (Jiang et al. 2018) and then 3.3 Å (He et al. 2021) for *Tetrahymena* telomerase in the absence and presence of telomeric DNA, respectively, and 8 Å (Nguyen et al. 2018) and 4 Å (Ghanim et al. 2021) for human telomerase associated with its DNA substrate. No ME structure of yeast telomerases has been reported to date.

Figure 8.3 shows the schematic structure of the human complex forming two lobes in which the RNA (hTERC) acts as a scaffold for different proteins. One lobe constitutes the catalytic core and contains the t/PK domain of hTERC and the TEN (DNA binding) and RT (reverse transcriptase activity) domains of TERT. The other lobe contains the STE domain of hTERC and the various proteins required for telomerase biogenesis (described below). The structure obtained in Cryo-EM in 2021 revealed the presence of a histone H2A-H2B dimer linked to the CR4/5 motif, suggesting a role for this dimer in the folding and activity of hTERC (Ghanim et al. 2021).

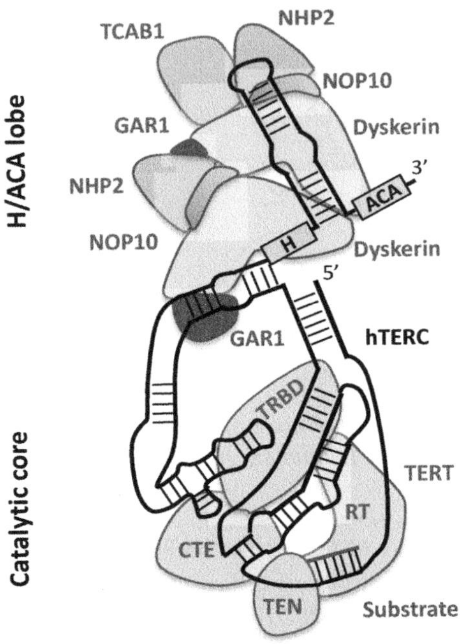

Figure 8.3. *Structure of the human telomerase holoenzyme. Adapted from Nguyen et al. (2018)*

Although Cryo-EM is a powerful method to obtain structures with sub-nanometric resolution, we hope to be able to obtain structures with atomic resolution in the near future in order to elucidate the molecular mechanisms involved. A recent NMR analysis study of RNA-protein interactions in the ciliate has allowed us to optimize the constructs for crystallographic study (Wang et al. 2019).

8.4. Biogenesis

Biogenesis is a term that groups together all the metabolic steps that lead to an active and correctly localized ribonucleoprotein complex. Telomerase biogenesis is a very complex process that requires a large number of maturation steps including the synthesis, splicing and chemical modification of telomerase RNA, its assembly with telomerase component proteins, and the transport of the different partners between different compartments in the nucleus. The proteins and molecular mechanisms involved in telomerase biogenesis are highly diverse across species (Egan and Collins 2012). In ciliates, RNA (TER) is transcribed by RNA polymerase III (Greider and Blackburn 1989), whereas it is transcribed by RNA polymerase II in vertebrates, yeast and plants. The different structural motifs present in RNA will participate in the recruitment of specific proteins of a given biogenesis pathway.

In yeast, it is the snRNA (small nuclear RNA) biogenesis pathway that is used, with different maturation mechanisms depending on the species (intervention of the Nrd1-Nab3-Sen1 complex for *S. cerevisiae* or of the spliceosome for *S. pombe*) (Tang et al. 2012; Noel et al. 2012). In ciliates, biogenesis follows the RNA polymerase III small RNA transcript biogenesis pathway, which requires La-family proteins such as the p65 protein in *Tetrahymena*, which is essential for folding RNA and stabilizing the TERT-TER interaction (Singh et al. 2012). In humans (Figure 8.1(B)), hTERC has the H/ACA motif (Mitchell et al. 1999; Massenet et al. 2017), specific to the snoRNA (small nucleolar RNA) biogenesis pathway, as well as a binding motif (CAB) of the TCAB1 protein, involved in specific localization to Cajal bodies (the nuclear compartment where post-transcriptional modifications and RNA splicing take place). Transcribed hTERC RNA initially binds two sets of H/ACA proteins (DKC1 (for Dyskerin), NHP2 and NOP10), as well as two NAF1 proteins (Figure 8.4). This complex thus formed leaves the transcription site to quickly localize in Cajal's bodies where the RNA will undergo a post-transcriptional modification, the 5' TMG cap(trimethylguanine). Subsequently, the two NAF1 proteins are replaced by two GAR1 proteins. This new complex is the one found in the holoenzyme and is essential for the proper functioning of telomerase. The location of the assembly of hTERC with the TERT subunit is not yet known, this assembly could be in the nucleolus, the nucleoplasm or in the Cajal bodies.

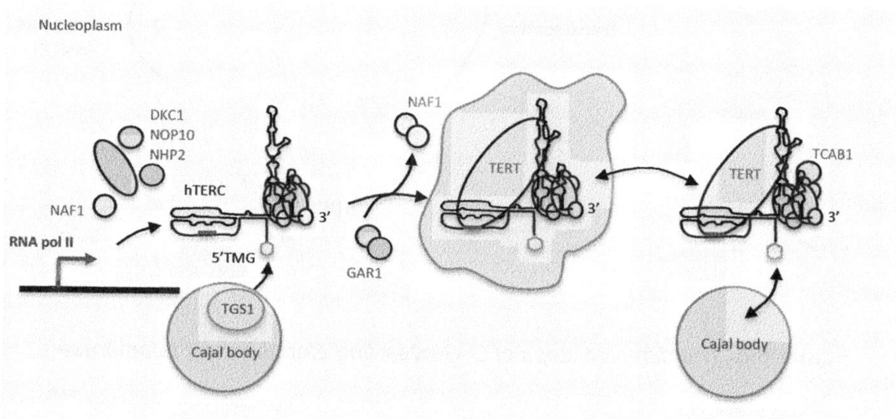

Figure 8.4. *The different stages of human telomerase biogenesis. Adapted from Wu et al. (2017)*

8.5. Mechanism

Greider and Blackburn (1989) were the first to propose a model of telomere size maintenance by telomerase. In this model, telomerase uses its RNA as a template to add nucleotides to the end of chromosomes, because of the catalytic subunit with a reverse transcriptase activity (TERT). The mechanism consists of three steps (Figure 8.5):

1) telomere–telomerase association step and binding of the RNA template to the 3' single-stranded end of the telomeric DNA by sequence complementarity;

2) elongation step by reverse transcription of the RNA template and addition of nucleotides to the 3' end of the telomeric DNA;

3) translocation step: once the reverse transcription of the RNA template sequence is completed, a translocation of the telomerase complex takes place to allow another cycle of synthesis (step 2).

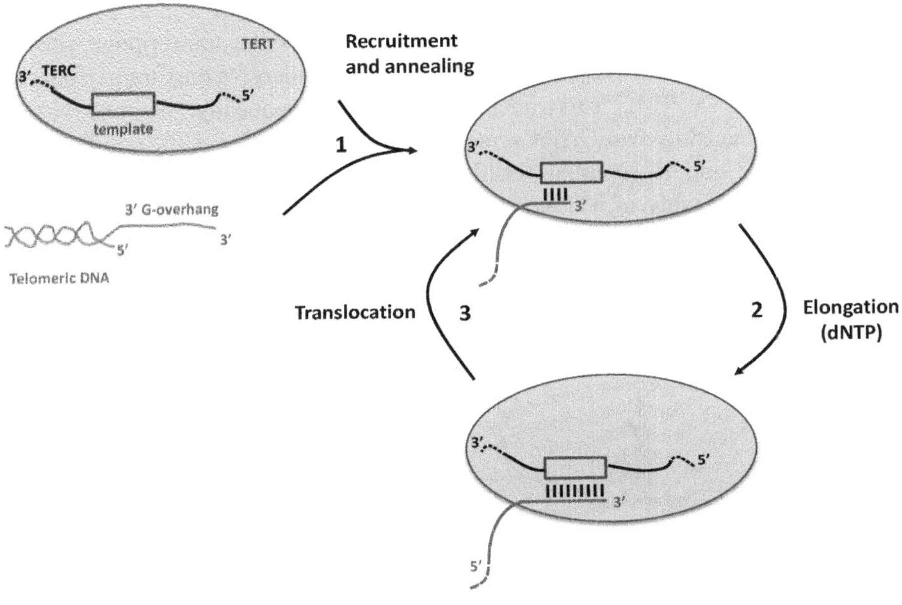

Figure 8.5. *The different steps of 3' G-overhang elongation by telomerase*

Numerous structural (mainly Cryo-EM) and functional studies have shown the critical role of the different RNA structural motifs and proposed models of molecular mechanism (for reviews, see Feigon et al. (2016); Wu et al. (2017);

Musgrove et al. (2018); Wang et al. (2019)). In addition, single-molecule resonance fluorescence energy transfer (smFRET) studies have revealed conformational changes in RNA during DNA synthesis (Berman et al. 2011; Parks and Stone 2014; Parks et al. 2017). In the case of *Tetrahymena*, an accordion RNA model has been proposed: extension and shortening of the regions flanking the template would facilitate template movement through the active site of the TERT RT domain (Berman et al. 2011). In humans, the dynamics of the PK motif would promote telomerase processivity (Parks et al. 2017).

All these data allow us to better understand the molecular mechanisms involved, however only the atomic structure of the holoenzyme with its substrate will enable the telomerase machinery to be fully understood.

8.6. Regulation

The regulation of telomerase activity is multiple. To have a good activity of telomerase, it is necessary to have it in quantity, well associated, in the right place and active.

This regulation can occur at the level of expression of the different TERT and RNA subunits of telomerase (Jie et al. 2019). It has been shown that the expression of the human TERT subunit (hTERT) can be regulated at several levels: at the transcriptional level, via a large number of transcriptional protein factors (activators, inhibitors or both), via mutations in the hTERT promoter or via epigenetic modifications, such as methylation of the hTERT promoter; at the post-transcriptional level, via alternative splicing and the intervention of non-coding RNAs such as microRNAs or long non-coding RNAs.

Post-translational modifications can modulate the stability and the subcellular localization of telomerase and consequently its concentration and activity. Two modifications have been mainly observed: phosphorylation that modulates its subcellular localization and ubiquitination that induces its degradation by the proteasome (Jie et al. 2019).

Because telomerase biogenesis is a key step in achieving active telomerase (see above), it is fully involved in the regulation of telomerase activity (Wu et al. 2017; Nguyen et al. 2019).

The telomerase recruitment step at telomeres is also a step that regulates telomerase activity. The diversity observed in the nature of telomeres and holoenzymes implies that the mode of telomerase recruitment is species-dependent

(Figure 8.2) (Nandakumar and Cech 2013; Nguyen et al. 2019). In humans, telomerase is recruited by the shelterin complex via the N-terminal domain of the TPP1 protein (Nandakumar et al. 2012; Sextonet al. 2012; Zhong et al. 2012); in yeast *S. cerevisiae*, the Sir4 and Cdc13 proteins recruit telomerasevia interactions with the Ku70/ku80 heterodimer and the telomerase Est1 subunit, respectively (Nugent et al. 1996; Hass and Zappulla 2015; Chen et al. 2018); in yeast, *S. pombe* recruitment occurs via the Ccq1 protein (Moyoshi et al. 2008; Moser et al. 2011; Yamazaki et al. 2012). In the ciliate *Tetrahymena*, telomerase is recruited to telomeres through the interaction of single-stranded telomeric DNA with the Teb1 subunit (subunit of the TEB complex), which has a high affinity and specificity for single-stranded telomeric DNA (Upton et al. 2014). However, by analogy with humans (and the yeast *S. pombe*), telomerase could be recruited by the Tpt1 protein (homolog of TPP1), which remains to be demonstrated. Furthermore, it has been observed that expression of TERRA (TElomeric repeat-containing RNA) positively modulates telomerase recruitment to telomeres (Farnung et al. 2012; Cusanelli et al. 2013; Moravec et al. 2016). However, the impact of TERRA on human telomerase activity *in vivo* is yet to be demonstrated (for more details, see Chapter 10).

In humans, enzymatic activity, and especially telomerase processivity, is regulated via the action of the shelterin proteins, POT1-TPP1 (Wang et al. 2007), and the CST complex (Chen et al. 2012, 2013). More recently, it has been proposed that the elongating telomeric DNA may regulate the processivity of telomerase by adopting G-quadruplex secondary structures (Jansson et al. (2019); see Chapter 1).

8.7. Disease and therapy

Telomerase plays a fundamental role in health by participating in longevity, as an anti-aging agent, and in tumorigenesis, as a pro-cancer agent; this role is discussed in detail in Chapter 11. In addition, mutations in telomerase-associated genes lead to degenerative constitutional diseases called telomeropathies, which are described in Chapter 12. In the medical context, telomerase can therefore be considered as a therapeutic tool to slow down or even stop aging, or as a target for anti-cancer therapy. Chapter 13 will discuss approaches that have been used to target telomerase.

8.8. Conclusion

Telomerase is a ribonucleoprotein complex whose structure and mechanisms of action and regulation have been extensively investigated since its discovery in 1985, and whose fundamental role in the maintenance of telomere size has made it a prime target for anti-aging and anti-cancer research. Obtaining telomerase structures with atomic resolution in the near future should allow for the development of new anti-telomerase drugs and the understanding of the effect of certain mutations of the telomerase complex in several diseases. However, using telomerase as a therapeutic target requires a better understanding of the other alternative roles it plays and their interactions with telomere elongation activity.

8.9. References

Artandi, S.E. and DePinho, R.A. (2010). Telomeres and telomerase in cancer. *Carcinogenesis*, 31(1), 9–18.

Berman, A.J., Akiyama, B.M., Stone, M.D., Cech, T.R. (2011). The RNA accordion model for template positioning by telomerase RNA during telomeric DNA synthesis. *Nat. Struct. Mol. Biol.*, 18(12), 1371–1375.

Blasco, M.A. (2003). Telomeres in cancer and aging: Lessons from the mouse. *Cancer Letters*, 194(2), 183–188.

Budarf, M.L. and Blackburn, E.H. (1986). Chromatin structure of the telomeric region and 3'-nontranscribes spacer of Tetrahymena ribosomal RNA genes. *The Journal of Biological Chemistry*, 261(5), 363–369.

Cash, D.D. and Feigon, J. (2017). Structure and folding of the Tetrahymena telomerase RNA pseudoknot. *Nucleic Acids Research*, 45, 482–495.

Chan, H., Wang, Y., Feigon, J. (2017). Progress in human and Tetrahymena telomerase structure determination. *Annu. Rev. Biophys.*, 46, 199–225.

Chen, L.-Y. and Lingner, J. (2013). CST for the grand finale of telomere replication. *Nucleus*, 4(4), 277–282.

Chen, L.-Y., Redon, S., Lingner, J. (2012). The human CST complex is a terminator of telomerase activity. *Nature*, 488, 540–544.

Chen H., Xue J., Churikov D., Hass E.P., Shi S., Lemon L.D., Luciano P., Bertuch A.A., Zappulla D.C., Géli V. et al. (2018). Structural insights into yeast telomerase recruitment to telomeres. *Cell*, 172, 331–343.

Counter, C.M., Meyerson, M., Ng Eaton, E., Weinberg, R.A. (1997). The catalytic subunit of yeast telomerase. *Proc. Natl. Acad. Sci. USA*, 94, 9202–9207.

Cusanelli, E., Perez Romero, C.A., Chartrand, P. (2013). Telomeric noncoding RNA TERRA is induced by telomere shortening to nucleate telomerase molecules at short telomeres. *Mol. Cell*, 51, 780–791.

Egan, E.D. and Collins, K. (2012). Biogenesis of telomerase ribonucleoproteins. *RNA*, 18, 1747–1759.

Farnung, B.O., Brun, C.M., Arora, R., Lorenzi, L.E., Azzalin, C.M. (2012). Telomerase efficiency elongates highly transcribing telomeres in human cancer cells. *PLoS One*, 7, e35714.

Feigon, J., Chan, H., Jiang, J. (2016). Integrative structural biology of *Tetrahymena* telomerase – Insights into catalytic mechanism and interaction at telomeres. *FEBS J.*, 283, 2044–2050.

Ghanim, G.E., Fountain, A.J., van Roon, A-M.M., Rangan, R., Das, R., Collins, K., Nguyen, T.H.D. (2021). Structure of human telomerase holoenzyme with bound telomeric DNA. *Nature*, 593, 449–453.

Gillis, A.J., Schuller, A.P., Skordalakes, E. (2008). Structure of the *Triboliumcastaneum* telomerase catalytic subunit TERT. *Nature*, 455, 633–637.

Greider, C. and Blackburn, E.H. (1985). Identification of a specific telomere terminal transferase activity in Tetrahymena extracts. *Cell*, 43, 405–413.

Greider, C. and Blackburn, E.H. (1987). The telomere terminal transferase of *Tetrahymena* is a ribonucleoprotein enzyme with two kinds of primer specificity. *Cell*, 51, 887–898.

Greider, C.W. and Blackburn, E.H. (1989). A telomeric sequence in the RNA of Tetrahymena telomerase required for telomere repeat synthesis. *Nature*, 337, 331–337.

Harkisheimer, M., Mason, M., Shuvaeva, E., Skordalakes, E. (2013). A motif in the vertebrate telomerase N-terminal linker of TERT contributes to RNA binding and telomerase activity and processivity. *Structure*, 21, 1870–1878.

Hass, E.P. and Zappulla, D.C. (2015). The Ku subunit of telomerase binds Sir4 to recruit telomerase to lengthen telomeres in *S. cerevisiae*. *eLife*, 4, e07750.

He, Y., Wang, Y., Liu, B., Helmling, C., Susac, L., Cheng, R., Zhou, Z.H., Feigon, J. (2021). Structures of telomerase at several steps of telomere repeat synthesis. *Nature*, 593, 454–459.

Huang, J., Brown, A.F., Wu, J., Xue, J., Bley, C.J., Rand, D.P., Wu, L., Zhang, R., Chen, J. J.-L., Lei, M. (2014). Structural basis for protein-RNA recognition in telomerase. *Nat. Struct. Mol. Biol.*, 21(6), 507–512.

Jacobs, S.A., Podell, E.R., Cech, T.R. (2006). Crystal structure of the essential N-terminal domain of telomerase reverse transcriptase. *Nat. Struct. Mol. Biol.*, 13, 218–225.

Jansson, L.I., Akiyama, B.M., Ooms, A., Lu, C., Rubin, S.M., Stone, M.D. (2015). Structural basis of template boundary definition in *Tetrahymena* telomerase. *Nat. Struct. Mol. Biol.*, 22, 883–888.

Jansson, L.I., Hentschel, J., Parks, J.W., Chang, T.R., Lu, C., Baral, R., Bagshaw, C.R., Stone, M.D. (2019). Telomere DNA G-quadruplex folding within actively extending human telomerase. *Proc. Natl. Acad. Sci. USA*, 116(19), 9350–9359.

Jiang, J., Miraco, E.J., Hong, K., Eckert, B., Chan, H., Cash, D.D, Min, B., Zhou, Z.H., Collins, K., Feigon, J. (2013). The architecture of *Tetrahymena* telomerase holoenzyme. *Nature*, 496(7444), 187–192.

Jiang, J., Chan, H., Cash, D.D., Miraco, E.J., Ogorzalek Loo, R.R., Upton, H.E., Cascio, D., Johnson, R.O'B., Collins, K., Loo, J.A. et al. (2015). Structure of Tetrahymena telomerase reveals previously unknown subunits, functions, and interactions. *Science*, 350(6260), aab4070.

Jiang, J., Wang, Y., Susac, L., Chan, H., Basu, R., Zhou, Z.H., Feigon, J. (2018). Structure of telomerase with telomeric DNA. *Cell*, 173(5), 1179–1190.

Jie, M.-M., Chang, X., Zeng, S., Liu, C., Liao, G.-B., Wu, Y.-R., Liu, C.-H., Hu, C.-J., Yang, S.-M., Li, X.-Z. (2019). Diverse regulatory manners of human telomerase reverse transcriptase. *Cell Commun. Signal.*, 17, 63–77.

Kapler, G.M. (1993). Developmentally regulated processing and replication of the *Tetrahymena* rDNA minichromosome. *Current Opinion in Genetics and Development,* 3(7), 730–735.

Lingner, J., Hughes, T.R., Shevchenko, A., Mann, M., Lundblad, V., Cech, T.R. (1997). Reverse transcriptase motifs in the catalytic subunit of telomerase. *Science*, 276, 561–567.

Lundblad, V. and Szostak, J.W. (1989). A mutant with a defect in telomere elongation leads to senescence in yeast. *Cell*, 57, 633–643.

Massenet, S., Bertrand, E., Verheggen, C. (2017). Assembly and trafficking of box C/D and H/ACA snoRNPs. *RNA Biol.*, 14, 680–692.

Mitchell, J.R., Cheng, J., Collins, K. (1999). A box H/ACA small nucleolar RNA-like domain at the human telomerase RNA 30 end. *Mol. Cell Biol.*, 19, 567–576.

Mitchell, M.T., Gillis, A., Futahashi, M., Fujiwara, H., Skordalakes, E. (2010). Structural basis for telomerase catalytic subunit TERT binding to RNA template and telomeric DNA. *Nat. Struct. Mol. Biol.*, 17, 513–518.

Miyoshi, T., Kanoh, J., Saito, M., Ishikawa, F. (2008). Fission yeast Pot1-Tpp1 protects telomeres and regulates telomere length. *Science*, 320, 1341–1344.

Moravec, M., Wischnewski, H., Bah, A., Hu, Y., Liu, N., Lafranchi, L., King, M.C., Azzalin, C.M. (2016). TERRA promotes telomerase-mediated telomere elongation in *Schizosaccharomyces pombe. EMBO Reports*, 17(7), 999–1012.

Morin, G.B. (1989). The human telomere terminal transferase enzyme is a ribonucleoprotein that synthesizes TTAGGG repeats. *Cell*, 59(3), 521–529.

Moser, B.A., Chang, Y.T., Kosti, J., Nakamura, T.M. (2011). Tel1ATM and Rad3ATR kinases promote Ccq1-Est1 interaction to maintain telomeres in fission yeast. *Nat. Struct. Mol. Biol.*, 18, 1408–1413.

Musgrove, C., Jansson, L.I., Stone, M.D. (2018). New perspectives on telomerase RNA structure and function. *Wiley Interdiscip. Rev. RNA*, 9(2). doi: 10.1002/wrna.1456.

Nakumara, T.M., Morin, G.B., Chapman, K.B., Weinrich, S.L., Andrews, W.H., Lingner, J., Harley, C.B., Cech, T.R. (1997). Telomerase catalytic subunit homologues from fission yeast and human. *Science*, 277, 955–959.

Nandakumar, J. and Cech, T.R. (2013). Finding the end: Recruitment of telomerase to the telomere. *Nature Reviews Molecular Cell Biology*, 14, 69–82.

Nandakumar, J., Bell, C.F., Weidenberg, I., Zaug, A.J., Leinwand, L.A., Cech, T.R. (2012). The TEL patch of telomere protein TPP1 mediates telomerase recruitment and processivity. *Nature*, 492(7428), 285–289.

Nguyen, T.H.D., Tam, J., Wu, R.A., Greber, B.J., Toso, D., Nogales, E., Collins, K. (2018). Cryo-EM structure of substrate-bound human telomerase holoenzyme. *Nature*, 557(7704), 190–195.

Nguyen, T.H.D., Collins, K., Nogales, E. (2019). Telomerase structures and regulation: Shedding light on the chromosome end. *Curr. Opin. Struct. Biol.*, 55, 185–193.

Noel, J.-F., Larose, S., Elela, S.A., Wellinger, R.J. (2012). Budding yeast telomerase RNA transcription termination is dictated by the Nrd1/Nab3 non-coding RNA termination pathway. *Nucleic Acids Research*, 40, 5625–5636.

Nugent, C.I., Hughes, T.R., Lue, N.F., Lundblad, V. (1996). Cdc13p: A single-strand telomeric DNA binding protein with a dual role in yeast telomere maintenance. *Science*, 274, 249–252.

Parks, J.W. and Stone, M.D. (2014). Coordinated DNA dynamics during the human telomerase catalytic cycle. *Nat. Commun.*, 5, 4146–4155.

Parks, J.W., Kappel, K., Das, R., Stone, M.D. (2017). Single-molecule FRET-Rosetta reveals RNA structural rearrangements during human telomerase catalysis. *RNA*, 23, 175–188.

Podlevsky, J.D. and Chen, J.J.-L. (2016). Evolutionary perspectives of telomerase RNA structure and function. *RNA Biology*, 13, 720–732.

Podlevsky, J.D., Li, Y., Chen, J.L. (2016). The functional requirement of two structural domains within telomerase RNA emerged early in eukaryotes. *Nucleic Acids Research*, 44, 9891–9901.

Rouda, S. and Skordalakes, E. (2007). Structure of the RNA-binding domain of telomerase: Implications for RNA recognition and binding. *Structure*, 15, 1403–1412.

Sarek, G., Marzec, P., Margalef, P., Boulton, S.J. (2015). Molecular basis of telomere dysfunction in human genetic diseases. *Nat. Struct. Mol. Biol.*, 22(11), 867–874.

Sauerwald, A., Sandin, S., Cristofari, G., Scheres, S.H.W., Lingner, J., Rhodes, D. (2013). Structure of active, dimeric human telomerase. *Nat. Struct. Mol. Biol.*, 20(4), 454–460.

Schmidt, J.C. and Cech, T.R. (2015). Human telomerase: Biogenesis, trafficking, recruitment and activation. *Genes and Development*, 29, 1095–1105.

Ségal-Bendirdjian, E. and Géli, V. (2019). Non-canonical roles of telomerase: Unraveling the imbroglio. *Frontiers in Cell and Developmental Biology*, 7, 332.

Sexton, A.N., Youmans, D.T., Collins, K. (2012). Specificity requirements for human telomere protein interaction with telomerase holoenzyme. *Journal of Biological Chemistry*, 287, 34455–34464.

Shay, J.W. (2016). Role of telomeres and telomerase in aging and cancer. *Cancer Discov.*, 6(6), 584–593.

Shippen-Lentz, D. and Blackburn, E.H. (1989). Telomere terminal transferase activity from *Euplotescrassus* adds large numbers of TTTTGGGG repeats onto telomeric primers. *Molecular and Cellular Biology*, 9(6), 2761–2764.

Singh, M., Wang, Z., Koo, B.K., Patel, A., Cascio, D., Collins, K., Feigon, J. (2012). Structural basis for telomerase RNA recognition and RNP assembly by the holoenzyme La family protein p65. *Mol. Cell*, 16–26.

Spangler, E.A., Ryan, T., Blackburn, E.H. (1988). Developmentally regulated telomere addition in *Tetrahymena thermophila*. *Nucleic Acids Research*, 16(12), 5569–5585.

Stewart, S.A. and Weinberg, R.A. (2006). Telomeres: Cancer to human Aging. *Annu. Rev. Cell Dev. Biol.*, 22, 531–557.

Tang, W., Kannan, R., Blanchette, M., Baumann, P. (2012). Telomerase RNA biogenesis involves sequential binding by Sm and Lsm complexes. *Nature*, 484, 260–264.

Townsley, D.M., Dumitriu, B., Young, N.S. (2014). Bone marrow failure and the telomeropathies. *Blood*, 124(18), 2775–2783.

Upton, H.E., Hong, K., Collins, K. (2014). Direct single-strand DNA binding by Teb1 mediates the recruitment of *Tetrahymena thermophila* telomerase to telomeres. *Mol. Cell. Biol.*, 34(22), 4200–4212.

Wang, F., Podell, E.R., Zaug, A.J., Yang, Y., Baciu, P., Cech, T.R., Lei, M. (2007). The POT1-TPP1 telomers complex is a telomerase processivity factor. *Nature*, 445, 506–510.

Wang, Y., Yesselman, J.D., Zhang, Q., Kang, M., Feigon, J. (2016). Structural conservation in the template/pseudoknot domain of vertebrate telomerase RNA from teleost fish and human. *Proc. Natl. Acad. Sci. USA*, 113(35), 5125–5134.

Wang, Y., Susac, L., Feigon, J. (2019). Structural biology of telomerase. *Cold Spring Harb. Perspect. Biol.*, 11(12), a032383.

Wu, R.A., Upton, H.E., Vogan, J.M., Collins, K. (2017). Telomerase mechanism of telomere synthesis. *Annu. Rev. Biochem.*, 86, 439–460.

Xi, L. and Cech, T.R. (2014). Inventory of telomerase components in human cells reveals multiple subpopulations of hTR and hTERT. *Nucleic Acids Research*, 42, 8565–8577.

Yamazaki, H., Tarumoto, Y., Ishikawa, F. (2012). Tel1(ATM) and Rad3(ATR) phosphorylate the telomere protein Ccq1 to recruit telomerase and elongate telomeres in fission yeast. *Genes Dev.*, 26, 241–246.

Zahler, A.M. and Prescott, D.M. (1988). Telomere terminal transferase activity in the hypotrichous ciliate *Oxytricha nova* and a model for replication of the ends of linear DNA molecules. *Nucleic Acids Research*, 16(4), 6953–6972.

Zheng, Q., Huang, J., Wang, G. (2019). Mitochondria, telomeres and telomerase subunits. *Frontiers in Cell and Developmental Biology*, 7, 274.

Zhong, F.L., Batista, L.F.Z., Freund, A., Pech, M.F., Venteicher, A.S., Artandi, S.E. (2012). TPP1 OB-fold domain controls telomeres maintenance by recruiting telomerase to chromosome ends. *Cell*, 150, 481–494.

9

Alternative Lengthening of Telomeres

Mathilde GAUCHIER
Institute of Human Genetics, CNRS,
University of Montpellier, France

All eukaryotes, as well as some prokaryotes, maintain their genome in the form of one or more linear DNA molecules. At each cell cycle, replication of linear DNA is incomplete at its ends, which poses a major problem for the faithful transmission of genetic material. Telomeres allow chromosomes to withstand progressive erosion at each division cycle while protecting genetic integrity (Watson 1972; Olovnikov 1973). Cells with too short telomeres are programmed to go into an irreversible cell cycle arrest, the replicative senescence and then stop proliferating. This process of telomere shortening, sometimes called the mitotic clock, is an essential barrier against genomic instability. However, certain mechanisms can be activated to compensate for this progressive erosion and maintain the size of telomeres beyond a critical size, thus favoring the proliferation of certain cell types. This is the case for stem cells or cancer cells in humans, which are characterized by their ability to divide indefinitely. These telomere maintenance mechanisms very often involve the reactivation of the telomerase, a specific reverse transcriptase that compensates for telomere erosion at each cell cycle (Chapter 8). But some cells are also able to maintain the size of their telomeres without detectable activity of this enzyme. These cells use alternative mechanisms to telomerase. These alternative mechanisms of telomere lengthening are called alternative lengthening of telomeres (ALT) and

For a color version of all the figures in this chapter, see www.iste.co.uk/saintome/telomeres.zip.

often use homologous recombination. In this chapter, we summarize the state of knowledge on the maintenance of telomere size by the ALT pathway, present in some human cancers, but also used by other organisms when telomerase is absent or dysfunctional.

9.1. ALT cell markers

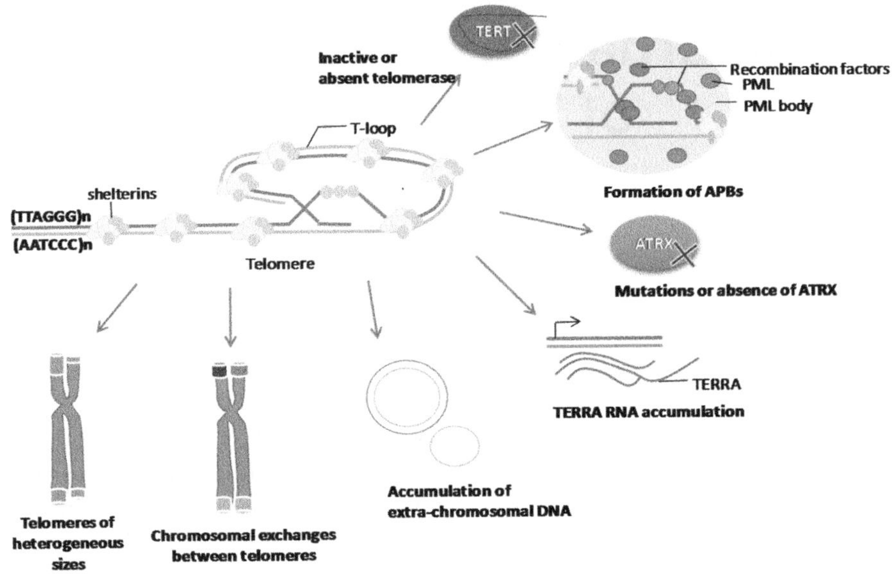

Figure 9.1. *Phenotypic markers indicating the presence of an active ALT pathway in the cell. Unlike telomerase maintenance, which is easily measured, none of the ALT markers quantify the activity of this pathway because telomere elongation by ALT is still poorly understood*

Historically, even before the discovery of telomerase, recombination between telomeric sequences had been proposed by Dancis and Holmquist as a mechanism involved in telomere maintenance (Dancis and Holmquist 1979). The existence of this mechanism was only later demonstrated as participating in the de novo synthesis of telomeres to escape senescence when telomerase is absent (Lundblad and Blackburn 1993). These telomerase-independent mechanisms have been termed alternative pathways of telomere elongation: ALT pathway(s). In humans, the ALT pathway was first described by studying cells in culture. Despite the absence of measurable telomerase activity, some cells are able to elongate their telomeres while retaining their ability to enter into a process of immortalization (Bryan et al. 1995,

1997). These telomeres, called ALTs, use a complementary DNA molecule as a template for the synthesis of new telomeric motif repeats at the end of chromosomes. This elongation mechanism has been shown to rely on homologous recombination. In effect, a transgene inserted into a single telomere becomes copied and amplified on the telomeres of different chromosomes. This phenomenon occurs only in human ALT cells, which do not use telomerase (Bryan et al. 1995, 1997). In the absence of a simple method to measure ALT activity, a set of cellular phenotypes that are almost all suggestive of the presence of this recombination mechanism have been described (Figure 9.1). These features can be identified by the complementary approaches detailed in the appendix to this chapter. However, an important unknown remains: unlike telomerase activity, whose values reflect telomere lengthening quite accurately and which is easily measured, the characteristics of the ALT pathway is not a measure that precisely quantifies this mechanism. In other words, these phenotypes are indicators of the presence of the ALT pathway but cannot measure its activity.

9.1.1. *ALT telomeres are associated with PML bodies: APBs*

Most ALT cells exhibit telomere co-localization with PML nuclear bodies. PML and SP100 sumoylated proteins are the main structural components of PML bodies (Chung et al. 2012). The number of PML nuclear bodies varies between 5 and 30 depending on cell type, stress conditions and cell cycle phase. These nuclear compartments consist of proteins and exclude the presence of nucleic acids. In contrast, in ALT cells APBs are formed, compartments that contain proteins such as PML, but also telomeric DNA sequences, and telomere-specific proteins such as shelterin (Yeager et al. 1999). A large number of factors involved in homologous recombination are also recruited to these APBs, suggesting that these nuclear bodies may serve as platforms for recombination (Henson and Reddel 2010). For this reason, it has been proposed that recombination between ALT telomeres leading to their elongation occurs in APBs. However, the mechanisms behind the formation of APBs are still not known, but may involve the chromatin status of telomeres (Jiang et al. 2009). DNA damage to telomeres is another source of APB formation (Cho et al. 2014).

9.1.2. *Telomeres of heterogeneous size*

Human cancer cells using telomerase have homogeneous telomere sizes, typically less than 10 kb. In contrast, the size distribution of ALT telomeres is more heterogeneous and ranges from less than 3 kb to more than 50 kb in the same cell. This heterogeneity in size is probably a sign of rapid loss or gain of telomeric motif repeats at the end of chromosomes during cell divisions, another sign of a recombination mechanism at work at telomeres (Murnane et al. 1994).

9.1.3. *The accumulation of extra-chromosomal circular telomeric DNA*

ALT cells accumulate different types of extra-chromosomal DNA of telomeric nature, including single-stranded DNA circles called C or G circles, double-stranded DNA circles or t-circles (Cesare and Reddel 2010). The origin of this extra-chromosomal DNA could come from an exchange between telomeres by recombination. It has also been proposed that this extra-chromosomal telomeric DNA could serve as a template for homologous recombination and would be involved in telomere elongation by a circular replication mechanism, such as those employed by bacteriophages (Natarajan and McEachern 2002).

9.1.4. *Variant telomeric DNA repeats*

Healthy cells and cancer cells using telomerase have variable telomeric repeats such as TGAGGG, TCAGGG and TTGGGG at the proximal region of the telomere only (Allshire et al. 1989). However, in ALT cells these variable repeat patterns are scattered along the telomere, certainly induced by telomere recombination in these cells (Varley et al. 2002). Telomere-specific proteins, forming the shelterin complex, bind with greater affinity to the canonical TTAGGG repeat motif. The presence of non-canonical motifs along ALT telomeres, presumably, disrupts the binding of the shelterin complex or the recruitment of other proteins to these telomeres. Among others, a family of nuclear receptors is recruited specifically to ALT telomeres, due to its affinity for the non-canonical TCAGGG motif (Dejardin and Kingston 2009; Conomos et al. 2012). It has been proposed that these variable repeat motifs and the binding of novel factors such as orphan receptors could change the chromatin status of telomeres, induce the recruitment of additional factors involved in homologous recombination or promote telomere clustering favoring recombination events.

9.1.5. *Sister chromatid exchange at telomeres*

Telomere sister chromatid exchange (T-SCE) tends to occur more frequently in cells using the ALT pathway than in primary or cancer cells positive for telomerase (Londoño-Vallejo et al. 2004). Sister chromatid exchange is a process that occurs during replication, whereby two sister chromatids break and join together, physically exchanging the parental strands of the duplicated chromosome (Wilson 3rd and Thompson 2007). T-SCE results from the invasion of the sister chromatid strand of a telomere and the formation of a secondary homologous recombination structure that can be resolved by a crossing-over event.

9.1.6. *The accumulation of TERRA*

Telomeres are transcribed by RNA polymerase II into long non-coding RNAs called TERRAs. TERRAs are transcribed from several telomeres from the upstream promoter region called subtelomere and this transcription is regulated during the cell cycle (Azzalin et al. 2007; Feretzaki et al. 2019). ALT cells show a significant accumulation of TERRA RNAs when compared to telomerase positive cells. This accumulation of TERRA could be induced by an increase in transcriptional activity at telomeres accompanied by the loss of cell cycle control (Arora and Azzalin 2015). In yeast, TERRA transcription can generate structures called R-loops with short telomeres. These structures, induced by the formation of DNA:RNA hybrids, have the potential to induce replicative stress and lead to the recruitment of recombination factors (Graf et al. 2017). In mammalian cells, deletion of RNAseH1, an endonuclease that degrades RNA from DNA:RNA hybrids, generates stabilization of R-loops at telomeres. This phenomenon correlates with a stabilization at telomeres of RPA, a marker of increased replicative stress in ALT cells. In contrast, overexpression of RNAseH1 in these cells compromises telomere maintenance by the ALT pathway (Arora et al. 2014). These pieces of evidence show that TERRA by forming secondary structures can stimulate homologous recombination at telomeres.

9.1.7. *Loss of function of the ATRX signaling pathway*

ALT tumors are very often correlated with the presence of mutations in ATRX or its partners involved in histone variant H3.3 deposition (Heaphy et al. 2011; Lovejoy et al. 2012; Schwartzentruber et al. 2012). ATRX is a chromatin remodeling factor that is found on regions of heterochromatin and in PML nuclear bodies. For example, ATRX is found at pericentromeres and at transposable elements which are repeat regions organized in a compact form of chromatin, the heterochromatin (McDowell et al. 1999; Sadic et al. 2015). Different modes of recruitment of ATRX have been described in the literature. Some evidence suggests that ATRX localizes to heterochromatin regions such as pericentromeres or telomeres through recognition of the tri-methylation of lysine 9 of histone H3 (H3K9me3) (Eustermann et al. 2011; Iwase et al. 2011; Gauchier et al. 2019). Indeed, ATRX has an ADD domain that specifically recognizes H3K9me3, allowing its correct localization on the genome for the deposition of the variant histone H3.3 in cooperation with the histone chaperone DAXX (Dhayalan et al. 2011; Eustermann et al. 2011; Iwase et al. 2011). However, other studies propose that

ATRX may recognize DNA structures such as G-quadruplexes or DNA:RNA structures found at telomeres (Law et al. 2010; Nguyen et al. 2017). As for the molecular function of ATRX, its mechanism of action is unclear but ATRX is crucial to telomeres in regulating the ALT pathway. Indeed, ATRX is the only factor known to be a suppressor of the ALT pathway. Loss of function of ATRX promotes activation of the ALT pathway in human fibroblasts, whereas transient expression of ATRX in ALT cells inhibits this pathway (Clynes et al. 2015; Napier et al. 2015). However, ATRX dysfunction is not always sufficient to activate the ALT pathway, suggesting that other determinants are required to initiate ALT (Bower et al. 2012; Lovejoy et al. 2012; Napier et al. 2015; Brosnan-Cashman et al. 2018).

9.2. ALT mechanisms

Strong evidence suggests a model whereby telomere elongation in ALT cells involves DNA substrate and recombination mechanisms. Telomerase-independent telomere lengthening mechanisms were first described in yeast through the isolation of surviving colonies from loss of telomerase function (Lundblad and Blackburn 1993; Teng and Zakian 1999). Genes encoding recombination proteins such as Rad51, Rad52 and Sgs1 are essential for proliferation of telomerase-deficient yeast, again indicating that the ALT pathway relies on homologous recombination to lengthen telomeres. Other recombination proteins have been implicated in the maintenance of the ALT pathway such as the MRN complex, SMC5-SMC6 (Potts and Yu 2007; Zhong et al. 2007). The loss of these factors leads to the shortening of ALT telomeres.

The addition of telomeric sequences to short telomeres in telomerase-negative cells was the first indication that recombination can also occur in human cells when telomerase is not active. Later, it has been observed that the insertion of a transgene on a telomere can be copied to the telomere of another chromosome in ALT cells (Dunham et al. 2000). This observation represents an indisputable argument that ALT telomeres recombine. However, the mechanism(s) leading to telomere elongation remains unclear (Figure 9.2). The study of telomere transgene amplification suggests that recombination occurs between pre-existing telomere sequences on a homologous or non-homologous chromosome (Dunham et al. 2000; Muntoni et al. 2009). In addition, during DNA replication, non-equal exchanges of sister chromatids at telomeres could participate in the formation of very long telomeres. To date, this phenomenon does not appear to be controlled in ALT cells. Despite this, random synthesis of long telomeres associated with preferential segregation, that is, chromosomes with long telomeres are found together in the same daughter cell during cell division, could be a mechanism for chromosome stabilization in ALT tumors. From this mechanism, only cells that inherited long telomeres would have a proliferative advantage over those that inherited short telomeres (Blagoev and Goodwin 2008). Finally, recombination events could involve extra-chromosomal telomeric DNA sequences. In human cells, it is not known whether telomeric DNA circles are the cause of ALT mechanisms or whether they might actively participate in recombination. However, it has been shown that this circular DNA is used as a substrate for recombination and telomere lengthening in the yeast *Kluyveromyces lactis* (Natarajan and McEachern 2002).

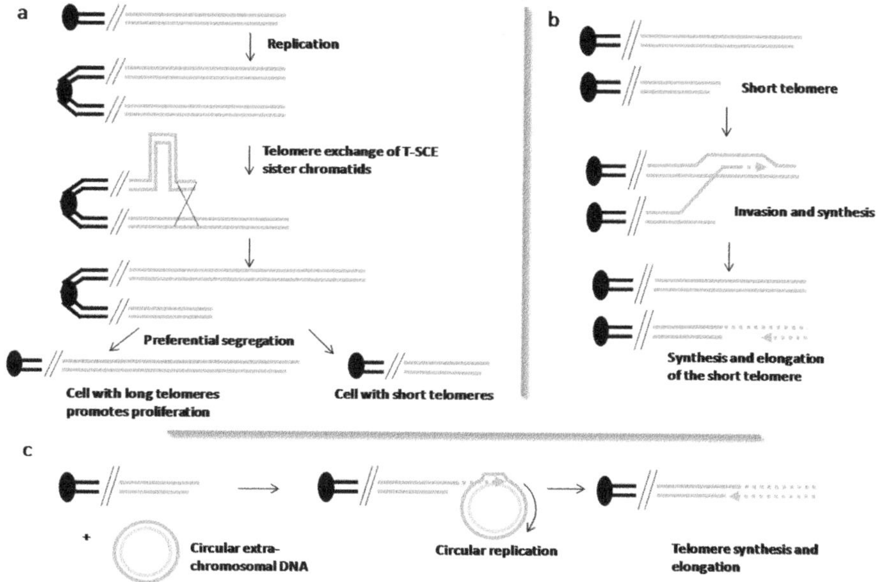

Figure 9.2. *Mechanisms of recombination and telomere lengthening that could be involved in the ALT mechanism in human cancer cells. (a) Telomere sister chromatid exchange (T-SCEs) can generate long telomeres that could provide a proliferative advantage to the cell when associated with unequal segregation. On the contrary, the daughter cell with short telomeres will have a more limited proliferative capacity. (b) Recombination combined with telomeric DNA synthesis using the sequence of an adjacent telomere as a substrate could be a mechanism participating in an ALT pathway. (c) Alternatively to (b), recombination allowing de novo telomere synthesis could use extrachromosomal DNA of telomeric nature such as circular DNA*

9.3. Replicative stress and DNA repair pathway at ALT telomeres

Telomeres are known to be difficult to replicate and therefore tend to accumulate DNA damage. This difficulty in replicating could be due to their repetitive nature, the formation of telomeric heterochromatin and/or their tendency to form secondary structures of DNA or RNA TERRA. Replicative stress and the accumulation of DNA damage at telomeres stimulate homologous recombination at telomeres. Different signaling may resolve replication fork blockage involving, for example, SMARCAL1 or FANCM helicases (Pan et al. 2017; Lu et al. 2019; Silva et al. 2019). The deletion of these factors leads to replicative fork blockage inducing double-strand breaks on telomeres and the recruitment of factors involved in resection events allowing repair by homologous recombination of ALT telomeres. It has been proposed that the BLM helicase would be involved in 5′-3′ resection by recruiting the EXO1 and DNA2 exonuclease, whereas the WRN helicase, which would have the same role, would rather recruit DNA2 (Mendez-Bermudez et al. 2012; Min et al. 2019). After resection, the single-stranded part of the DNA is covered by RPA, then RAD51 or RAD52 which will subsequently make it possible to initiate homologous strand search and invasion, and thus activate inter- or intra-telomeric homologous recombination and repair. The replisome consisting of Polδ-PCNA-RFC1 then enables synthesis of the invasive strand of the telomere. Within the BTR complex (comprising TOP3α, RMI1 and RMI2), BLM enables the dissolution of recombination intermediates during homologous recombination facilitating telomere elongation by Polδ (Bizard and Hickson 2014). This mechanism can be inhibited by the SMX complex containing SLX4, which promotes the resolution of recombination intermediates (Sobinoff et al. 2017).

9.4. The structure of chromatin at ALT telomeres

When an active gene is placed near a telomere, its expression tends to be repressed. More precisely, the transcription of this gene juxtaposed to a telomere is stochastic and fluctuates between a repressed and active state. This phenomenon called telomere position effect (TPE) reflects the presence of a dynamic structure in the chromatin that has the ability to propagate along the telomere and silence the initiation of transcription in its proximity (Tham and Zakian 2002). How chromatin accessibility and remodeling might play a role in the recruitment of recombination factors or DNA repair pathways to ALT telomeres is an unresolved question. It has been proposed that accessible or euchromatic chromatin may facilitate genome transactions such as transcription or recombination, whereas a closed or heterochromatic form of chromatin would have an inhibitory effect. However, despite the great efforts that have been made to characterize and differentiate between ALT and non-ALT telomere-associated chromatin, it is still difficult to understand how the structure of telomeric chromatin could initiate or maintain the ALT pathway. Indeed, there are many inconsistencies between studies. Human telomeres have few nucleosomes, especially at short telomeres (Makarov et al. 1993; Tommerup et al. 1994; Nikitina and Woodcock 2004; Dejardin and Kingston 2009).

Enzymatic digestions with micrococcal nuclease (MNase) show that the nucleosome density is more reduced at the telomeres of ALT cells than cells using telomerase to lengthen their telomeres (Episkopou et al. 2014). It has therefore been proposed that chromatin modifying factors may be involved in the activation of the ALT pathway. In agreement, ALT can be initiated by disrupting chromatin assembly mechanisms. For example, dysfunction of ASF1, a chaperone of canonical histones H3 and H4, is sufficient to activate the ALT pathway in telomerase cancer cells (O'Sullivan et al. 2014). However, no mutations in the ASF1 gene have been associated with ALT tumors, suggesting that the causes of the development of these tumors are other. We have previously written that most ALT tumors correlate with loss-of-function mutations in *ATRX*, which is a marker of heterochromatin. This simple observation suggests that defects in heterochromatin may initiate alternative mechanisms of telomere maintenance. Telomeres can be heterochromatic and marked by tri-methylation of histone H3 on lysine 9, more commonly known as H3K9me3. At telomeres, SETDB1 is the enzyme essential for the deposition of the H3K9me3 mark, which is itself recognized by the heterochromatin protein HP1 and ATRX (Gauchier et al. 2019). It has been proposed that ALT cells may be characterized by the loss of H3K9me3 accompanied by a global decompaction of chromatin at telomeres suggesting that heterochromatin protects telomeres from recombination (Episkopou et al. 2014). However, contradictory studies show that heterochromatin stimulates recombination and is associated with the development of ALT phenotypes. Indeed, it was shown that the NuRD complex was recruited by ZNF827 to ALT telomeres. NuRD allows chromatin remodeling and histone deacetylation thus inducing a heterochromatin environment (Conomos et al. 2014). Furthermore, it was observed that ectopic heterochromatin formation by SETDB1 at telomeres is sufficient to result in the appearance of ALT markers (Gauchier et al. 2019). The ALT pathway appears to be initiated by two consecutive events, the formation of heterochromatin at telomeres associated with the ATRX loss of function. In accordance, meta-analysis of chromatin markings in telomerase-positive human cancer cells revealed that their telomeres were not heterochromatic but had histone active marks that are usually found on regions that stimulate gene transcription (Cubiles et al. 2018; Vaquero-Sedas and Vega-Palas 2019). However, this same analysis noted that at least two ALT pathway cell lines, U2OS (osteosarcoma) and a neuroblastoma-derived line, show significant enrichment of the H3K9me3 marker, again suggesting that ALT telomeres are characterized by their heterochromatic status. Similar observations were reported in a clinical study seeking to characterize ALT markers in neuroblastoma (Hartlieb et al. 2021). It is also surprising to note that the presence of repeated sequences of different natures as well as the process of heterochromatinization on this repeated DNA are criteria required for the maintenance of telomeres in the absence of telomerase in several organisms such as yeast or Drosophila.

Recent advances have identified new factors involved in chromatin formation at telomeres; further studies are needed to better understand their potential roles in the initiation and maintenance of this pathway in human cells. Systematic characterization of ALT tumors from large cohorts supports a recurrent heterogeneity of ALT pathway-associated markers between patients. Telomere chromatin and its dysfunction is an exciting field of study to explore to better understand ALT telomere biology.

9.5. ALT and cancer

ALT is detected in 10–20% of human tumors and is over-represented in tumors of mesenchymal origin such as bone and brain (Henson and Reddel 2010). Very few ALT tumors have an epithelial origin. This finding could be explained by a better repression of the telomerase gene in mesenchymal cells, thus inducing an over-representation of the ALT pathway in these tissues. In addition, the presence or absence of unknown signaling pathways in mesenchyme cells could provide conditions for activation of the ALT pathway.

The prognosis associated with activation of the ALT pathway is complex and varies between tumor types in humans. Indeed, the ALT pathway is associated with a poor prognosis when activated in neuroblastoma and some soft tissue liposarcomas (Onitake et al. 2009; Costa et al. 2006). In contrast, activation of the ALT pathway in glioblastoma multiforme appears to be associated with improved survival (Hakin-Smith et al. 2003).

It is generally accepted that immortalized cell lines exclusively use either telomerase or recombination as a telomere maintenance mechanism for telomere elongation. Indeed, telomerase activity appears to be able to inhibit the ALT pathway (Perrem et al. 1999). However, several examples have been described where telomerase activity coexists with the ALT pathway, and this is particularly true in tumors (Bryan et al. 1997; Perrem et al. 2001; Pezzolo et al. 2015). However, it is not clear whether these two mechanisms of telomere lengthening occur in the same cells or form heterogeneous populations within the tumor.

9.6. The ALT pathway: a non-pathological function?

While the ALT pathway is most commonly associated with immortalization of human cancer cells, there is evidence to suggest that homologous recombination may also have a role in rapid telomere lengthening during early mammalian development (Liu et al. 2007). Indeed, oocytes and early embryos show a very low activity of telomerase which increases only later at the morula and blastocyst stages (Wright et al. 1996; Schaetzlein et al. 2004; Liu et al. 2007). However, rapid telomere elongation occurs at the two-cell zygotic stage, suggesting the involvement of an alternative telomerase-independent mechanism. This hypothesis is likely, as telomerase-deficient mice still show a 10 kb extension of their telomeres at the 1- and 2-cell embryonic stages (Liu et al. 2007). These early embryonic stages are characterized by the accumulation of T-SCEs, suggesting that recombination may be the mechanism leading to telomere elongation. While recombination produces telomeres of very heterogeneous size in tumors, the early embryo has telomeres of homogeneous size. This homogeneity in size could be explained by the coexistence of recombination and telomerase activity at the end of the pre-implantation stage (Tardat and Dejardin 2018). The molecular mechanisms involved in massive lengthening during early development require further investigation. In addition, amplification of a telomeric transgene between telomeres in murine somatic cells suggests that an ALT mechanism may also occur in adult tissues (Neumann et al. 2013). These physiological aspects of the ALT pathway will have to be studied in more detail in order to better understand the mechanisms of initiation of the ALT pathway in a pathological context.

9.7. Alternative pathways not involving recombination: evolutionary aspects

In contrast to the ALT pathway, telomerase in eukaryotes generates telomeres that are both homogeneous and stable, especially in terms of sequence. For these reasons, telomerase activity can be considered a major advantage in solving the problem of chromosome end replication and certainly explains the conservation of this enzyme during evolution. The origin of its enzymatic activity remains enigmatic, however its mode of operation is similar to retrotransposons, which may lead to the idea that telomerase derives from the domestication of a retro-element. Despite this obvious evolutionary success, many alternative mechanisms to telomerase exist to maintain telomere size in different organisms. Some of them do not have telomerase, as in some diptera. To compensate for the absence of telomerase, different alternative pathways have replaced the function of this enzyme, involving recombination events of course, but also chromosomal rearrangements and even retrotransposition. These examples are detailed below.

Retrotransposons are mobile genetic units contained in the eukaryotic genome itself (the host) and are the ancestral traces of invasion by retroviruses. Retrotransposons, often considered as irrelevant DNA, are capable of self-replicating through an RNA intermediate, by a step called reverse transcription, to integrate in several copies in the host genome. These so-called retrotransposition events remain rare, because they can induce deleterious mutations, particularly for the fertility and reproduction of the species. However, these mobile elements have been and still remain a driver of genetic innovation during evolution. It has been proposed that telomerase itself, the enzyme that uses the reverse transcription mechanism to lengthen telomeres, could be the product of evolution from a retrotransposon (Eickbush 1997). Equally intriguing, some organisms compensate for the loss of telomerase activity by maintaining the size of their telomeres through the activity of retrotransposons. This is true for some dipterans, of which Drosophila melanogaster, deficient in telomerase, is one of the best studied examples. In this particular case, it is repeated sequences of retrotransposons called HTT from their names HeTA, TAHRE and TART, which allow the maintenance of the end of the chromosomes by accumulating by retrotransposition (Casacuberta and Pardue 2005; Casacuberta 2017) (Figure 9.3). The structure of chromatin, especially the formation of heterochromatin on these telomere-reporting retrotransposons, is important for the recruitment of factors involved in chromosome end protection.

The domestication of retrotransposons for telomere maintenance in Drosophila is not an exception. Indeed, other organisms have retrotransposon insertions in their telomere sequences. This is the case of other insects such as *Bombyx mori* and *Tribolium castaneum*, which show a poor telomerase activity, but also of the protozoan parasite *Girardia lambia*, the green alga *Clorella vulgaris*, the fungus *Magnaporthe oryzae* and the rotifer *bdelloid* (Fujiwara 2015). Other alternative mechanisms have been observed in yeast to compensate for telomere erosion. In *Schizosaccharomyces pombe*, some colonies surviving telomerase loss and deficient in the recombination pathway are characterized by the formation of large palindromes that compensate for telomere erosion (Figure 9.3). This mechanism, called PAL, does not rely on homologous recombination but is sufficient to protect essential genes and lead to cell immortalization (Maringele and Lydall 2004). Also, other types of "survivors" called HAATI are able to maintain their linear chromosomal DNA by replacing their telomeres with blocks of heterochromatin from ribosomal regions to protect their linear ends. Despite the loss of telomeric sequences, this recombinant heterochromatin allows the recruitment of POT1, a protein of the shelterin complex that allows the preservation of chromosome end identity (Jain et al. 2010).

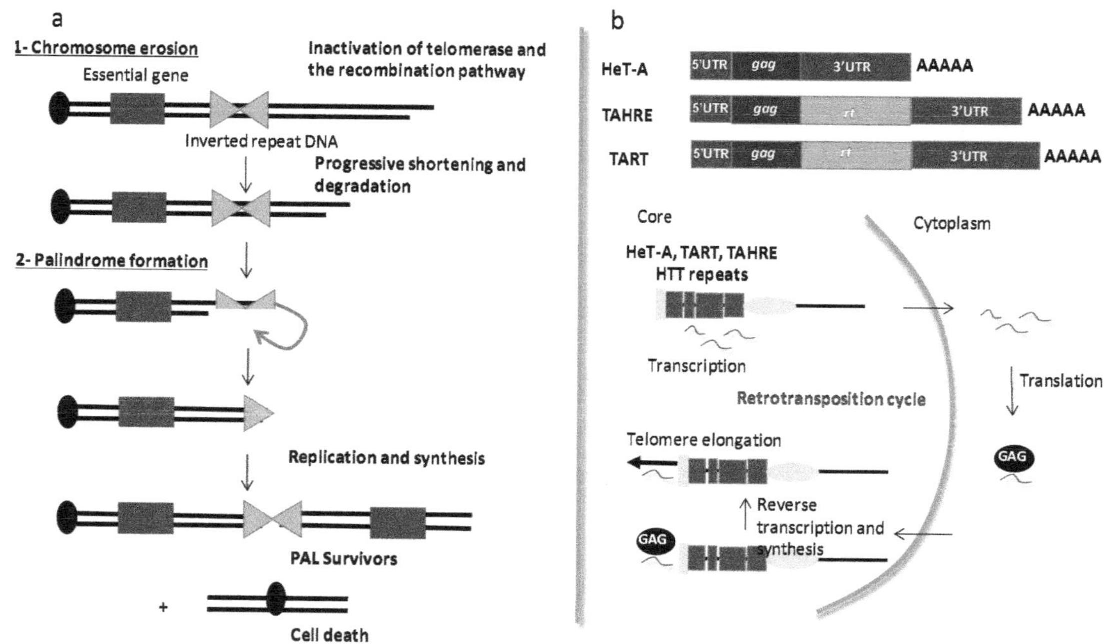

Figure 9.3. *Alternative mechanisms of telomere maintenance independent of homologous recombination in yeast and Drosophila. (a) Proposed mechanism to explain the formation of large palindromes at the end of chromosomes in yeast to maintain essential genes and provide a proliferative advantage to the cell. (b) Representation of the molecular structure composed of different genes allowing the retrotransposition of HeT-A, TAHRE and TART elements found at telomeres in Drosophila. A retrotransposition model that allows telomere elongation has been proposed*

Evolution has multiplied solutions to the problem of chromosome end replication and telomere maintenance. Telomere elongation necessary for cell proliferation often involves telomerase but also alternatives such as homologous recombination, retrotransposition, rearrangement of repeated sequences at the end of chromosomes. Some of these alternative mechanisms (retrotransposition, PAL) have been described in the organisms mentioned above, but what about in humans? For example, it has been observed that hamster cells deficient in the NHEJ repair pathway have dysfunctional telomeres that become target sites for retrotransposition of LINE1 elements (Morrish et al. 2007). On the other hand, palindrome formation is over-represented in the genome of some human cancer cells (Tanaka et al. 2005). A better characterization of telomeric sequences with the growing progress in long molecule sequencing will probably be a technological key to reveal the presence (or absence) of such alternative mechanisms. Finally, although the ALT pathway in humans is rather associated with tumorigenesis processes, the question arises whether alternative mechanisms could occur and be regulated in a physiological context as already discussed during early development. In this case, the ALT phenotypes observed in cancers could be the result of the deregulation of these mechanisms. Further study of alternative pathways of telomere elongation could lead to new targeted therapies for human diseases and will undoubtedly provide many answers to more fundamental questions in biology.

9.8. Appendix: experimental methods that allow the identification of the ALT phenotype in human cancer cells

Telomeric restriction fragment analysis: This approach is most often used to visualize and measure telomere size. Genomic DNA extracted from cells is digested with restriction enzymes that cut the entire genome into small fragments except for telomeres. The digested DNA is then placed in an agarose gel and the DNA is separated by electrophoresis. The migrated DNA is then transferred to a nitrocellulose membrane and hybridized with a radioactive probe complementary to the telomeric sequence.

Amplification of extra-chromosomal rings of telomeric DNA: The DNA circles are amplified using a DNA polymerase called phi29. The amplified DNA circles are then deposited on a loaded nylon membrane under non-denaturing conditions. The telomeric DNA circles are then detected by hybridization of a telomeric radioactive probe.

CO-FISH: This fluorescence in situ hybridization (FISH) method is used to detect T-SCEs. The approach consists of digesting the newly synthesized DNA strand with an exonuclease after the incorporation of BrdU during a replication cycle and the introduction of cuts in this fragment by UV radiation. In this process, the template strand remains intact and becomes single stranded. FISH is performed under non-denaturing conditions to detect these matrix strands, with hybridization of two telomeric fluorescent probes, one color for each strand. If a chromosomal exchange at the telomeres occurs after replication, the signals of the two probes will co-localize.

IF and FISH: Immunofluorescence (IF) combined with FISH of telomeres allows visualization of whether protein factors localize to telomeres in the cell.

Northern-blot: This method allows analysis of the level and size of TERRA transcripts. RNA is extracted from cells, and then RNAs are migrated into a denaturing agarose gel containing formaldehyde, transferred onto a loaded nylon membrane before hybridization of a radioactive telomeric probe.

9.9. References

Allshire, R.C., Dempster, M., Hastie, N.D. (1989). Human telomeres contain at least three types of G-rich repeat distributed non-randomly. *Nucleic Acids Res.*, 17, 4611–4627.

Arora, R. and Azzalin, C.M. (2015). Telomere elongation chooses TERRA ALTernatives. *RNA Biol.*, 12, 938–941.

Arora, R., Lee, Y., Wischnewski, H., Brun, C.M., Schwarz, T., Azzalin, C.M. (2014). RNaseH1 regulates TERRA-telomeric DNA hybrids and telomere maintenance in ALT tumour cells. *Nat Commun*, 5, 5220.

Azzalin, C.M., Reichenbach, P., Khoriauli, L., Giulotto, E., Lingner, J. (2007). Telomeric repeat containing RNA and RNA surveillance factors at mammalian chromosome ends. *Science*, 318, 798–801.

Bizard, A.H. and Hickson, I.D. (2014). The dissolution of double Holliday junctions. *Cold Spring Harb. Perspect. Biol.*, 6, a016477.

Blagoev, K.B. and Goodwin, E.H. (2008). Telomere exchange and asymmetric segregation of chromosomes can account for the unlimited proliferative potential of ALT cell populations. *DNA Repair (Amst)*, 7, 199–204.

Bower, K., Napier, C.E., Cole, S.L., Dagg, R.A., Lau, L.M., Duncan, E.L., Moy, E.L., Reddel, R.R. (2012). Loss of wild-type ATRX expression in somatic cell hybrids segregates with activation of alternative lengthening of telomeres. *PLoS One*, 7, e50062.

Brosnan-Cashman, J.A., Yuan, M., Graham, M.K., Rizzo, A.J., Myers, K.M., Davis, C., Zhang, R., Esopi, D.M., Raabe, E.H., Eberhart, C.G. et al. (2018). ATRX loss induces multiple hallmarks of the alternative lengthening of telomeres (ALT) phenotype in human glioma cell lines in a cell line-specific manner. *PLoS One*, 13, e0204159.

Bryan, T.M., Englezou, A., Gupta, J., Bacchetti, S., Reddel, R.R. (1995). Telomere elongation in immortal human cells without detectable telomerase activity. *The EMBO Journal*, 14, 4240–4248.

Bryan, T.M., Englezou, A., Dalla-Pozza, L., Dunham, M.A., Reddel, R.R. (1997). Evidence for an alternative mechanism for maintaining telomere length in human tumors and tumor-derived cell lines. *Nat. Med.*, 3, 1271–1274.

Casacuberta, E. (2017). Drosophila: Retrotransposons making up telomeres. *Viruses*, 19, 9(7), 192.

Casacuberta, E. and Pardue, M.L. (2005). HeT-A and TART, two Drosophila retrotransposons with a bona fide role in chromosome structure for more than 60 million years. *Cytogenet. Genome Res.*, 110, 152–159.

Cesare, A.J. and Reddel, R.R. (2010). Alternative lengthening of telomeres: Models, mechanisms and implications. *Nat. Rev. Genet.*, 11, 319–330.

Cho, N.W., Dilley, R.L., Lampson, M.A., Greenberg, R.A. (2014). Interchromosomal homology searches drive directional ALT telomere movement and synapsis. *Cell*, 159, 108–121.

Chung, I., Osterwald, S., Deeg, K.I., Rippe, K. (2012). PML body meets telomere: The beginning of an ALTernate ending? *Nucleus*, 3, 263–275.

Clynes, D., Jelinska, C., Xella, B., Ayyub, H., Scott, C., Mitson, M., Taylor, S., Higgs, D.R., Gibbons, R.J. (2015). Suppression of the alternative lengthening of telomere pathway by the chromatin remodelling factor ATRX. *Nat. Commun.*, 6, 7538.

Conomos, D., Stutz, M.D., Hills, M., Neumann, A.A., Bryan, T.M., Reddel, R.R., Pickett, H.A. (2012). Variant repeats are interspersed throughout the telomeres and recruit nuclear receptors in ALT cells. *J. Cell. Biol.*, 199, 893–906.

Conomos, D., Reddel, R.R., Pickett, H.A. (2014). NuRD-ZNF827 recruitment to telomeres creates a molecular scaffold for homologous recombination. *Nat. Struct. Mol. Biol.*, 21, 760–770.

Costa, A., Daidone, M.G., Daprai, L., Villa, R., Cantu, S., Pilotti, S., Mariani, L., Gronchi, A., Henson, J.D., Reddel, R.R., Zaffaroni, N. (2006). Telomere maintenance mechanisms in liposarcomas: Association with histologic subtypes and disease progression. *Cancer Res.*, 66, 8918–8924.

Cubiles, M.D., Barroso, S., Vaquero-Sedas, M.I., Enguix, A., Aguilera, A., Vega-Palas, M.A. (2018). Epigenetic features of human telomeres. *Nucleic Acids Res.*, 46, 2347–2355.

Dancis, B.M. and Holmquist, G.P. (1979). Telomere replication and fusion in eukaryotes. *Journal of Theoretical Biology*, 78, 211–224.

Dejardin, J. and Kingston, R.E. (2009). Purification of proteins associated with specific genomic loci. *Cell*, 136, 175–186.

Dhayalan, A., Tamas, R., Bock, I., Tattermusch, A., Dimitrova, E., Kudithipudi, S., Ragozin, S., Jeltsch, A. (2011). The ATRX-ADD domain binds to H3 tail peptides and reads the combined methylation state of K4 and K9. *Hum. Mol. Genet.*, 20, 2195–2203.

Dunham, M.A., Neumann, A.A., Fasching, C.L., Reddel, R.R. (2000). Telomere maintenance by recombination in human cells. *Nat. Genet.*, 26, 447–450.

Eickbush, T.H. (1997). Telomerase and retrotransposons: Which came first? *Science*, 277, 911–912.

Episkopou, H., Draskovic, I., van Beneden, A., Tilman, G., Mattiussi, M., Gobin, M., Arnoult, N., Londono-Vallejo, A., Decottignies, A. (2014). Alternative lengthening of telomeres is characterized by reduced compaction of telomeric chromatin. *Nucleic Acids Res.*, 42, 4391–4405.

Eustermann, S., Yang, J.C., Law, M.J., Amos, R., Chapman, L.M., Jelinska, C., Garrick, D., Clynes, D., Gibbons, R.J., Rhodes, D. et al. (2011). Combinatorial readout of histone H3 modifications specifies localization of ATRX to heterochromatin. *Nat. Struct. Mol. Biol.*, 18, 777–782.

Feretzaki, M., Renck Nunes, P., Lingner, J. (2019). Expression and differential regulation of human TERRA at several chromosome ends. *RNA*, 25(11), 1470–1480.

Fujiwara, H. (2015). Site-specific non-LTR retrotransposons. *Microbiol Spectr.*, 3, MDNA3-0001-2014.

Gauchier, M., Kan, S., Barral, A., Sauzet, S., Agirre, E., Bonnell, E., Saksouk, N., Barth, T. K., Ide, S., Urbach, S. et al. (2019). SETDB1-dependent heterochromatin stimulates alternative lengthening of telomeres. *Science Advances*, 5, eaav3673.

Graf, M., Bonetti, D., Lockhart, A., Serhal, K., Kellner, V., Maicher, A., Jolivet, P., Teixeira, M.T., Luke, B. (2017). Telomere length determines TERRA and R-loop regulation through the cell cycle. *Cell*, 170, 72–85.e14.

Hakin-Smith, V., Jellinek, D.A., Levy, D., Carroll, T., Teo, M., Timperley, W.R., Mckay, M.J., Reddel, R.R., Royds, J.A. (2003). Alternative lengthening of telomeres and survival in patients with glioblastoma multiforme. *Lancet*, 361, 836–838.

Hartlieb, S.A., Sieverling, L., Nadler-Holly, M., Ziehm, M., Toprak, U.H., Herrmann, C., Ishaque, N., Okonechnikov, K., Gartlgruber, M., Park, Y.G. et al. (2021). Alternative lengthening of telomeres in childhood neuroblastoma from genome to proteome. *Nat. Commun.*, 12, 1269.

Heaphy, C.M., De Wilde, R.F., Jiao, Y., Klein, A.P., Edil, B.H., Shi, C., Bettegowda, C., Rodriguez, F.J., Eberhart, C.G., Hebbar, S. et al. (2011). Altered telomeres in tumors with ATRX and DAXX mutations. *Science*, 333, 425.

Henson, J.D. and Reddel, R.R. (2010). Assaying and investigating alternative lengthening of telomeres activity in human cells and cancers. *FEBS Lett.*, 584, 3800–3811.

Iwase, S., Xiang, B., Ghosh, S., Ren, T., Lewis, P.W., Cochrane, J.C., Allis, C.D., Picketts, D.J., Patel, D.J., Li, H. et al. (2011). ATRX ADD domain links an atypical histone methylation recognition mechanism to human mental-retardation syndrome. *Nat. Struct. Mol. Biol.*, 18, 769–776.

Jain, D., Hebden, A.K., Nakamura, T.M., Miller, K.M., Cooper, J.P. (2010). HAATI survivors replace canonical telomeres with blocks of generic heterochromatin. *Nature*, 467, 223–227.

Jiang, W.Q., Zhong, Z.H., Nguyen, A., Henson, J.D., Toouli, C.D., Braithwaite, A.W., Reddel, R.R. (2009). Induction of alternative lengthening of telomeres-associated PML bodies by p53/p21 requires HP1 proteins. *J. Cell Biol.*, 185, 797–810.

Law, M.J., Lower, K.M., Voon, H.P., Hughes, J.R., Garrick, D., Viprakasit, V., Mitson, M., de Gobbi, M., Marra, M., Morris, A. et al. (2010). ATR-X syndrome protein targets tandem repeats and influences allele-specific expression in a size-dependent manner. *Cell*, 143, 367–378.

Liu, L., Bailey, S.M., Okuka, M., Munoz, P., Li, C., Zhou, L., Wu, C., Czerwiec, E., Sandler, L., Seyfang, A. et al. (2007). Telomere lengthening early in development. *Nat. Cell Biol.*, 9, 1436–1441.

Londoño-Vallejo, J.A., Der-Sarkissian, H., Cazes, L., Bacchetti, S., Reddel, R.R. (2004). Alternative lengthening of telomeres is characterized by high rates of telomeric exchange. *Cancer Research*, 64, 2324–2327.

Lovejoy, C.A., Li, W., Reisenweber, S., Thongthip, S., Bruno, J., de Lange, T., De, S., Petrini, J.H., Sung, P.A., Jasin, M. et al. (2012). Loss of ATRX, genome instability, and an altered DNA damage response are hallmarks of the alternative lengthening of telomeres pathway. *PLoS Genet*, 8, e1002772.

Lu, R., O'Rourke, J.J., Sobinoff, A.P., Allen, J.A.M., Nelson, C.B., Tomlinson, C.G., Lee, M., Reddel, R.R., Deans, A.J., Pickett, H.A. (2019). The FANCM-BLM-TOP3A-RMI complex suppresses alternative lengthening of telomeres (ALT). *Nat. Commun.*, 10, 2252.

Lundblad, V. and Blackburn, E.H. (1993). An alternative pathway for yeast telomere maintenance rescues est1-senescence. *Cell*, 73, 347–360.

Makarov, V.L., Lejnine, S., Bedoyan, J., Langmore, J.P. (1993). Nucleosomal organization of telomere-specific chromatin in rat. *Cell*, 73, 775–787.

Maringele, L. and Lydall, D. (2004). Telomerase- and recombination-independent immortalization of budding yeast. *Genes Dev.*, 18, 2663–2675.

McDowell, T.L., Gibbons, R.J., Sutherland, H., O'Rourke, D.M., Bickmore, W.A., Pombo, A., Turley, H., Gatter, K., Picketts, D.J., Buckle, V.J. et al. (1999). Localization of a putative transcriptional regulator (ATRX) at pericentromeric heterochromatin and the short arms of acrocentric chromosomes. *Proc. Natl. Acad. Sci. USA*, 96, 13983–13988.

Mendez-Bermudez, A., Hidalgo-Bravo, A., Cotton, V.E., Gravani, A., Jeyapalan, J.N., Royle, N.J. (2012). The roles of WRN and BLM RecQ helicases in the alternative lengthening of telomeres. *Nucleic Acids Res.*, 40, 10809–10820.

Min, J., Wright, W.E., Shay, J.W. (2019). Clustered telomeres in phase-separated nuclear condensates engage mitotic DNA synthesis through BLM and RAD52. *Genes Dev.*, 33, 814–827.

Morrish, T.A., Garcia-Perez, J.L., Stamato, T.D., Taccioli, G.E., Sekiguchi, J., Moran, J.V. (2007). Endonuclease-independent LINE-1 retrotransposition at mammalian telomeres. *Nature*, 446, 208–212.

Muntoni, A., Neumann, A.A., Hills, M., Reddel, R.R. (2009). Telomere elongation involves intra-molecular DNA replication in cells utilizing alternative lengthening of telomeres. *Hum. Mol. Genet.*, 18, 1017–1027.

Murnane, J.P., Sabatier, L., Marder, B.A., Morgan, W.F. (1994). Telomere dynamics in an immortal human cell line. *EMBO J.*, 13, 4953–4962.

Napier, C.E., Huschtscha, L.I., Harvey, A., Bower, K., Noble, J.R., Hendrickson, E.A., Reddel, R.R. (2015). ATRX represses alternative lengthening of telomeres. *Oncotarget*, 6, 16543–16558.

Natarajan, S. and Mceachern, M.J. (2002). Recombinational telomere elongation promoted by DNA circles. *Mol. Cell. Biol.*, 22, 4512–4521.

Neumann, A.A., Watson, C.M., Noble, J.R., Pickett, H.A., Tam, P.P., Reddel, R.R. (2013). Alternative lengthening of telomeres in normal mammalian somatic cells. *Genes Dev.*, 27, 18–23.

Nguyen, D.T., Voon, H.P.J., Xella, B., Scott, C., Clynes, D., Babbs, C., Ayyub, H., Kerry, J., Sharpe, J.A., Sloane-Stanley, J.A. et al. (2017). The chromatin remodelling factor ATRX suppresses R-loops in transcribed telomeric repeats. *EMBO Rep.*, 18, 914–928.

Nikitina, T. and Woodcock, C.L. (2004). Closed chromatin loops at the ends of chromosomes. *J. Cell. Biol.*, 166, 161–165.

O'Sullivan, R.J., Arnoult, N., Lackner, D.H., Oganesian, L., Haggblom, C., Corpet, A., Almouzni, G., Karlseder, J. (2014). Rapid induction of alternative lengthening of telomeres by depletion of the histone chaperone ASF1. *Nat. Struct. Mol. Biol.*, 21, 167–174.

Olovnikov, A.M. (1973). A theory of marginotomy: The incomplete copying of template margin in enzymic synthesis of polynucleotides and biological significance of the phenomenon. *Journal of Theoretical Biology*, 41, 181–190.

Onitake, Y., Hiyama, E., Kamei, N., Yamaoka, H., Sueda, T., Hiyama, K. (2009). Telomere biology in neuroblastoma: Telomere binding proteins and alternative strengthening of telomeres. *J. Pediatr. Surg.*, 44, 2258–2266.

Pan, X., Drosopoulos, W.C., Sethi, L., Madireddy, A., Schildkraut, C.L., Zhang, D. (2017). FANCM, BRCA1, and BLM cooperatively resolve the replication stress at the ALT telomeres. *Proc. Natl. Acad. Sci. USA*, 114, E5940–E5949.

Perrem, K., Bryan, T.M., Englezou, A., Hackl, T., Moy, E.L., Reddel, R.R. (1999). Repression of an alternative mechanism for lengthening of telomeres in somatic cell hybrids. *Oncogene*, 18, 3383–3390.

Perrem, K., Colgin, L.M., Neumann, A.A., Yeager, T.R., Reddel, R.R. (2001). Coexistence of alternative lengthening of telomeres and telomerase in hTERT-transfected GM847 cells. *Mol. Cell. Biol.*, 21, 3862–3875.

Pezzolo, A., Pistorio, A., Gambini, C., Haupt, R., Ferraro, M., Erminio, G., De Bernardi, B., Garaventa, A., Pistoia, V. (2015). Intratumoral diversity of telomere length in individual neuroblastoma tumors. *Oncotarget*, 6, 7493–7503.

Potts, P.R. and Yu, H. (2007). The SMC5/6 complex maintains telomere length in ALT cancer cells through SUMOylation of telomere-binding proteins. *Nat. Struct. Mol. Biol.*, 14, 581–590.

Sadic, D., Schmidt, K., Groh, S., Kondofersky, I., Ellwart, J., Fuchs, C., Theis, F.J., Schotta, G. (2015). Atrx promotes heterochromatin formation at retrotransposons. *EMBO Rep.*, 16, 836–850.

Schaetzlein, S., Lucas-Hahn, A., Lemme, E., Kues, W.A., Dorsch, M., Manns, M.P., Niemann, H., Rudolph, K.L. (2004). Telomere length is reset during early mammalian embryogenesis. *Proc. Natl. Acad. Sci. USA*, 101, 8034–8038.

Schwartzentruber, J., Korshunov, A., Liu, X.Y., Jones, D.T., Pfaff, E., Jacob, K., Sturm, D., Fontebasso, A.M., Quang, D.A., Tonjes, M. et al. (2012). Driver mutations in histone H3.3 and chromatin remodelling genes in paediatric glioblastoma. *Nature*, 482, 226–231.

Silva, B., Pentz, R., Figueira, A.M., Arora, R., Lee, Y.W., Hodson, C., Wischnewski, H., Deans, A.J., Azzalin, C.M. (2019). FANCM limits ALT activity by restricting telomeric replication stress induced by deregulated BLM and R-loops. *Nat. Commun.*, 10, 2253.

Sobinoff, A.P., Allen, J.A., Neumann, A.A., Yang, S.F., Walsh, M.E., Henson, J.D., Reddel, R.R., Pickett, H.A. (2017). BLM and SLX4 play opposing roles in recombination-dependent replication at human telomeres. *EMBO J.*, 36, 2907–2919.

Tanaka, H., Bergstrom, D.A., Yao, M.C., Tapscott, S.J. (2005). Widespread and nonrandom distribution of DNA palindromes in cancer cells provides a structural platform for subsequent gene amplification. *Nat. Genet.*, 37, 320–327.

Tardat, M. and Dejardin, J. (2018). Telomere chromatin establishment and its maintenance during mammalian development. *Chromosoma*, 127, 3–18.

Teng, S.C. and Zakian, V.A. (1999). Telomere-telomere recombination is an efficient bypass pathway for telomere maintenance in *Saccharomyces cerevisiae. Mol. Cell. Biol.*, 19, 8083–8093.

Tham, W.H. and Zakian, V.A. (2002). Transcriptional silencing at Saccharomyces telomeres: Implications for other organisms. *Oncogene*, 21, 512–521.

Tommerup, H., Dousmanis, A., de Lange, T. (1994). Unusual chromatin in human telomeres. *Mol. Cell. Biol.*, 14, 5777–5785.

Vaquero-Sedas, M.I. and Vega-Palas, M.A. (2019). Targeting cancer through the epigenetic features of telomeric regions. *Trends Cell. Biol.*, 29, 281–290.

Varley, H., Pickett, H.A., Foxon, J.L., Reddel, R.R., Royle, N.J. (2002). Molecular characterization of inter-telomere and intra-telomere mutations in human ALT cells. *Nat. Genet.*, 30, 301–305.

Watson, J. (1972). Origin of concatemeric T7DNA. *Nature New Biology*, 239, 197–201.

Wilson 3rd, D.M., and Thompson, L.H. (2007). Molecular mechanisms of sister-chromatid exchange. *Mutat. Res.*, 616, 11–23.

Wright, W.E., Piatyszek, M.A., Rainey, W.E., Byrd, W., Shay, J.W. (1996). Telomerase activity in human germline and embryonic tissues and cells. *Dev. Genet.*, 18, 173–179.

Yeager, T.R., Neumann, A.A., Englezou, A., Huschtscha, L.I., Noble, J.R., Reddel, R.R. (1999). Telomerase-negative immortalized human cells contain a novel type of promyelocytic leukemia (PML) body. *Cancer Res.*, 59, 4175–4179.

Zhong, Z.H., Jiang, W.Q., Cesare, A.J., Neumann, A.A., Wadhwa, R., Reddel, R.R. (2007). Disruption of telomere maintenance by depletion of the MRE11/RAD50/NBS1 complex in cells that use alternative lengthening of telomeres. *J. Biol. Chem.*, 282, 29314–29322.

10

TERRA, Telomeric Non-coding RNAs

Anabelle DECOTTIGNIES
Genetic and Epigenetic Alterations of Genomes, Duve Institute, UCLouvain, Brussels, Belgium

10.1. Introduction

Telomeres have long been recognized as heterochromatic structures that, unlike euchromatin regions, are less favorable to transcription. The first time that the transcription of these terminal regions of chromosomes was observed was when Rudenko and van der Ploeg (1989) demonstrated transcription of *Trypanosoma brucei* telomeres into non-coding RNAs. A few years later, transcriptional activity was detected at the telomeres of bird "brush" chromosomes (Solovei et al. 1994). These chromosomes are specifically observed during the diplotene stage of prophase I in meiosis and are characterized by intense transcriptional activity. It was not until 2007–2008 that the first evidence of telomere transcription in mammals was obtained. In humans and mice, telomeric transcription from the C-rich strand, to form G-rich telomeric RNA molecules, can be detected with $(CCCTAA)_n$ (Azzalin et al. 2007; Schoeftner and Blasco 2008). It was at this time that the term "TERRA" (TElomeric repeat-containing RNA) appeared in the literature (Azzalin et al. 2007) and has since been widely adopted by the scientific community. The presence of TERRA has also been demonstrated in yeast, *Saccharomyces cerevisiae* (Luke et al. 2008) or *Schizosaccharomyces pombe* (Bah et al. 2012; Greenwood and Cooper

Telomeres,
coordinated by Carole SAINTOMÉ.

2012), zebrafish (Schoeftner and Blasco 2008) or plants (Vrbsky et al. 2010). In recent years, however, the transcription of mouse telomeres has been challenged by the discovery that PAR-TERRA, a non-coding RNA containing $(UUAGGG)_n$ sequences from the pseudoautosomal locus (PAR) located in the subtelomeric regions of the X and Y sex chromosomes, appeared to be the major constituent of TERRAs produced in murine embryonic stem cells (Chu et al. 2017a). It is currently unknown whether these pseudoautosomal locus-derived PAR-TERRA molecules play a similar role to that of telomere-derived TERRAs.

In this chapter, I will develop the state of the art related to human and murine TERRAs as well as the known roles for these non-coding RNAs so far.

10.2. The human TERRAs

10.2.1. *The majority of human telomeres are transcribed*

The first studies carried out on human cells revealed the presence of TERRA at the ends of chromosomes via a fluorescent in situ hybridization (FISH) approach under non-denaturing conditions, that is, without a denaturing step that separates the DNA strands, coupled with immunofluorescence (IF) against telomere proteins (Azzalin et al. 2007). TERRA molecules were detected using fluorescent sequence probes $(CCCTAA)_n$ directed against the $(UUAGGG)_n$ motifs of TERRA molecules. These observations in cell culture were then validated in human tissue using similar RNA-FISH/IF approaches (Diman et al. 2016). Importantly, while these experiments suggested that the TERRA molecules detected at telomeres originated from transcription of the telomeres themselves, they did not formally demonstrate this. Similarly, they did not show whether all human telomeres were transcribed or whether TERRAs originated mainly from a single locus – telomere or intra-chromosome – before localizing, in *trans*, on the other telomeres of the nucleus. Additional experiments were therefore required at this stage to provide evidence for transcription of human telomeres. Through quantitative PCR approaches using primers located in subtelomeric regions of chromosomes, directly upstream of telomeric repeats, it was possible to show that human telomeres were, for the most part, transcribed into TERRA from promoters located within 1 kb of telomeres (Nergadze et al. 2009; Deng et al. 2012; Porro et al. 2014a; Diman et al. 2016; Feretzaki and Lingner 2017; Feretzaki et al. 2019). However, it is possible that other subtelomeric promoters, located further upstream, that is, approximately 5–10 kb from telomeres, may also initiate TERRA production in human cells (Porro et al. 2014a).

However, not all telomeres are transcribed with the same intensity, due in particular to the nucleotide sequence and chromatin status of the different subtelomeric promoters. The transcriptional activity of these promoters is indeed, in part, controlled by the level of DNA methylation (Nergadze et al. 2009). Because DNA methylation occurs on cytosine residues that precede guanine residues and negatively influence transcription factor binding, the CpG island content of subtelomeric promoters has a strong influence on the accessibility of chromatin to these factors (Yehezkel et al. 2008; Nergadze et al. 2009; Diman et al. 2016; Sagie et al. 2017; Feretzaki et al. 2019; Le Berre et al. 2019). However, the presence of these CpG islands is required for transcriptional activity since subtelomeres lacking such motifs, such as the p-terminus of chromosome 7 for example, are unable to initiate transcription of the corresponding telomere (Nergadze et al. 2009; Diman et al. 2016; Sagie et al. 2017; Feretzaki et al. 2019). Consistent with a role for CpG islands on telomeric transcript production, the transcription factor NRF1 (nuclear respiratory factor 1), which binds to these subtelomeric DNA motifs, is a key regulator of TERRA expression (Diman et al. 2016; Le Berre et al. 2019). Regulation of TERRA by NRF1 at least partially explains the impact of DNA methylation since NRF1 binding to promoters is sensitive to the latter. Another study showed that the transcription factor CTCF also binds to subtelomeric CpG islands (Deng et al. 2012). Additional chromatin immunoprecipitation (ChIP) experiments confirmed the presence of the active form of RNA polymerase II at a range of subtelomeric promoters, thereby providing further evidence for the existence of transcriptional activity in these chromosome terminal regions (Nergadze et al. 2009). Finally, two other transcription factors, HSF1 (heat shock factor 1), induced under heat shock conditions, or p53, activated in response to DNA-damaging treatments, such as etoposide, have also been shown to induce transcription of some human telomeres by binding to their subtelomeric promoters (Tutton et al. 2016; Koskas et al. 2017).

Together, these different studies therefore helped to demonstrate that the vast majority of human telomeres are transcribed from subtelomeric promoters rich in CpG islands (Figure 10.1). Note that a recent study showed that TERRA molecules, produced from a plasmid, could localize in *trans* to several human telomeres (Feretzaki et al. 2020), suggesting that, even if not all human telomeres are transcribed, they could be linked by TERRA molecules produced from other chromosome ends.

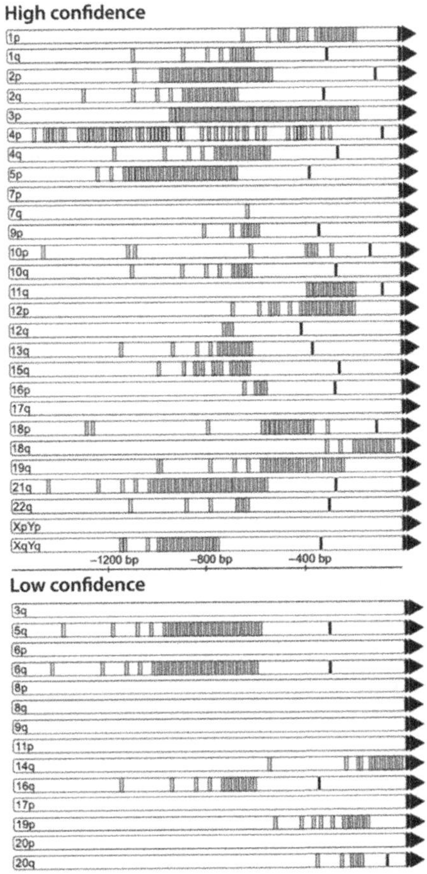

	Ilot CpG	NRF1	HSF1	CTCF	p53
High confidence					
1p	+	+	+	-	-
1q	+	+	-	+	+
2p	+	+	-	+	-
2q	+	+	-	+	+
3p	+	+	+	-	-
4p	+	+	+	+	+
4q	+	+	+	+	+
5p	+	+	+	+	+
7p	-	-	-	-	-
7q	-	+	+	-	-
9p	+	+	-	+	+
10p	+	+	+	-	-
10q	+	+	-	+	+
11q	+	+	-	+	-
12p	+	+	-	+	+
12q	+	+	+	-	-
13q	+	+	-	+	+
15q	+	+	-	+	+
16p	+	+	-	+	+
17q	-	-	+	-	+
18p	+	+	+	+	-
18q	-	+	+	-	+
19q	+	+	-	+	+
21q	+	+	-	+	+
22q	+	+	-	+	+
Xp/Yp	-	-	-	-	-
Xq/Yq	+	+	-	+	+
Low confidence					
3q	-	-	-	-	-
5q	+	+	-	-	+
6p	-	-	-	-	-
6q	+	+	-	+	+
8p	-	-	-	-	+
8q	-	-	+	-	-
9q	-	-	+	-	-
11p	-	-	-	-	-
14q	+	+	-	+	-
16q	+	+	-	-	-
17p	-	-	+	+	+
19p	+	+	-	+	+
20p	-	-	+	-	-
20q	+	+	+	+	+

Figure 10.1. *Overview of human subtelomeres. Human subtelomeric sequences available in the databases are shown graphically and are taken from Diman et al. (2016). "High confidence" lists sequences for which subtelomeric localization is highly probable. "Low confidence" shows sequences for which subtelomeric origin is not certain in view of assembly difficulties during sequencing. Predicted NRF1 binding sites, based on in silico analysis, are indicated by thin gray rectangles (Diman et al. 2016). Black rectangles indicate the position of putative transcriptional starts based on the study by Nergadze et al. (2009). The presence of putative binding sites for HSF1, shown in the third column, is based on the study by Koskas et al. (2017). Note that the position of the HSF1 binding sites is, for the most part, upstream of the region shown in the graph. The presence of binding sites for CTCF and p53 is based on analysis of ChIP-seq data by Lieberman's group (Deng et al. 2012; Tutton et al. 2015)*

10.2.2. *Telomere length influences human TERRAs*

The very first Northern blot analyses of TERRAs revealed the presence of molecules with size heterogeneity (Azzalin et al. 2007), although it was not known at the time whether this heterogeneity was the consequence of the presence of telomeres of different lengths. The first evidence for a correlation between telomere length and TERRA length came in 2008 from telomerase overexpression experiments in short telomere cells (Yehezkel et al. 2008). The study revealed that the length of TERRA molecules mirrored telomere lengthening in these cells. Similar experiments were then performed in a series of other human cell lines, recapitulating the previous results and providing further evidence for the telomeric origin of TERRA molecules (Arnoult et al. 2012; Van Beneden et al. 2013). However, it is possible that shorter TERRA molecules are also produced from long telomeres, perhaps following early transcriptional arrest (Porro et al. 2010).

Telomere length not only affects the length of TERRA molecules, but also the amount of molecules produced. Thus, it appears that long telomeres, associated with more heterochromatin marks, have reduced transcription compared to shorter telomeres in the same cellular context (Arnoult et al. 2012; Van Beneden et al. 2013). The presence of short telomeres in a cell is therefore associated with an increased TERRA molecule number. Similar observations in yeast, coupled with evidence of interaction between TERRA and the telomerase complex, led to the proposal that increased production of TERRA at short telomeres may facilitate telomerase recruitment and thereby enable preferential elongation of short telomeres (Cusanelli et al. 2013). The ability of TERRA to recruit human telomerase to telomeres, however, is still controversial, as discussed later in this chapter.

10.2.3. *Induction of human TERRA expression by different types of stress: metabolic, oxidative stress, heat shock or DNA damage*

The discovery of NRF1 binding sites on human subtelomeric promoters and the demonstration of its impact on TERRA production was the first indication of a possible link between energy stress and/or mitochondrial function and telomere transcription (Diman et al. 2016). Indeed, NRF1 is a transcription factor that, when co-activated by PGC-1α (peroxisome proliferator-activated receptor gamma co-activator 1α), enables the transcription of genes involved in mitochondrial biogenesis and defense against oxidative stress (Ventura-Clapier et al. 2008). We have proposed that transcription of TERRA by NRF1 could be part of the cell's antioxidant response and, in this way, contribute to the protection of telomeres from reactive oxygen species (ROS) (Diman et al. 2016). One possibility – not yet demonstrated – would be that TERRA could act as a protective shield for telomeres which, rich in guanine residues, are particularly sensitive to oxidation. In this scenario, single-stranded $(UUAGGG)_n$ species would be oxidized more readily than double-stranded telomere sequences, thus protecting $(TTAGGG)_n$ sequences from oxidation. This mechanism could have an advantage in the context of non-mitotic cells in which new telomeric DNA synthesis no longer occurs, whereas replacement of oxidized TERRA molecules by induction of new telomere transcription would be possible. Although this hypothesis remains to be demonstrated, a recent study indicates that *in vitro* oxidative stress resulting from H_2O_2 treatment of human cells in culture induces an increase in TERRA expression, supporting an antioxidant function of these non-coding RNAs (Galigniana et al. 2020). Furthermore, we had shown that intensive endurance exercise, known to be associated with a transient increase in ROS, leads to activation of human telomere transcription *in vivo* in skeletal muscle (Diman et al. 2016).

Other types of stress, such as heat shock, are associated with a risk of telomere dysfunction since, under these conditions, the telomeric protein TRF2 partially dissociates from telomeres, inducing their deprotection (Koskas et al. 2017). The transcription factor HSF1 (heat shock factor 1) plays a major role in response to a variety of endogenous or exogenous cellular stresses, such as heat shock, especially in tumors (Wang et al. 2020). Artificial induction of heat shock, by incubating cells at 43°C, promotes the recruitment of HSF1 to human subtelomeric promoters to induce telomere transcription (Koskas et al. 2017). Cells depleted for HSF1 show more telomere damage after heat shock, suggesting that TERRAs may, perhaps through their binding to TRF2, contribute to the re-association of TRF2 to telomeres after heat shock (Koskas et al. 2017).

Finally, another study showed that, in response to DNA-damaging treatments, such as etoposide, the transcription factor p53 is recruited to human subtelomeres and induces chromatin remodeling (increased H3K9 and H3K27 acetylation) that stimulates telomere transcription (Tutton et al. 2016). Although the role of TERRA induction in etoposide-treated cells could not be formally demonstrated, as etoposide resistance was not tested under TERRA depletion conditions, the authors were able to show that in the absence of p53, telomeres showed signs of DNA damage.

Overall, it appears that human TERRA induction may play multiple roles in telomere protection under different stress conditions.

10.2.4. *Human TERRAs form RNA:DNA hybrids at telomeres to form R-loops*

Gradually, it has become apparent that human TERRAs are, at least in part, engaged in RNA:DNA hybrid structures at telomeres. The formation of these particular structures involves the displacement of a strand of DNA to allow the formation of so-called "R-loops". Although in the context of ectopic expression from a plasmid, TERRAs can localize in *trans* to multiple telomeres through the action of the DNA recombinase RAD51 (Feretzaki et al. 2021), the current assumption is that R-loops form primarily in a co-transcriptional manner. However, since not all human telomeres appear to be transcribed (see above), it remains to be shown that R-loop formation can occur in *trans* in a natural context. It is conceivable that this *trans* formation may also be facilitated by the presence of the telomeric protein TRF2 which, *in vitro*, is able to promote R-loop formation by allowing TERRA invasion of double-stranded telomeric DNA (Lee et al. 2018). Regardless of *cis*- or *trans*-formation, the amount of RNA:DNA hybrids detected by monoclonal antibodies specific for these structures correlates with the level of telomere transcription (Arora et al. 2014; Sagie et al. 2017; Toubiana and Selig 2018).

The presence of R-loops is not inconsequential to telomere biology and can influence their maintenance, as well as genomic stability in a variety of ways, including homologous recombination activation following the presence of replication fork blocks, followed by replication fork collapses (reviewed by Toubiana and Selig (2018); Bettin et al. (2019); Feretzaki et al. (2020)). R-loops are therefore a source of telomere danger and must be properly regulated. To this end, helicases, such as Pif1 or BLM, as well as the RNAse H1/H2, play important roles and help eliminate RNA:DNA hybrids (Toubiana and Selig 2018; Feretzaki et al. 2020).

R-loops are particularly important in the context of cancer cells using an alternative telomerase-independent mechanism for maintaining their telomeres based on telomere homologous recombination events (ALT cells) (reviewed by Toubiana and Selig (2018)). In this case, the presence of R-loops may have a catalytic effect for telomere homologous recombination (Arora et al. 2014). But here too, ensuring the balance between R-loop formation and dissolution is crucial, and RNAse H1 plays an important role in maintaining this balance. Indeed, the depletion of RNAse H1 in ALT cells leads to the accumulation of R-loops accompanied by replication stress at telomeres and telomeric DNA excision phenomena, whereas overexpression of the same enzyme induces progressive telomere shortening by reducing the efficiency of homologous recombination (Arora et al. 2014). Another regulatory element for the presence of R-loops at ALT telomeres is related to the loss of expression of the chromatin remodeling factor ATRX in the majority of these cells. Indeed, it appears that loss of function of ATRX in ALT cells, generally following the acquisition of deleterious somatic mutations in the coding phase of the gene, is associated with greater persistence of telomeric R-loops, which likely promotes homologous recombination events (reviewed by Toubiana and Selig (2018)). Of note, the observations related to the favorable role of R-loops in human ALT telomere recombination are similar to those made in yeast, in which accumulation of R-loops at short telomeres of telomerase-deficient cells activates homologous recombination and delays entry into senescence (Graf et al. 2017).

The presence of R-loops is also of particular importance in the context of patients with ICF (centromeric instability-immunodeficiency-dysmorphic) syndrome, in whom a loss of function of the DNA methyltransferase DNMT3B results in increased telomere transcription. In cells from these patients, telomeres show extensive damage related to the presence of R-loops and increased collapse of replication forks at telomeres, a phenomenon that, *in vitro*, can be mitigated by overexpression of RNase H1 (reviewed by Toubiana and Selig (2018)).

Besides these two pathological situations – ALT cancer cells or cells from ICF patients – the deleterious effect of R-loops to telomeres of normal cells remains to be validated since cancer cells relying on telomerase for telomere maintenance do not seem to be sensitive to changes in the expression level of RNAse H1 (reviewed by Toubiana and Selig (2018)).

10.2.5. *Other known functions of human TERRAs*

In addition to the likely protective functions in response to oxidative stress or heat shock and an impact on the telomere via the formation of R-loops, TERRA plays diverse roles in the human cell, particularly through its ability to interact with a range of proteins (Figure 10.2).

10.2.5.1. *Regulation of telomeric chromatin*

Early on, a role for TERRA in chromatin regulation was proposed following the observation that TERRA interacts with proteins such as H3K9me3, HP1, the histone methyltransferase Suv39h1, the nucleolar remodeling complex NoRC, or other proteins involved in chromatin remodeling identified by *in vitro* screens (reviewed by Bettin et al. (2019)). The first study in this sense, published in 2009, proposed a model in which TERRA, through its interaction with the telomeric protein TRF2, on the one hand, and the ORC complex, on the other hand, allows the formation of H3K9me3 at telomeres (Deng et al. 2009). Similarly, TERRA has also been proposed to act as a guide RNA for the recruitment of the NoRC complex to telomeres, thus participating in its heterochromatinization and the decreased rate of telomeric recombination (Postepska-Igielska et al. 2013). TERRA was also identified as interacting *in vitro* with the PRC2 complex responsible for the deposition of the H3K27me3 repressive mark (Montero et al. 2018). The impact of this interaction on the density of H3K27me3 marks at human telomeres, however, remains to be confirmed because, to date, no effective tool could be developed to remove TERRA content from human cells. In general, the current lack of depletion tools is a hindrance to the study of human TERRA functions.

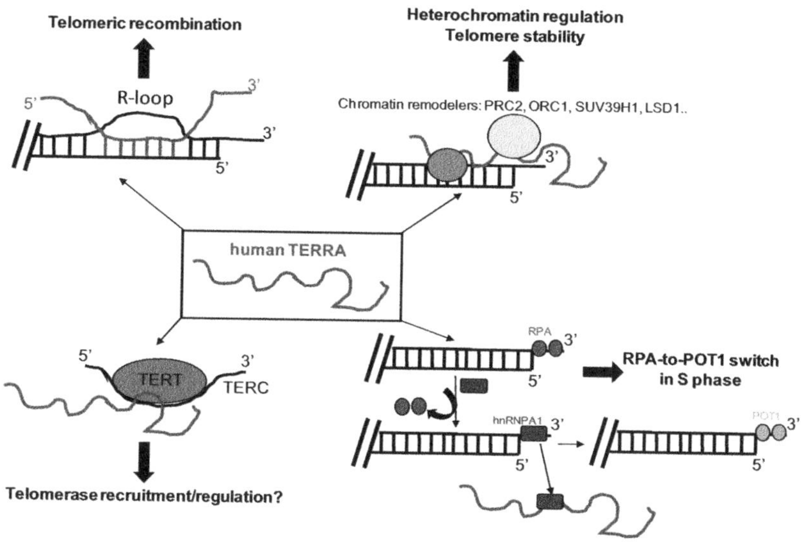

Figure 10.2. *Possible roles of human TERRAs. Because of their ability to bind to a range of proteins, human TERRAs likely play multiple roles at the telomere. Although the function of regulating the switch between RPA and POT1 at telomeres after replication has been clearly demonstrated (Flynn et al. 2011), the impact of human TERRAs on telomerase regulation and recruitment at telomeres is less clear. Evidence is also accumulating in favor of an important role for TERRA in telomere chromatin regulation (see text). On the other hand, the formation of R-loops by hybridization between TERRA and telomeric DNA also impacts telomere biology by promoting replication fork progression problems, a phenomenon that helps to induce recombination events at the telomere. For a color version of this figure, see www.iste.co.uk/saintome/telomeres.zip*

10.2.5.2. *Response to deprotected telomeres and protective role*

An interaction of TERRA with the lysine demethylase LSD1 was demonstrated in cells of the HeLa cancer line (Porro et al. 2014b). The study showed that, in cells depleted for the telomeric protein TRF2, telomere transcription is increased and accompanied by the additional recruitment of LSD1 to the deprotected telomeres. This recruitment of LSD1, in turn, allows the nuclease MRE11 to be delivered to telomeres to cleave single-stranded G-rich telomeric DNA. Although the study showed a role for LSD1 in the response to telomere damage induced by the loss of TRF2, the role of TERRA in the recruitment of LSD1 to telomeres could not be formally demonstrated due to the lack of effective TERRA depletion tools.

In a second study by the same group of Lingner, TERRA induction in cells depleted for TRF2 correlated with an increased recruitment of the histone methylase SUV39H1 to telomeres, accompanied by an increase in the H3K9me3 mark and telomere fusions (Porro et al. 2014a).

Overall, although there is ample evidence pointing to an increase in telomere transcription in response to telomere deprotection, the demonstration of a role for TERRA in this response has, as mentioned above, been hampered by the lack of effective tools for depletion of these non-coding RNAs. However, a recent study in which the q telomere of endogenous chromosome 15 (Tel15q) was tagged showed that depletion of this single Tel15q in unstressed cells induces multiple DNA damage signals in the nucleus (Avogaro et al. 2018). These observations thus suggest a role for TERRA in protecting, in *cis*, the telomere and, in *trans*, other genomic loci.

10.2.5.3. *Regulation of the control of RPA and POT1 binding to single-stranded telomeric DNA*

TERRA, through its interaction with the RNA-binding protein hnRNPA1, participates in the detachment of the RPA (replication protein A) protein from G-rich single-stranded telomeric DNA to allow the switch between RPA and POT1 at the exit from S phase (Flynn et al. 2011). It is indeed important that the RPA protein, deposited on single-stranded DNA during the S phase, does not remain bound to telomeres in order to avoid the initiation of the DNA damage response via activation of ATR kinase. The study elegantly showed that this switch between RPA and POT1 at the end of the S phase was a consequence of fluctuations in TERRA levels during the cell cycle. Indeed, the cellular amounts of TERRA decrease during the S phase to reach their minimum level at the end of the replication phase. When TERRA levels are low, the hnRNPA1 protein detaches from its TERRA RNA substrate and binds to telomeres where it displaces the RPA protein. In turn, POT1 displaces hnRNPA1 from the telomeres and binds to it to prevent activation of the ATR kinase (Flynn et al. 2011).

10.2.5.4. *Modulation of telomerase activity and its recruitment to telomeres*

Very soon after the discovery of TERRAs and the presence in their sequence of motifs complementary to the template region of the telomerase h*TR* subunit, the idea arose that these non-coding RNA molecules may negatively regulate telomerase (reviewed by Cusanelli and Chartrand (2015)). Although oligonucleotides with a sequence matching that of TERRA are indeed capable, *in vitro*, of inhibiting telomerase activity in human cell extracts (Redon et al. 2010), *in vivo* hybridization between h*TR* and TERRA has not been demonstrated to date. Another study by the group of Lingner then proposed that, in cells, there may be regulation of telomerase activity that involves three partners: hnRNPA1, TERRA and telomerase (Redon et al. 2013). In this model, the authors propose that the negative impact of TERRA on telomerase – observed *in vitro* – could be compensated by the interaction between TERRA and hnRNPA1 *in vivo*. In this scenario, TERRA and hnRNPA1 would therefore act in concert to positively or negatively regulate telomerase. Furthermore, the demonstration that telomerase is able to lengthen telomeres of ICF patient-derived cells in which telomere transcription is very high (Farnung et al. 2012) counters the idea that TERRAs inhibit telomerase activity at telomeres. This observation is, in contrast, rather consistent with the positive impact of TERRAs on telomerase recruitment to short telomeres of budding yeast (Cusanelli et al. 2013; Moravec et al. 2016).

To date, the impact of TERRA on human telomerase activity *in vivo* is still uncertain and will require further investigation.

10.3. The mouse TERRAs

10.3.1. *The majority of murine telomeres are probably not transcribed*

The first study related to murine TERRAs was published almost 15 years ago (Azzalin et al. 2007). Since then, the genomic origin of murine TERRAs is still hotly debated (reviewed by Diman and Decottignies (2018)). The debate about the telomeric or intrachromosomal origin of mouse TERRAs was initiated by the observation by FISH-RNA/IF that, unlike what is observed in human cells, murine TERRA foci are, for the most part, not localized at telomeres. Instead of this telomere labeling, murine cell nuclei typically have one or two large TERRA foci associated with the q-ends of sex chromosomes (Schoeftner and Blasco 2008; Zhang et al. 2009). In addition to these two large TERRA foci, smaller foci are visible that in rare cases (two to three events per nucleus) co-localize with telomeres (Lopez de Silanes et al. 2014).

Two studies have contributed to a better understanding of the murine TERRA repertoire. The first, published in 2014, suggested that the subtelomere and/or telomere at the 18q end is the major contributor to the TERRA pool in mice (Lopez de Silanes et al. 2014). Depletion experiments of TERRA 18q seemed to indicate an important role for this molecule in protecting a range of other mouse telomeres, suggesting a role in *trans* for this TERRA. Although this transcript was also detected in another study to describe the murine TERRA repertoire in embryonic stem cells, another TERRA-like RNA was identified that is more than 200-fold more abundant than the 18q TERRA transcript (Chu et al. 2017a). This RNA, named PAR-TERRA, is transcribed from the pseudoautosomal subtelomeric region of the X and Y sex chromosomes, a region characterized by the presence of intra-chromosomal $(TTAGGG)_n$ motifs. According to this study, the TERRA pool of embryonic stem cells is over 99% PAR-TERRA (Chu et al. 2017a). Chromosomal localization search screens of chromatin-bound PAR-TERRA subsequently revealed that it binds to some telomeres, but also to many intra-chromosomal loci (Chu et al. 2017a). Consistent with at least partial co-localization of PAR-TERRA with mouse telomeres, depletion of PAR-TERRA in embryonic stem cells results in telomere damage (Chu et al. 2017b). However, it cannot be formally excluded that telomere damage in cells treated with antisense oligonucleotides directed against $(UUAGGG)_n$ sequences was not caused by the binding of these oligonucleotides to the G-rich single-stranded portion of the telomere, and the impact of PAR-TERRA on the protection of mouse telomeres therefore remains to be formally demonstrated.

Our team recently showed that PAR-TERRA is also the major contributor to the pool of $(UUAGGG)_n$ in mouse embryonic fibroblasts, as well as in a range of transformed murine cell lines, while 18q TERRA contributes only very weakly to the pool of TERRAs in these different cell types (Viceconte et al. 2021). It therefore appears that mice behave differently from humans – and other eukaryotic species – in terms of telomere transcription. The reasons for these differences are unknown at this time, but could be related to the higher level of heterochromatinization of mouse telomeres. Indeed, their heterochromatinization state may not be compatible with transcription (reviewed by Diman and Decottignies (2018)). An additional argument for the very low transcription of mouse telomeres comes from ChIP experiments that revealed the absence of the activating mark H3K4me3 at mouse telomeres, whereas this mark is well present at human telomeres (Rosenfeld et al. 2009). If TERRA plays a role in telomere heterochromatinization, as proposed in human cells, then it is conceivable that murine telomeres may have evolved another TERRA-independent mechanism to achieve this.

It is also currently unknown whether murine telomeres exhibit R-loops. Although most R-loops appear to form co-transcriptionally, as mentioned above, it cannot be excluded that these hybrid DNA:RNA structures may also result from transcription-independent invasion events. Again, further investigations will be required in the future to answer this question.

10.3.2. *PAR-TERRA: a crucial role in homologous sex chromosome pairing and in the regulation of ATRX target gene expression*

Early observations of co-localization of mouse *(UUAGGG)*$_n$ molecules with the inactive X chromosome or Y chromosome were followed by other equivalent observations that gradually led to the question of a possible role for TERRA in mouse sex chromosome biology. It was in 2017 that this evidence was provided when J.T. Lee's group showed that PAR-TERRA plays a crucial role in sex chromosome pairing in mouse embryonic stem cells, which in turn allows for X chromosome inactivation in female cells (Chu et al. 2017a). The authors proposed that PAR-TERRA could serve as a bridge molecule, allowing sex chromosome pairing upon X-inactivation. Sex chromosome pairing has also been observed in male cells that are not subject to X-inactivation, but nevertheless require an X-counting system to ensure the absence of a second X. To date, a role for TERRA in sex chromosome pairing has not been reported in any other species, although the requirement for sex chromosome pairing for X inactivation appears to be conserved among eukaryotic species (Lee 2005). It remains to be seen whether the presence of *(UUAGGG)*$_n$ repeats in murine PAR-TERRA plays, per se, a role in chromosome pairing.

PAR-TERRA also appears to play other functions across the genome by modulating the expression of several hundred genes (Chu et al. 2017b). Thus, PAR-TERRA competes with ATRX at many genomic loci to activate gene transcription, while ATRX has a repressive effect on these same genes. However, the reasons why the same non-coding RNA plays so many different roles in the nucleus remain to be elucidated.

10.3.3. *Known or predicted functions of murine TERRAs*

Apart from what has been described above about PAR-TERRA, research on the functions of murine TERRAs is less advanced than that dedicated to understanding human TERRAs. When identifying the TERRA proteome by an iDRIP approach (identification of proteins interacting directly with RNA) coupled with mass spectrometry in murine ES cells, J.T. Lee's group nonetheless identified potential new protein partners, suggesting a role for *(UUAGGG)*$_n$ motif RNAs in the processes of DNA double-strand break repair by homologous recombination, DNA replication, or cell cycle control (Chu et al. 2017b). Interactions with various chromatin remodeling factors, including the PRC2 complex, were also found, reinforcing the already known role of TERRA in chromatin regulation. A more recent study in mouse pluripotent cells also correlated TERRA induction, induced by TRF1 depletion, with differential expression of target genes regulated by the PRC2 complex and H3K27me3 (Marión et al. 2019). Although TERRA is physically associated with many of these genes, the study did not formally demonstrate that transcriptional deregulation of PRC2 target genes in cells depleted for TRF1 was dependent on TERRA interaction with the PRC2 complex.

Together, the various studies on the functions of mouse TERRAs point to a role for these non-coding RNAs in the general regulation of chromatin. However, as discussed above, a role for murine TERRAs in telomeric chromatin regulation remains unclear, suggesting that, although functions may be common between humans and mice, the genomic loci regulated by TERRAs may be distinct in these two species.

10.4. Conclusion

Since the discovery of TERRAs in humans and mice 15 years ago, numerous studies have been performed to understand the mechanisms of regulation of the production of these non-coding RNAs and their roles in the cell. Although much evidence has accumulated in favor of important roles for TERRAs, the formal demonstration of the functions of these non-coding RNAs is still hampered by the lack of efficient and specific depletion tools. The differences observed between humans and mice do not facilitate the work of understanding either. However, one element that seems to be common to human and mouse TERRAs is their ability to interact with a whole series of proteins, particularly those linked to the regulation of heterochromatin. It is therefore possible that, in a manner similar to what has already been described for other non-coding RNAs, TERRAs could play an important role in the regulation of telomeric and/or chromosomal heterochromatin.

The field is waiting for new powerful tools that will allow the further study of the functions of these enigmatic TERRAs in mammalian cells.

10.5. References

Arnoult, N., Van Beneden, A., Decottignies, A. (2012). Telomere length regulates TERRA levels through increased trimethylation of telomeric H3K9 and HP1α. *Nat. Struct. Mol. Biol.*, 19(9), 948–956.

Arora, R., Lee, Y., Wischnewski, H., Brun, C.M., Schwarz, T., Azzalin, C.M. (2014). RNaseH1 regulates TERRA-telomeric DNA hybrids and telomere maintenance in ALT tumour cells. *Nat. Commun.,* 5, 5220.

Avogaro, L., Querido, E., Dalachi, M., Jantsch, M.F., Chartrand, P., Cusanelli, E. (2018). Live-cell imaging reveals the dynamics and function of single-telomere TERRA molecules in cancer cells. *RNA Biol.*, 15(6), 787–796.

Azzalin, C.M., Reichenbach, P., Khoriauli, L., Giulotto, E., Lingner, J. (2007). Telomeric repeat containing RNA and RNA surveillance factors at mammalian chromosome ends. *Science,* 318(5851), 798–801.

Bah, A., Wischnewski, H., Shchepachev, V., Azzalin, C.M. (2012). The telomeric transcriptome of *Schizosaccharomyces pombe*. *Nucl. Acids Res.,* 40(7), 2995–3005.

Bettin, N., Oss Pegorar, C., Cusanelli, E. (2019). The emerging roles of TERRA in telomere maintenance and genome stability. *Cells*, 8(3). doi: 10.3390/cells8030246.

Chu, H.P., Froberg, J.E., Kesner, B., Oh, H.J., Ji, F., Sadreyev, R., Pinter, S.F., Lee, J.T. (2017a). PAR-TERRA directs homologous sex chromosome pairing. *Nat. Struct. Mol. Biol.,* 24(8), 620–631.

Chu, H.P., Cifuentes-Rojas, C., Kesner, B., Aeby, E., Lee, H.G., Wei, C., Oh, H.J., Boukhali, M., Haas, Lee, J.T. (2017b). TERRA RNA antagonizes ATRX and protects telomeres. *Cell,* 170(1), 86–101.

Cusanelli, E. and Chartrand, P. (2015). Telomeric repeat-containing RNA TERRA: A noncoding RNA connecting telomere biology to genome integrity. *Front Genet,* 6, 143.

Cusanelli, E., Romero, C.A., Chartrand, P. (2013). Telomeric noncoding RNA TERRA is induced by telomere shortening to nucleate telomerase at short telomeres. *Mol. Cell,* 51(6), 780–791.

Deng, Z., Norseen, J., Wiedmer, A., Riethman, H., Lieberman, P.M. (2009). TERRA RNA binding to TRF2 facilitates heterochromatin formation and ORC recruitment at telomeres. *Mol. Cell,* 35(4), 403–413.

Deng, Z., Wang, Z., Stong, N., Plasschaert, R., Moczan, A., Chen, H.S., Hu, S., Wikramasinghe, P., Davuluri, R.V., Bartolomei, M.S. et al. (2012). A role for CTCF and cohesin in subtelomere chromatin organization, TERRA transcription, and telomere end protection. *EMBO J.*, 31(21), 4165–4178.

Diman, A. and Decottignies, A. (2018). Genomic origin and nuclear localization of TERRA telomeric repeat-containing RNA: From darkness to dawn. *FEBS J.*, 285(8), 1389–1398.

Diman, A., Boros, J., Poulain, F., Rodriguez, J., Purnelle, M., Episkopou, H., Bertrand, L., Francaux, M., Deldicque, L., Decottignies, A. (2016). Nuclear respiratory factor 1 and endurance exercise promote human telomere transcription. *Sci. Adv.,* 2(7), e1600031.

Farnung, B.O., Brun, C.M., Arora, R., Lorenzi, L.E., Azzalin, C.M. (2012). Telomerase efficiently elongates highly transcribing telomeres in human cancer cells. *PLoS One*, 7(4), e35714.

Feretzaki, M. and Lingner, J. (2017). A practical qPCR approach to detect TERRA, the elusive telomeric repeat-containing RNA. *Methods*, 114, 39–45.

Feretzaki, M., Renck Nunes, P., Lingner, J. (2019). Expression and differential regulation of human TERRA at several chromosome ends. *RNA*, 25(11), 1470–1480.

Feretzaki, M., Pospisilova, M., Valador Fernandes, R., Lunardi, T., Krejci, L., Lingner, J. (2020). RAD51-dependent recruitment of TERRA lncRNA to telomeres through R-loops. *Nature*, 587(7833), 303–308.

Flynn, R., Centore, R.C., O'Sullivan, R.J., Rai, R., Tse, A., Songyang, Z., Chang, S., Karlseder, J., Zou, L. (2011). TERRA and hnRNPA1 orchestrate an RPA-to-POT1 switch on telomeric single-stranded DNA. *Nature*, 471(7339), 532–536.

Galigniana, N.M., Charó, N.L., Uranga, R., Cabanillas, A.M., Piwien-Pilipuk, G. (2020). Oxidative stress induces transcription of telomeric repeat-containing RNA (TERRA) by engaging PKA signaling and cytoskeleton dynamics. *Biochim. Biophys. Acta Mol. Cell Res.*, 1867(4), 118643.

Graf, M., Bonetti, D., Lockhart, A., Serhal, K., Kellner, V., Maicher, A., Jolivet, P., Teixeira, M.T., Luke, B. (2017). Telomere length determines TERRA and R-loop regulation through the cell cycle. *Cell*, 170(1), 72–85.

Greenwood, J. and Cooper, J.P. (2012). Non-coding telomeric and subtelomeric transcripts are differentially regulated by telomeric and heterochromatin assembly factors in fission yeast. *Nucleic Acids Res.*, 40(7), 2956–2963.

Koskas, S., Decottignies, A., Dufour, S., Pezet, M., Verdel, A., Vourc'h, C., Faure, V. (2017). Heat shock factor 1 promotes TERRA transcription and telomere protection upon heat stress. *Nucleic Acids Res.*, 45(11), 6321–6333.

Le Berre, G., Hossard, V., Riou, J.F., Guieysse-Peugeot, A.L. (2019). Repression of TERRA expression by subtelomeric DNA methylation is dependent on NRF1 binding. *Int. J. Mol. Sci.*, 20(11). doi: 10.3390/ijms20112791.

Lee, J.T. (2005). Sex chromosome inactivation: The importance of pairing. *Current Biol.*, 15(7), R249–R252.

Lee, Y.W., Arora, R., Wischnewski, H., Azzalin, C.M. (2018). TRF1 participates in chromosome end protection by averting TRF2-dependent telomeric R loops. *Nat. Struct. Mol. Biol.*, 25(2), 147–153.

López de Silanes, I., Graña, O., Luigia De Bonis, M., Dominguez, O., Pisano, D.G., Blasco, M.A. (2014). Identification of TERRA locus unveils a telomere protection role through association to nearly all chromosomes. *Nat. Commun.*, 5, 4723.

Luke, B., Panza, A., Redon, S., Iglesias, N., Li, Z., Lingner, J. (2008). The Rat1p 5' to 3' exonuclease degrades telomeric repeat-containing RNA and promotes telomere elongation in *Saccharomyces cerevisiae*. *Mol. Cell,* 32(4), 465–477.

Marión, R.M., Montero, J.J., López de Silanes, I., Graña-Castro, O., Martinez, P., Schoeftner, S., Palacios-Fábrega, J.A., Blasco, M.A. (2019). TERRA regulate the transcriptional landscape of pluripotent cells through TRF1-dependent recruitment of PRC2. *Elife*, 8. doi: 10.7554/eLife.44656.

Montero, J.J., López Silanes, I., Megias, D., Fraga, M.F., Castells-Garcia, A., Blasco, M.A. (2018). TERRA recruitment of polycomb to telomeres is essential for histone trimethylation marks at telomeric heterochromatin. *Nat. Commun.*, 9, 1548.

Moravec, M., Wischnewski, H., Bah, A., Hu, Y., Liu, N., Lafranchi, L., King, M.C., Azzalin, C.M. (2016). TERRA promotes telomerase-mediated telomere elongation in *Schizosaccharomyces pombe*. *EMBO Rep.*, 17(7), 999–1012.

Nergadze, S.G., Farnung, B.O., Wischnewski, H., Khoriauli, L., Vitelli, V., Chawla, R., Giulotto, E., Azzalin, C.M. (2009). CpG-island promoters drive transcription of human telomeres. *RNA,* 15(12), 2186–2194.

Porro, A., Feuerhahn, S., Reichenbach, P., Lingner, J. (2010). Molecular dissection of telomeric repeat-containing RNA biogenesis unveils the presence of distinct and multiple regulatory pathways. *Mol. Cell. Biol.,* 30(20), 4808–4817.

Porro, A., Feuerhahn, S., Delafontaine, J., Riethman, H., Rougemont, J., Lingner, J. (2014a). Functional characterization of the TERRA transcriptome at damaged telomeres. *Nat. Commun.,* 5, 5379.

Porro, A., Feuerhahn, S., Lingner, J. (2014b). TERRA-reinforced association of LSD1 with MRE11 promotes processing of uncapped telomeres. *Cell Rep.*, 6(4), 765–776.

Postepska-Igielska, A., Krunic, D., Schmitt, N., Greulich,-Bode, K.M., Boukamp, P., Grummt, I. (2013). The chromatin remodeling complex NoRC safeguards genome stability by heterochromatin formation at telomeres and centromeres. *EMBO Rep.*, 14(8), 704–710.

Redon, S., Reichenbach, P., Lingner, J. (2010). The non-coding RNA TERRA is a natural ligand and direct inhibitor of human telomerase. *Nucleic Acids Res.*, 38(17), 5797–5806.

Redon, S., Zemp, I., Lingner, J. (2013). A three-state model for the regulation of telomerase by TERRA and hnRNPA1. *Nucleic Acids Res.*, 41(19), 9117–9128.

Rosenfeld, J.A., Wang, Z., Schones, D.E., Zhao, K., DeSalle, R., Zhang, M.Q. (2009). Determination of enriched histone modifications in non-genic portions of the human genome. *BMC Genomics,* 10, 143.

Rudenko, G. and Van der Ploeg, L.H. (1989). Transcription of telomere repeats in protozoa. *EMBO J.*, 8(9), 2633–2638.

Sagie, S., Toubiana, S., Hartono, S.R., Katzir, H., Tzur-Gilat, A., Havazelet, S., Francastel, C., Velasco, G., Chédin, F., Selig, S. (2017). Telomeres in ICF syndrome cells are vulnerable to DNA damage due to elevated DNA:RNA hybrids. *Nat. Commun.*, 8, 14015.

Schoeftner, S. and Blasco, M.A. (2008). Developmentally regulated transcription of mammalian telomeres by DNA-dependent RNA polymerase II. *Nat. Cell Biol.*, 10(2), 228–236.

Solovei, I., Gaginskaya, E.R., Macgregor, H.C. (1994). The arrangement and transcription of telomere DNA sequences at the ends of lampbrush chromosomes of birds. *Chromosome Res.*, 2, 460–470.

Toubiana, S. and Selig, S. (2018). DNA:RNA hybrids at telomeres – When it is better to be out of the (R) loop. *FEBS J.*, 285(14), 2552–2566.

Tutton, S., Azzam, G.A., Stong, N., Vladimirova, O., Wiedmer, A., Monteith, J.A., Beishline, K., Wang, Z., Deng, Z., Riethman, H. et al. (2016). Subtelomeric p53 binding prevents accumulation of DNA damage at human telomeres. *EMBO J.*, 35(2), 193–207.

Van Beneden, A., Arnoult, N., Decottignies, A. (2013). Telomeric RNA expression: Length matters. *Front. Oncol.*, 3,178.

Ventura-Clapier, R., Garnier, A., Veksler, V. (2008). Transcriptional control of mitochondrial biogenesis: The central role of PGC-1α. *Cardiovasc. Res.*, 79, 208–217.

Viceconte, N., Loriot, A., Lona Abreu, P., Scheibe, M., Fradera Sola, A., Butter, F., De Smet, C., Arnoult, N., Decottignies, A. (2021). PAR-TERRA is the main contributor to telomeric repeat-containing RNA transcripts in normal and cancer mouse cells. *RNA*, 27(1), 106–121.

Vrbsky, J., Akimcheva, S., Watson, J.M., Turner, T.L., Daxinger, L., Vyskot, B., Aufsatz, W., Riha, K. (2010). siRNA-mediated methylation of *Arabidopsis telomeres*. *PLoS Genet.*, 6(6), e1000986.

Wang, G., Cao, P., Fan, Y., Tan, K. (2020). Emerging roles of HSF1 in cancer: Cellular and molecular episodes. *Biochm. Biophys. Acta*, 1874(1), 188390.

Yehezkel, S., Segev, Y., Viegas-Péquignot, E., Skorecki, K., Selig, S. (2008). Hypomethylation of subtelomeric regions in ICF syndrome is associated with abnormally short telomeres and enhanced transcription from telomeric regions. *Hum. Mol. Genet.*, 17(18), 2776–2789.

Zhang, L.F., Ogawa, Y., Ahn, J.Y., Namekawa, S.H., Silva, S.S., Lee, J.T. (2009). Telomeric RNAs mark sex chromosomes in stem cells. *Genetics,* 182(3), 685–698.

11

Telomeres and Cancer

Arturo LONDOÑO-VALLEJO
Telomere and Cancer Lab, Institut Curie, CNRS, PSL Research University, Paris, France

11.1. Introduction

Cancer represents a large collection of different diseases in which transformed cells proliferate without control. Virtually any tissue in the human organism can be affected by a cancerous process. Although the mechanisms at the origin of cell transformation greatly differ between tumor types, many features ("hallmarks") are shared among cancer cells, such as accumulation of mutations, chromosome instability (CIN), modification of the chromatin landscape, altered metabolism, cell dedifferentiation and, of course, the ability to proliferate indefinitely (Hanahan and Weinberg 2011). A key aspect to the latter hallmark is telomere biology. Indeed, the acquisition of a telomere maintenance mechanism (TMM) is absolutely required for a cancer cell to become immortal. In addition, telomeres exert crucial roles before and during cell transformation, above all by participating in CIN, another cancer hallmark (Hanahan and Weinberg 2011). This chapter will describe how telomere biology impacts the development of cancer.

For a color version of all the figures in this chapter, see www.iste.co.uk/saintome/telomeres.zip.

11.2. Short telomeres and telomere-induced senescence are barriers to oncogenic development

As explained in previous chapters, telomeres are essential to protect chromosome ends from unwanted repair reactions that could result in end-to-end fusions or in chromosome end degradation. A minimum number of telomeric repeats, as well as the presence of all components of the telomere-specific protein complex shelterin, are required to ensure telomere protection. However, as seen elsewhere, telomeres are very dynamic structures subjected to shortening and lengthening events (telomere length homeostasis).

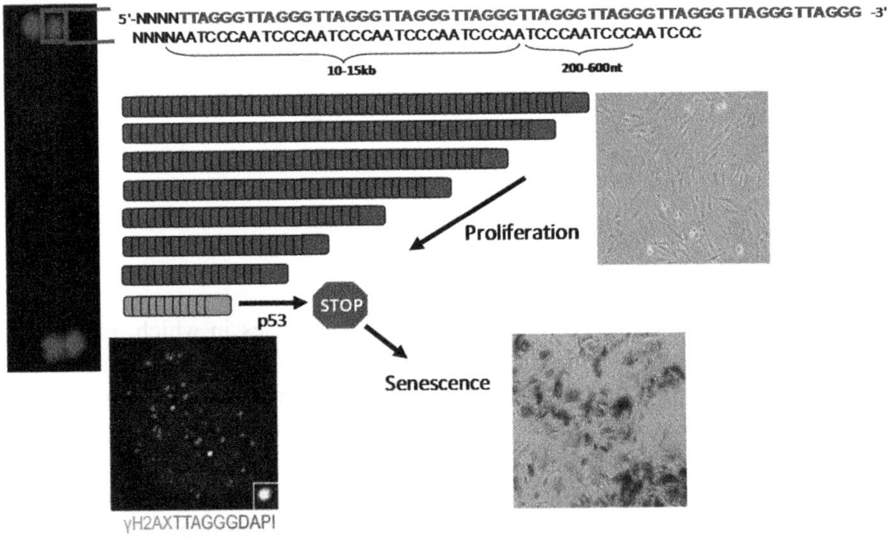

Figure 11.1. *Mitotic senescence as an anti-tumor mechanism in human cells*

NOTES ON FIGURE 11.1.– *During the S-phase of somatic cells, which do not express telomerase, the extremities of chromosomes are not completely replicated and therefore telomeres shorten at each cell division. The kinetics of telomere shortening depends on the efficiency of replicating mechanisms, the extent of C-strand degradation to recreate the 3′ overhang in chromatids as well as the occurrence of replication blocks or the activity of repair mechanisms that deal with eventual DNA lesions (resulting from, for instance, oxidative damage). When telomeres reach a critical length, the chromosome extremities are no longer protected and are detected by the cell as double strand breaks (which can be visualized by the co-localization of a telomere signal – revealed by the red-labeled telomeric probe – and a DNA damage marker such as the phosphorylation of the histone H2AX). The fact that, as shown in the picture, only one telomere appears as "deprotected" is explained by the inherited heterogeneity in telomere lengths amongst the different chromosome arms within the cell. Several deprotected telomeres are required for the cell to activate the p53 pathway, which will lead to a block in the progression of the cell cycle. If the block persist, the cell will enter the senescence program, which leads to a permanent cell-cycle arrest and includes large metabolic and morphological changes (as illustrated in the lower right picture where senescent cells are stained for beta-galactosidase expression, the canonical senescent marker. The staining is specific since it is negative in proliferating cells as shown in the upper right picture). Senescence is a very strong anti-tumor mechanism since it prevents the uncontrolled proliferation of cells. However, the senescent microenvironment may also promote tumorigenesis by the secretion of many factors (the senescence-associated secretory phenotype or SASP) into the extracellular space, including cytokines and other inflammatory factors, growth factors and extracellular matrix components.*

At the single cell level, and in the absence of a telomere elongation mechanism, telomeres shorten progressively with every cell cycle (Figure 11.1) (Shay and Wright 2000). After a variable number of cell divisions, telomeres inevitably become critically short and unable to protect chromosome ends. At this moment, the cell detects the unprotected telomere(s) and a local DNA damage response is elicited (Cesare et al. 2013), followed by an attempt to repair, for instance, by fusing the unprotected end to another available chromosome end (Cleal et al. 2018). If the DNA damage persists, a more global DNA damage response is triggered in which p53 activation plays a central role (Herbig et al. 2004). The p53-mediated transcriptional response that follows blocks the initiation of another cell cycle, thus creating a permanent state of cell cycle arrest called senescence (Hahn 2002). The senescence state induced by telomere shortening thus constitutes a strong barrier to uncontrolled proliferation and therefore acts as one of the strongest mechanisms against cancer (Reddel 2000).

At the organismal level, and since telomerase activity is suppressed after embryogenesis in most differentiated somatic cells, telomeres shorten in all tissues as humans age (Figure 11.2) (Demanelis et al. 2020). With age, then, cells exhaust their proliferation capacity and enter senescence, thus undergoing profound changes in their transcriptomic, metabolic and structural landscape (Lopez-Otin et al. 2013). It is believed that the accumulation of senescent cells with age, in particular in organs with high regeneration capacity, is responsible for the decrease in performance and fitness that affect aged organisms (Campisi 2011). Indeed, individuals affected by age-associated pathological manifestations tend to have shorter telomeres than same-age class healthy individuals (Herrmann et al. 2018).

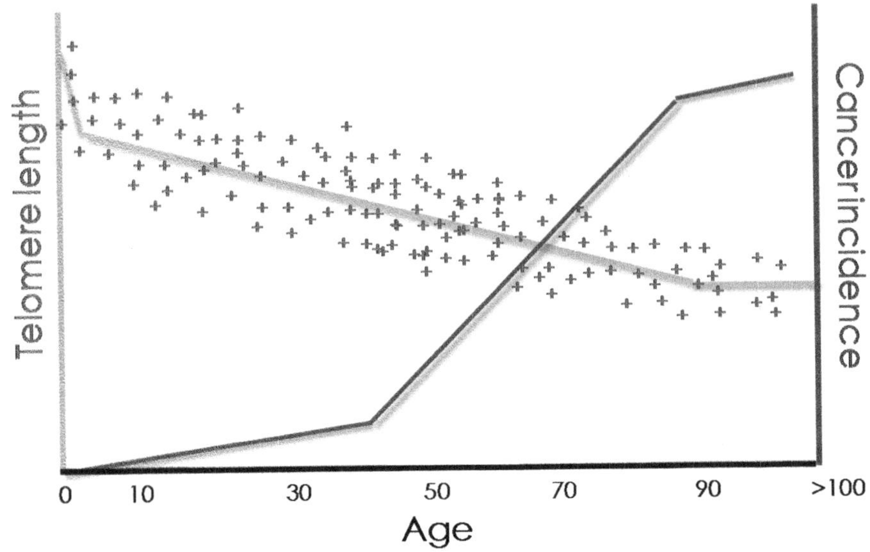

Figure 11.2. *Telomere shortening and cancer incidence during aging. The yellow solid line represents a hypothetical telomere length mean obtained through a hypothetical population study measuring telomere lengths in peripheral blood white cells from individuals (+) of different ages. Telomeres are long at birth (15–18 kb) and shorten rapidly in the first years of life because of intense cell proliferation and in spite of the presence of telomerase activity in bone marrow cells progenitors. Afterwards, telomere lengths shorten very progressively with age (23–47 bp/year) to reach a lower level, around 6-8 kb. The important dispersion of measurements within age classes mainly reflects the genetic heterogeneity amongst individuals and the fact that telomere lengths and (likely) the strength of telomerase activity are genetically determined. However, sex and exogenous factors (such as smoking) have a significant impact on telomere lengths. The blue line represents the incidence of all types of carcinomas in the whole population, with a strong increase around 40–50 years old. Initial studies pointed to the association between individuals with short telomeres and a higher risk for them of developing cancer. However, more recent studies using a genetic approach (Mendelian randomization, see the text) discovered that individuals with a genetic predisposition to have short telomeres were also less predisposed to cancer*

Interestingly, individuals affected with cancer display shorter telomeres than individuals of the same age without cancer (Willeit et al. 2010; Wentzensen et al. 2011). Furthermore, cancer patients with the shortest telomeres have a higher risk of dying from a metastatic disease than those with longer telomeres (Willeit et al. 2010). These observations, together with the fact that the frequency of carcinomas increases almost exponentially after the fourth decade of life (Figure 11.2)

(Jemal et al. 2008), supported for a long time the view that individuals bearing short telomeres had a higher risk of developing a cancer. However, this conclusion contrasted sharply with strong experimental evidence in mouse models which indicated that, in an otherwise wild type context, tumor progression is impeded by the presence of short telomeres (Gunes and Rudolph 2013). Furthermore, while longitudinal studies in which healthy individuals were followed for at least two decades showed that individuals with shorter telomeres bore a higher risk of premature death, this death was mostly due to cardiovascular and inflammatory diseases, but not to cancer (Cawthon et al. 2003). It follows that longer telomeres are associated with better health but short telomeres do not predispose to cancer, thus questioning the observed association between short telomeres and cancer. This intriguing conundrum was recently solved by elegant epidemiological studies that used a Mendelian randomization approach (Rode et al. 2015). In this type of study, rather than measuring telomere lengths in the population, individuals are instead given a score calculated by the presence of single nucleotide polymorphisms (SNPs) that had been previously found strongly associated with short telomeres and physically linked to TERT, TERC and OBFC1 genes (Codd et al. 2013). As mentioned in other chapters, TERT and TERC are key components of the telomerase enzyme, while OBFC1 is involved in telomere replication. Remarkably, it was observed that individuals with a maximum score of genetic association with short telomeres were in fact protected against cancer in comparison to the rest of the population (Rode et al. 2015), even though the same individuals were predisposed to cardiovascular diseases. This extraordinary result perfectly fits with the *in vitro* and *in vivo* experimental evidence and provides a very strong support to the contention that short telomeres do act as barriers to uncontrolled cell proliferation and therefore to cancer.

11.3. Excessive shortening of telomeres promotes CIN in the early stages of cancer transformation

If shortening of telomeres protects against oncogenic development, this is only true in the context where cell cycle checkpoints, particularly the p53-dependent pathway, are intact. Notably, this pathway is directly or indirectly inactivated in the great majority of cancers (Müller and Vousden 2013), and therefore the detection of unprotected chromosome ends no longer triggers growth arrest in pre-tumoral cells (Frias et al. 2012). Instead, cells keep dividing and the shortening of telomeres continues, thus deprotecting an increasing number of chromosome ends (Figure 11.3) (Muraki et al. 2012). These unprotected ends then undergo repair through end-to-end fusions, either between sister chromatids, at the end of the

S-phase, or between individual chromosomes, at any point of the cell cycle. In the former case, sister chromatids will segregate into different daughter cells and the point of fusion will form a bridge which will eventually break when the cells drift away thus creating new chromosome extremities ready to undergo a new fusion. The point of break may be far from the point of the initial fusion and therefore could be associated with loss of genetic material in one daughter cell and gain of the same material in the other, thus creating genetic imbalances (Londono-Vallejo 2008). A similar situation occurs when a dicentric chromosome is formed by a fusion between different chromosomes, an event that can also lead to aneuploidy, that is, the gain or loss of a whole chromosome (or chromosome arm) in a cell. Repeated cycles of breakage, fusion and bridge (BFB) formation during mitosis, as well as whole chromosome mis-segregations (leading to micronuclei formation and to chromothripsis), as well as the failure to undergo mitosis (leading to tetraploidization), all can result from telomere-mediated CIN (Figure 11.3) (Davoli et al., 2010; Maciejowski et al. 2015). Experimentally, during cell proliferation, it is possible to observe a point where so many chromosome extremities become deprotected that a cell can no longer divide without losing essential genetic material. At this point, called telomere crisis, most if not all cells in the culture die, thus creating a second barrier to cell proliferation (Counter et al. 1992). Unfortunately, this point also generates a very strong bottle neck that favors the emergence of unique cells that manage to reactivate a mechanism that will stabilize telomeres and therefore chromosome ends (Counter et al. 1992) (Figure 11.3). The qualitative and quantitative CIN characteristics of post-crisis cells obtained *in vitro* closely phenocopy the characteristics of many cancer cells *in vivo*, suggesting that telomere-mediated chromosome fusions are an important motor for genome instability during transformation (Nowak et al. 2002; Gisselsson 2003; Der-Sarkissian et al. 2004; Swiggers et al. 2006; Zheng et al. 2009; Roger et al. 2013; Suraweera et al. 2016). Indeed, robust evidence accumulated during recent years indicate that telomere shortening is prevalent in pre-neoplastic lesions *in vivo* and that telomere–telomere fusions are quite common during the early stages of tumor development (Shih et al. 2001; Lantuejoul et al. 2005; Raynaud et al. 2008; Chene et al. 2013; Danforth 2016). However, many chromosome rearrangements observed in late tumors are the result of secondary events that did not necessarily involve a telomere or a chromosome extremity, making it difficult to associate them with a telomere-driven initiation event. Recent deep genome-wide short-read sequencing approaches have allowed us to identify fusion events between chromosomes that can only be explained by end-to-end fusions (Alexandrov et al. 2013). However, only the long-read sequence approaches or long fiber mapping technologies will provide a true appreciation of the prevalence of telomere-driven CIN in human cancer.

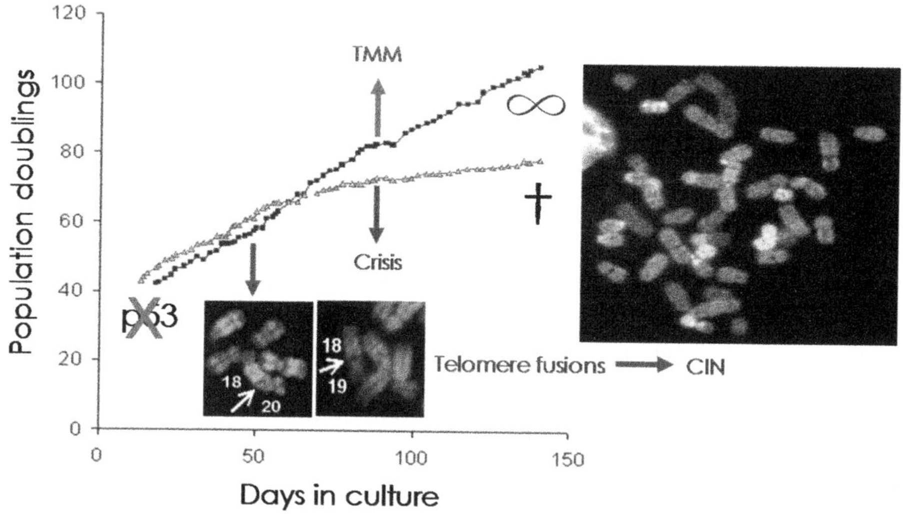

Figure 11.3. *Telomere-driven chromosome instability contributes to tumor cell transformation. On the graph, the two curves represent the growth kinetics of two independent human cell lines growing in vitro after inactivation of the p53 pathway (thus preventing senescence). In these cells, the shortest telomeres will reach their critical length after around 50 population doublings. As illustrated below the curves, the ends of the chromosome arms carrying unprotected telomeres will be fused to each other. These chromosome fusions will be at the origin of chromosome segregation conflicts during the next mitosis, which will likely be resolved by introducing new chromosome breaks. With further cell divisions, more telomeres will become deprotected and many other chromosome ends will be affected by cycles of breakage and fusion, thus leading to spread chromosome instability (CIN), as revealed on the right using multicolor FISH karyotyping analysis where color changes on the same chromosome indicate a translocation event. With time, cells approaching 100 population doublings will undergo massive CIN and will be unable to divide without losing genetic material, and therefore die, leading to "crisis" (defined operationally by the inflexion on the growth curve). On the other hand, and fueled by CIN and the selective pressure imposed by the crisis, a rare event may lead to the reactivation of a telomere maintenance mechanism (TMM), which most often corresponds to the re-expression of the catalytic subunit of telomerase (TERT). This event will allow the emergence of an immortal cell (∞), which will bear an abnormal karyotype (such as the one represented in the picture) and which is very similar to the type of rearrangements frequently detected in carcinomatous cells. However, these post-crisis cells are not completely transformed as they are unable to form tumors in the absence of a senescent microenvironment*

11.4. The contribution of telomere-driven CIN to tumor progression

How telomere-driven CIN contributes to tumor progression is still a matter of intense research. Local rearrangement events as a consequence of BFBs leading to gains or losses of genes directly involved in cell cycle control or differentiation are often revealed in cancer genomes but those invariably retrieved in all tumors of the same type or else proven to be a driver event in cancer progression are rare. For instance, the short arm of chromosome 17, whose telomere tends to be one of the shortest in a human cell, is often lost in many types of cancers (Martens et al. 1998). Interestingly, close to this chromosome extremity is the gene coding for p53, and therefore such telomere-mediated rearrangements immediately provide a clear advantage to the tumor cell. However, more often than not, lost or gained regions identified in cancer genomes do not carry well-known or even suspected tumor suppressor or cancer driver genes.

In striking contrast with this lack of direct connection with cancer gene pathways, it has been shown in multiple cancer types that the degree of CIN directly correlates with the aggressiveness of the disease (i.e. the occurrence of metastases) (Rajagopalan and Lengauer 2004; Campbell et al. 2010; Taylor et al. 2010; Brennan et al. 2013; Watanabe et al. 2012; Tijhuis et al. 2019). Efforts to demonstrate that metastatic tumor cells have gained new, key mutations in cancer driver genes have often failed and instead studies have shown that most of the driver mutations are already present in the primary tumor (Campbell et al. 2010). Interestingly, recent experimental studies have demonstrated that CIN constitute a totally independent factor that contributed to the metastatic behavior of tumors (Matano et al. 2015; Bakhoum et al. 2018; Tijhuis et al. 2019). In this context, work in the Telomere and Cancer lab at the Curie Institute has experimentally demonstrated that telomere-driven CIN in epithelial cells initiates a general response that comprises global epigenetic rearrangements and profound transcriptomic changes affecting coding and non-coding RNAs, including an extensive deregulation of the micro-RNA landscape (Castro-Vega et al. 2013) (Figure 11.4). Among the latter, micro-RNAs from the miR-200 family are strongly repressed in CIN+ cells, almost immediately after the initiation of CIN, an event directly connected with the induction of an epithelial-to-mesenchymal (EMT) transition. EMT is a cell transcription reprogramming phenomenon widely recognized as associated with the acquisition of metastatic properties by the epithelial cells (Pastushenko and Blanpain 2019). Concomitantly, HOTAIR, a long non-coding RNA also associated by multiple studies with metastatic potential, is up-regulated in cells undergoing CIN and it may also contribute to changes in the transcriptome landscape during transformation (Jarroux et al. 2021). Strikingly, most if not all transcription deregulation events detected in the telomere-driven CIN model are independent of copy number variations or local events (Castro-Vega et al. 2013). Of note, the CIN-associated reprograming of epithelial cells is stable and persists in post-crisis after spontaneous immortalization. Post-crisis cells display many chromosome rearrangements but they are no longer undergoing active CIN because telomeres are now maintained by the expression of endogenous telomerase. These cells display, though, a higher level of endogenous DNA damage, which seems to be required to maintain EMT (Castro-Vega et al. 2013). Our observations demonstrate that a transient destabilization of the genome may have a lasting impact on the genetic program of a transformed cell. Interestingly, this new program is still under the influence of the microenvironment, and in particular the presence of senescent cells for reasons that will be explained in the next section.

11.5. The senescent microenvironment promotes cancer development

There is no greater risk factor for developing a cancer, particularly a carcinoma, than advanced age (DePinho 2000). Indeed, and as mentioned above, the incidence of cancer increases almost exponentially after the fourth or fifth decade of life. Also, as explained, telomeres shorten with age and in normal cells they act as a barrier to tumor development as this shortening induces cell senescence. Nevertheless, the accumulation of senescent cells impacts not only the performance of the affected organs but also influences their microenvironment through the secretion of numerous factors, a phenomenon called the senescence-associated secretory phenotype (SASP) (Coppe et al. 2010). SASP composition varies following the type of cell that enters senescence but it is generally enriched in inflammatory cytokines, growth factors and enzymes capable of profoundly modifying the extracellular matrix (Basisty et al. 2020). It has been shown that the presence of senescent cells next to tumor cells has a profound impact in tumor cell behavior, making it prone to proliferation and invasion (Ortiz-Montero et al. 2017). Other experiments have demonstrated that telomere-driven CIN+ post-crisis cells were able to form tumors only when in the presence of telomere-driven senescent fibroblasts (Castro-Vega et al. 2015) (Figure 11.4). Remarkably, these post-crisis cells reversed their EMT phenotype when in contact with senescence-conditioned media and the cells issued from the tumors they formed in the presence of senescent fibroblasts had undergone a stable MET transition (Castro-Vega et al. 2015) (Figure 11.4). Interestingly, these MET cells were now endowed of all the characteristics ascribed to cancer stem cells (self-renewal capacity, high tumor initiation capacity and capacity to differentiate into other cell types). The characterization of these MET cells did not reveal further evolution in the genetic landscape with respect to the EMT state, whereas the micro-RNA transcriptome landscape almost completely reverted to its pre-CIN state, strongly suggesting an epigenetic mechanism involved in the response to the senescent microenvironment (Castro-Vega et al. 2015). These experiments perfectly illustrate the plasticity of CIN+ cells and suggest how the combination of CIN in tumor cells at a primary site (thus fueling the acquisition of migration properties through EMT) and the accumulation of senescent cells with age at distant places (thus pushing EMT cells into MET) may result in an increased risk of developing metastases. Indeed, it has been calculated that the natural evolution of a tumor, for instance, a colocarcinoma, can take 15-20 years before the occurrence of metastasis (Jones et al. 2008). Given that there is no further genetic evolution in the tumor between the last stage at the primary site and the metastatic site, it is highly likely that the acquisition of metastatic potential by the tumor predates the actual occurrence of metastasis by several years. It is then tempting to conclude that it is the accumulation of senescent cells in the target organ that makes it possible for EMT migrant cells to colonize, undergo MET and develop a full metastatic tumor. Supporting this scenario are findings suggesting that circulating cancer cells can be detected very soon after tumor initiation (Hosseini et al. 2016). In all, telomere shortening not only constitutes a critical facet of the evolution of CIN in carcinomas but is also an important aspect of the tumor microenvironment through the induction of senescence in stroma cells.

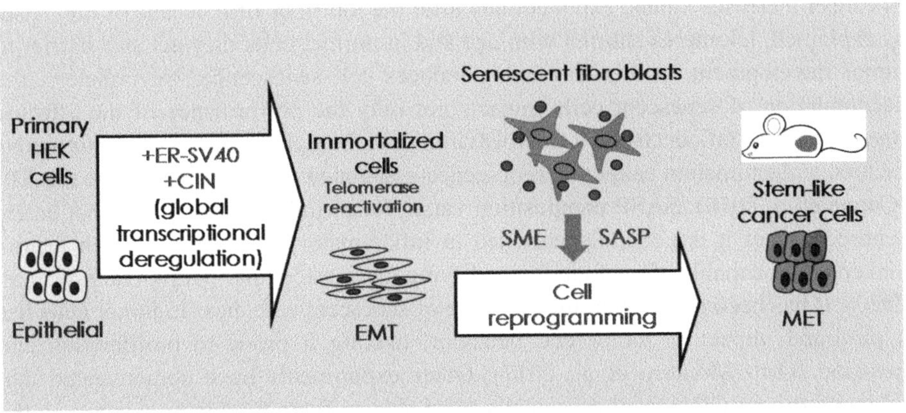

Figure 11.4. *Role of telomere-driven CIN in the acquisition of tumorigenic capacity by cancer cells*

NOTES ON FIGURE 11.4.– *In vitro models of telomere-driven CIN (such as the one presented in Figure 11.3) have allowed the dissection of some of the steps that contribute to the tumor transformation of cells undergoing a telomere-driven CIN. Cells in which the p53/Rb pathways have been abolished by the introduction of the early region of SV40 (ER-SV40), and that enter a period of telomere-driven CIN, undergo massive transcriptional deregulation that comprises the induction of an epithelial-to-mesenchymal transition (EMT). This EMT phenomenon has been associated with the acquisition of stemness and metastatic properties by cancer cells.*

The new transcriptomic landscape is fixed very rapidly after the initiation of CIN and persists after spontaneous or exogenous immortalization (telomerase reactivation). These CIN-positive post-crisis EMT cells are unable to form tumors in immunosuppressed mice unless they are co-injected with senescent fibroblasts. The senescent microenvironment (SME) favors, by the secretion of numerous factors (SASP), the growth of tumors composed of fully transformed cells that have undergone a mesenchymal-to-epithelial transition (MET). These cells now bear all the characteristics of stem-like cancer cells: high tumor initiation capacity (in the absence of SME), self renewal (sphere formation) potential and cell differentiation capacity (into adipocytes, for instance).

11.6. Mechanisms of cell immortalization

In order for a transformed cell to increase its proliferation potential and form a mass, it almost invariably needs to reactivate a mechanism of telomere maintenance. The most frequent TMM is the reactivation of telomerase activity (Kim et al. 1994). As explained elsewhere, telomerase is a reverse transcriptase that uses a specific RNA to add TTAGGG repeats to the 3′ end of telomeres. Telomerase activity is present at high levels during embryonic development and is detectable in most progenitor cells after birth (Wright et al. 1996). However, during cell differentiation, the expression of TERT, the catalytic subunit of the enzyme, is repressed, while expression of TERC, the RNA subunit, is generally maintained (Wright et al. 1996). It has been shown that forcing the expression of TERT in normal human diploid cells is sufficient to induce cell immortalization (Bodnar et al. 1998). In agreement with that, most cancer cells spontaneously re-express TERT at some point during tumor evolution, after a period of prolonged proliferation that leads to telomere shortening (Shay and Wright 1996). The mechanisms of re-activation of TERT expression are diverse and may result from CIN-associated events like locus amplification or translocation (Nowak et al. 2006; Zhu et al. 2006; Peifer et al. 2015; Dwight et al. 2018), overexpression of transcriptional activators (like Myc) (Khattar and Tergaonkar 2017; Baena-Del Valle et al. 2018), epigenetic modifications, including promoter methylation or demethylation (thus allowing the binding of activators or repressors) (Guilleret and Benhattar 2003; Lee et al. 2019; Lee et al. 2020), and the occurrence of promoter mutations, which create new binding sites for activators or destroy binding sites for suppressors (Borah et al. 2015; Huang et al. 2015; Stern et al. 2015) (Figure 11.5). The latter mechanism is predominant in particular tumors like bladder carcinoma, glioblastoma, hepatocellular carcinoma, etc. (Nault et al. 2013; Vinagre et al. 2013). Also, in hepatitis B virus-related hepatocellular carcinoma, the integration of virus at the level of the *TERT* promoter is rather frequent (Sung et al. 2012), leading to recruitment of host transcriptional factors that activate *TERT* expression (Sze et al. 2021). However, in the majority of the most frequent types of carcinomas (breast, prostate, colon, etc.), the mechanism of reactivation of telomerase expression is likely highly variable between tumors and in any case it is less well understood.

Although telomerase activity tends to be high in tumor cells, telomere lengths are in general quite short. This is likely the result of an aberrant homeostasis where mechanisms of length control are lost or only partially effective. Both telomerase recruitment at chromosome ends and its activity toward the 3′ overhang may be affected by alterations in the cell cycle checkpoints or by an abnormal composition in the shelterin complex, which may also lead to defects in telomere capping, thus allowing some degree of telomere degradation. The relative importance of these aspects for tumor progression remains unknown. In any case, mutations in shelterin components, such as POT1, have been associated with the emergence and progression of some types of cancers of great variety (Wu et al. 2020), some of which are familial (most notably melanoma (Potrony et al. 2015), chronic lymphocytic leukemia (Speedy et al. 2016), angiosarcoma (Calvete et al. 2015) and glioma (Bainbridge et al. 2015)), thus underlining the importance of telomere biology in cancer. Interestingly, and contrary to what it is usually observed in most carcinomas, POT1-mutated cancers are associated with long telomeres.

If over 90% of all human tumors reactivate telomerase, the majority of the remaining 10% utilizes a different mechanism based on homologous recombination, collectively named ALT (alternative lengthening of telomeres) (Reddel 2003) (Figure 11.5). Most ALT positive tumors are from a mesenchymal origin and, conversely, up to 50% of these tumors may be ALT (Heaphy et al. 2011). In the ALT context, telomeres are recombinogenic, allowing the G-rich 3′ overhang to invade the double strand portion of another telomere and to prime the extension of that 3′ end by a semiconservative replication-mediated mechanism akin to "Break Induced Replication" (BIR) (Cesare and Reddel 2010; Dilley et al. 2016; Roumelioti et al. 2016). After replication, which proceeds until the end of the invaded telomere is reached, the extended overhang is converted into double strand by standard filling-up mechanisms (Dilley et al. 2016; Zhang et al. 2019). The number of added repeats will depend on the length of the invaded telomere and may represent several kilobases. At the same time, ALT telomeres are the target of damage/repair mechanisms that may lead to abrupt telomere loss (Yang et al. 2020). Sudden gains and losses of long telomere repeat tracts constitute a major characteristic of ALT activity, a feature reflected in a Southern blot or in a telomere FISH image by a pronounced heterogeneity of telomere lengths (Murnane et al. 1994). Another characteristic of ALT cells is the presence of modified PML nuclear bodies or APBs (ALT-associated PML bodies) (Yeager et al. 1999) (Figure 11.5). In these bodies, we find all components of PML bodies together with telomere (chromosomal and extrachromosomal) DNA material as well as DNA damage response components. It is likely that APBs favor inter-telomere recombination activities by facilitating telomere–telomere interactions, mainly after replication (Draskovic et al. 2009; Zhang et al. 2020). The formation of APBs is quite dynamic and likely depends on epigenetic modifications, which may be quite pronounced in ALT telomeres, as well as in the presence of DNA damage and recombination mediating factors, which may influence the movements and directionality of damaged telomeres (Potts and Yu 2007;Cho et al. 2014; Conomos et al. 2014) (Figure 11.5).

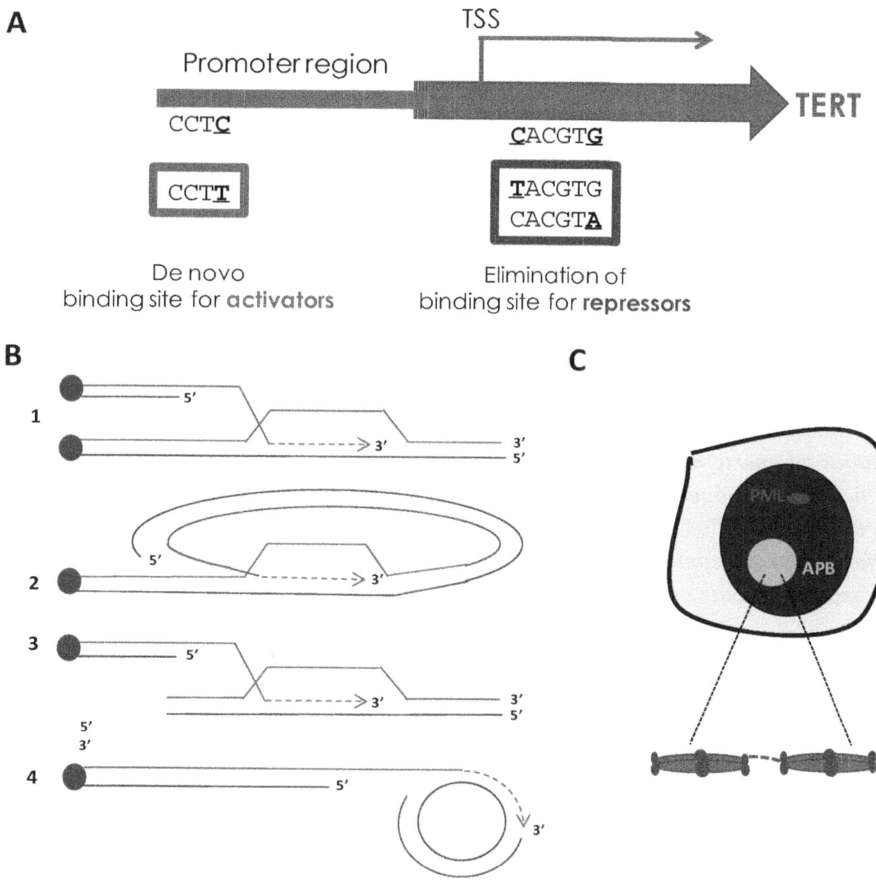

Figure 11.5. *Mechanisms of cell immortalization*

NOTES ON FIGURE 11.5.– *In order to become immortal, cancer cells should reactivate a mechanism of telomere maintenance. A) Most often, tumor cells reactivate the expression of TERT, the gene coding for the catalytic subunit of telomerase. This reactivation may be connected to CIN through amplification of the locus or translocation events that put active enhancers in proximity to the TERT promoter or that induce changes in the chromatin environment. In certain types of cancer, reactivation of TERT is associated with mutations in the promoter, some of which create binding sites for transcriptional activators (green box), others may eliminate binding sites for transcriptional repressors (red box). Very often, the mechanisms of TERT reactivation remain unknown. B) Another mode of immortalization is the maintenance of telomeres using a recombination-based mechanism (ALT). In this mechanism, the G-rich 3' overhang of a telomere invades another telomere sequence found on another chromosome and uses the complementary strand (in red) as a template to synthesize new repeats (broken blue line) thus lengthening the overhang (1). This very long G-rich overhang will then be converted into double stranded telomere by a fill-in reaction (not shown). It is also possible that the T-loop structure formed by the invasion of proximal sequences by the 3' extension of the same telomere provides an opportunity for this 3' end to be extended (2). In other modes of telomere synthesis, the 3' overhang will use extrachromosomal templates, which are present in abundance in ALT cells either in linear (3) or circular (4) forms. Indeed, the presence of covalently closed C-rich circles (in red) is considered pathognomonic of ALT. C) Another characteristic of ALT cells is the presence of APBs (ALT-associated PML bodies, yellow circle) in their nuclei (large brown oval). APBs are PML bodies (small red oval) that also contain telomeric material as well as Shelterin components and proteins belonging to the DNA damage/repair pathway. It has been shown that telomere exchanges can occur within APBs (illustrated by the broken red line connecting two chromatids on different chromosomes) after replication. High levels of telomere exchange between chromatids are also a characteristic of ALT cells.*

Finally, examples of malignant tumors where no TMM was detected have been described (Dagg et al. 2017; Claude and Decottignies 2020). In this case, one may argue that telomeres were long enough in the initially transformed cell to support cell growth all along the formation of the tumor. Indeed, even normal, diploid telomere-negative cells are capable of dozens of cell divisions, which may be sufficient, if kept unchecked, to form a tumor. In the case of TMM-negative tumor cells, some of which were reported to divide >200 times *in vitro*, clearly no TMM would be required for the tumor to reach a large size. However, it is likely that in the great majority of cases, cell transformation results in high cell fragility and mortality, hence the need for a TMM in order to ensure positive growth. Also, telomere dynamics in cancer cells is likely more complex that just the loss of telomere repeats due to the end replication problem (for instance, telomere damage may be permanently induced by the increased oxidative damage), thus requiring constant healing by telomerase or another TMM.

11.7. Therapeutic potential of telomeres and telomerase and prognostic value of telomere-related evaluations

Given their key roles both in cell viability and cell immortalization, enormous research efforts have been deployed during the last decades to develop therapeutic approaches targeting telomeres and telomerase to fight cancer (Figure 11.6). The discovery that telomerase was a reverse transcriptase opened the possibility that inhibitors similar to those developed against viral diseases could interfere with the telomerase enzymatic activity (Strahl and Blackburn 1996). Facilitated by the development of *in vitro* PCR-based telomerase activity tests, very large molecular screenings were also carried out to find new telomerase inhibitors. Strikingly, such efforts have remained largely fruitless, with the identification of very few molecules, of which only one (GRN163L or Imetelstat) has reached Phase III for restricted therapeutic indications, including myelofibrosis (https://clinicaltrials.gov/ct2/show/NCT04576156). Another obstacle encountered by telomerase inhibitors, including dominant-negative approaches (where dysfunctional forms of telomerase that interfere with telomere maintenance are expressed), is tumor cell resistance acquired through overexpressing of the endogenous telomerase (Incles et al. 2003; Gomez et al. 2003; Bojovic and Crowe 2011) or the activation of an alternative TMM (Queisser et al. 2013). During the search for telomerase inhibitors, a particular class of small molecules elicited a great interest: the G-quadruplex ligands. These molecules bind and stabilize secondary structures formed by G-rich sequences and were initially expected to prevent telomere elongation by telomerase (Guittat et al. 2004; Carvalho et al. 2020). In fact, it was soon realized that these ligands could target telomere sequences within cells and provoke DNA damage or inhibit their replication (Guittat et al. 2004; Carvalho et al. 2020) (Figure 11.6). Such ligands can also recognize G4s formed elsewhere in the genome and thus elicit damage responses unrelated to telomeres as well as transcriptional responses both in relation to both the stabilization of G4s found in promoters, the replication-related DNA damage response and other metabolic effects, including at the level of mitochondria (Hampel et al. 2013; Hasegawa et al. 2016; Falabella et al. 2019; Gray et al. 2019). In all, they have appeared to have antitumor cell activities and currently constitute good candidates for the development of anticancer strategies, in combination with other therapeutic approaches, such as radiotherapy (Merle et al. 2011) (Figure 11.6).

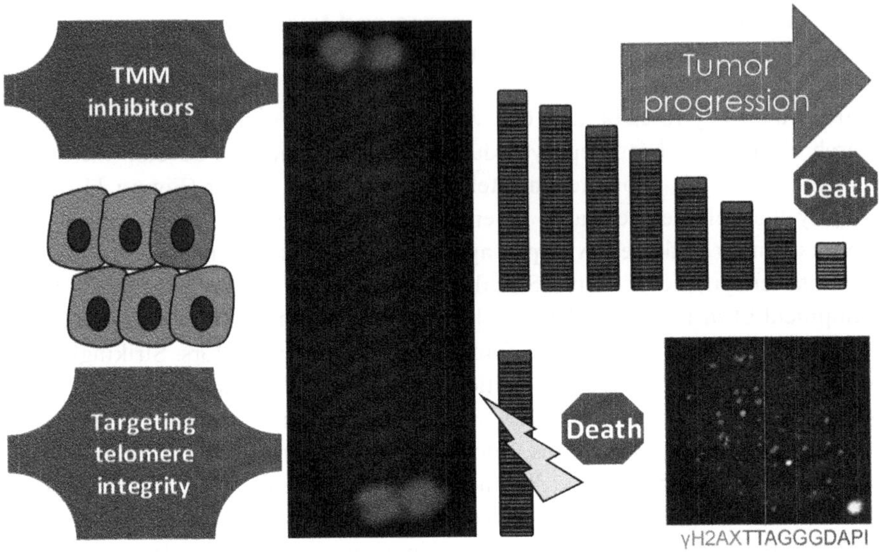

Figure 11.6. *Telomeres as a therapeutic target to fight cancer. Given the essential role played by telomeres in genome stability and the predicted consequences of telomere dysfunction in cell survival, telomeres have emerged as a potential target to prevent cancer cells from proliferating. Since the great majority of cancer cells have reactivated a telomere maintenance mechanism (TMM), which is absent in most normal somatic cells, interfering with that mechanism will lead to proliferation-associated telomere shortening. So far, only one inhibitor of telomerase (which is active in 90% of cancers) has reached clinical trials and it is still under evaluation. No inhibitors of ALT have been described up today. In any case, it is predicted that such approaches will take some time to exert their effects on cancer cells (thus allowing further tumor progression) since telomeres will have to shorten enough to trigger chromosome end deprotection, rampant genome instability and cell death. Also, when targeting one type of TMM, cells may develop resistance by activating the other type. Therefore, anti-TMM therapeutic approaches are likely to be combined with other therapies. On the other hand, therapeutic approaches that target the integrity of telomeres (by interrupting replication, inducing damage or disrupting telomere structure), such as G-quadruplex ligands, are expected to have an immediate effect on cancer cells. However, such agents may also affect normal cells in healthy tissues and also have non-telomeric effects because of the presence of similar targets in other parts of the genome*

Along with the search for therapeutic applications, a great effort has also been invested to better understand the clinical implications of TMMs used by tumor cells. For instance, high levels of telomerase have been found to be associated with more aggressive cancers and the absence of telomerase may indicate a better response to classic anti-cancer therapies (Shay et al. 1997; Carey et al. 1999; Gonzalez-Quevedo et al. 2002; Tomoda et al. 2002; Sanz-Casla et al. 2005; Kulic et al. 2016). At the same time, the absence of telomerase activity may indicate the presence of an alternative TMM, which can carry prognostic information (Hakin-Smith et al. 2003; Chen et al. 2006; Matsuo et al. 2010; Lee et al. 2012; Lee et al. 2015; Job et al. 2019; Kent et al. 2019; Koneru et al. 2020). For instance, in osteosarcoma, the presence of telomerase signifies a worse prognosis (Ulaner et al. 2003), while in liposarcomas the detection of an alternative TMM is associated with a worse outcome (Venturini et al. 2010; Lee et al. 2015). Nevertheless, measuring telomerase activity requires specialized equipment and reagents and, above all, fresh tumor tissue samples, thus making it quite difficult to realize. Also, absence of telomerase activity does not necessarily mean that an alternative TMM is at play (Claude and Decottignies 2020). The latter needs to be evidenced by the presence of APBs or of a high heterogeneity of telomere lengths in tumor cells, which can be done in fixed specimens (Heaphy et al. 2011), or else by the presence of C-circles, detected in tumor DNA by PCR (Henson et al. 2017). Another aspect of telomere biology that can be studied in fixed specimens is the level of expression of shelterin components, whose alterations can be associated with better or worse prognosis and with the response to therapy (Augereau et al. 2011; Ozden et al. 2014; Panero et al. 2014). Finally, as mentioned previously, the degree of genome instability of a tumor is often associated with the clinical outcome. In all, although determining the degree of genome instability, the type of TMM, the level of telomerase activity or the level of telomere damage can provide clinically relevant information, such approaches are far from being used in the common practice and remain associated with research activities.

11.8. Conclusion

Telomere maintenance is tightly linked to genome stability and cell proliferation. Telomere dysfunction is intimately associated with major aspects of tumor development. Thus, interfering with telomere maintenance continues to be a major objective in anticancer research. Nevertheless, much effort still needs to be deployed to better understand the nuances of cancer-associated telomere metabolism in order to improve the therapeutic margin and preserve the telomere status of non-tumor cells.

11.9. Acknowledgments

Work in the Telomeres and Cancer laboratory has been supported along the years with multiple grants from ANR, INCa, Fondation ARC, La Ligue contre le

Cancer, La Fondation de France, La Fondation pour les Maladies Rares, as well as internal grant calls from the Curie Institute. I thank all former and present members of the lab as well as all collaborators. I apologize to all colleagues in the telomere field whose original works were not cited due to space limitations.

11.10. References

Alexandrov, L.B., Nik-Zainal, S., Wedge, D.C., Aparicio, S.A., Behjati, S., Biankin, A.V., Bignell, G.R., Bolli, N., Borg, A., Borresen-Dale, A.L. et al. (2013). Signatures of mutational processes in human cancer. *Nature*, 500, 415–421.

Augereau, A., T'kint De Roodenbeke, C., Simonet, T., Bauwens, S., Horard, B., Callanan, M., Leroux, D., Jallades, L., Salles, G., Gilson, E. et al. (2011). Telomeric damage in early stage of chronic lymphocytic leukemia correlates with shelterin dysregulation. *Blood*, 118, 1316–1322.

Baena-Del Valle, J.A., Zheng, Q., Esopi, D.M., Rubenstein, M., Hubbard, G.K., Moncaliano, M.C., Hruszkewycz, A., Vaghasia, A., Yegnasubramanian, S., Wheelan, S.J. et al. (2018). MYC drives overexpression of telomerase RNA (hTR/TERC) in prostate cancer. *J. Pathol.*, 244, 11–24.

Bainbridge, M.N., Armstrong, G.N., Gramatges, M.M., Bertuch, A.A., Jhangiani, S.N., Doddapaneni, H., Lewis, L., Tombrello, J., Tsavachidis, S., Liu, Y. et al. (2015). Germline mutations in shelterin complex genes are associated with familial glioma. *J. Natl. Cancer Inst.*, 107, 384.

Bakhoum, S.F., Ngo, B., Laughney, A.M., Cavallo, J.A., Murphy, C.J., Ly, P., Shah, P., Sriram, R.K., Watkins, T.B.K., Taunk, N.K. et al. (2018). Chromosomal instability drives metastasis through a cytosolic DNA response. *Nature*, 553, 467–472.

Basisty, N., Kale, A., Jeon, O.H., Kuehnemann, C., Payne, T., Rao, C., Holtz, A., Shah, S., Sharma, V., Ferrucci, L. et al. (2020). A proteomic atlas of senescence-associated secretomes for aging biomarker development. *PLoS Biol*, 18, e3000599.

Bodnar, A.G., Ouellette, M., Frolkis, M., Holt, S.E., Chiu, C.P., Morin, G.B., Harley, C.B., Shay, J.W., Lichtsteiner, S., Wright, W.E. (1998). Extension of life-span by introduction of telomerase into normal human cells. *Science*, 279, 349–352.

Bojovic, B. and Crowe, D.L. (2011). Resistance to telomerase inhibition by human squamous cell carcinoma cell lines. *Int. J. Oncol.*, 38, 1175–1181.

Borah, S., Xi, L., Zaug, A.J., Powell, N.M., Dancik, G.M., Cohen, S.B., Costello, J.C., Theodorescu, D., Cech, T.R. (2015). Cancer. TERT promoter mutations and telomerase reactivation in urothelial cancer. *Science*, 347, 1006–1010.

Brennan, C.W., Verhaak, R.G., Mckenna, A., Campos, B., Noushmehr, H., Salama, S.R., Zheng, S., Chakravarty, D., Sanborn, J.Z., Berman, S.H. et al. (2013). The somatic genomic landscape of glioblastoma. *Cell*, 155, 462–477.

Calvete, O., Martinez, P., Garcia-Pavia, P., Benitez-Buelga, C., Paumard-Hernandez, B., Fernandez, V., Dominguez, F., Salas, C., Romero-Laorden, N., Garcia-Donas, J. et al. (2015). A mutation in the POT1 gene is responsible for cardiac angiosarcoma in TP53-negative Li-Fraumeni-like families. *Nat. Commun.*, 6, 8383.

Campbell, P.J., Yachida, S., Mudie, L.J., Stephens, P.J., Pleasance, E.D., Stebbings, L.A., Morsberger, L.A., Latimer, C., Mclaren, S., Lin, M.L. et al. (2010). The patterns and dynamics of genomic instability in metastatic pancreatic cancer. *Nature*, 467, 1109–1113.

Campisi, J. (2011). Cellular senescence: Putting the paradoxes in perspective. *Curr. Opin. Genet. Dev.*, 21, 107–112.

Carey, L.A., Kim, N.W., Goodman, S., Marks, J., Henderson, G., Umbricht, C.B., Dome, J.S., Dooley, W., Amshey, S.R., Sukumar, S. (1999). Telomerase activity and prognosis in primary breast cancers. *J. Clin. Oncol.*, 17, 3075–3081.

Carvalho, J., Mergny, J.L., Salgado, G.F., Queiroz, J.A., Cruz, C. (2020). G-quadruplex, friend or foe: The role of the G-quartet in anticancer strategies. *Trends Mol. Med.*, 26, 848–861.

Castro-Vega, L.J., Jouravleva, K., Liu, W.Y., Martinez, C., Gestraud, P., Hupe, P., Servant, N., Albaud, B., Gentien, D., Gad, S. et al. (2013). Telomere crisis in kidney epithelial cells promotes the acquisition of a microRNA signature retrieved in aggressive renal cell carcinomas. *Carcinogenesis*, 34, 1173–1180.

Castro-Vega, L.J., Jouravleva, K., Ortiz-Montero, P., Liu, W.Y., Galeano, J.L., Romero, M., Popova, T., Bacchetti, S., Vernot, J.P., Londono-Vallejo, A. (2015). The senescent microenvironment promotes the emergence of heterogeneous cancer stem-like cells. *Carcinogenesis*, 36, 1180–1192.

Cawthon, R.M., Smith, K.R., O'Brien, E., Sivatchenko, A., Kerber, R.A. (2003). Association between telomere length in blood and mortality in people aged 60 years or older. *Lancet*, 361, 393–395.

Cesare, A.J. and Reddel, R.R. (2010). Alternative lengthening of telomeres: Models, mechanisms and implications. *Nat. Rev. Genet.*, 11, 319–330.

Cesare, A.J., Hayashi, M.T., Crabbe, L., Karlseder, J. (2013). The telomere deprotection response is functionally distinct from the genomic DNA damage response. *Mol. Cell*, 51, 141–155.

Chen, Y.J., Hakin-Smith, V., Teo, M., Xinarianos, G.E., Jellinek, D.A., Carroll, T., Mcdowell, D., MacFarlane, M.R., Boet, R., Baguley, B.C. et al. (2006). Association of mutant TP53 with alternative lengthening of telomeres and favorable prognosis in glioma. *Cancer Res.*, 66, 6473–6476.

Chene, G., Tchirkov, A., Pierre-Eymard, E., Dauplat, J., Raoelfils, I., Cayre, A., Watkin, E., Vago, P., Penault-Llorca, F. (2013). Early telomere shortening and genomic instability in tubo-ovarian preneoplastic lesions. *Clin Cancer Res.*, 19, 2873–2882.

Cho, N.W., Dilley, R.L., Lampson, M.A., Greenberg, R.A. (2014). Interchromosomal homology searches drive directional ALT telomere movement and synapsis. *Cell*, 159, 108–121.

Claude, E. and Decottignies, A. (2020). Telomere maintenance mechanisms in cancer: Telomerase, ALT or lack thereof. *Curr. Opin. Genet. Dev.*, 60, 1–8.

Cleal, K., Norris, K., Baird, D. (2018). Telomere length dynamics and the evolution of cancer genome architecture. *Int. J. Mol. Sci.*, 19(2):482.

Codd, V., Nelson, C.P., Albrecht, E., Mangino, M., Deelen, J., Buxton, J.L., Hottenga, J.J., Fischer, K., Esko, T., Surakka, I. et al. (2013). Identification of seven loci affecting mean telomere length and their association with disease. *Nat. Genet.*, 45, 422–427, 427e1-2.

Conomos, D., Reddel, R.R., Pickett, H.A. (2014). NuRD-ZNF827 recruitment to telomeres creates a molecular scaffold for homologous recombination. *Nat. Struct. Mol. Biol.*, 21, 760–770.

Coppe, J.P., Desprez, P.Y., Krtolica, A., Campisi, J. (2010). The senescence-associated secretory phenotype: The dark side of tumor suppression. *Annu. Rev. Pathol.*, 5, 99–118.

Counter, C.M., Avilion, A.A., Lefeuvre, C.E., Stewart, N.G., Greider, C.W., Harley, C.B., Bacchetti, S. (1992). Telomere shortening associated with chromosome instability is arrested in immortal cells which express telomerase activity. *EMBO J.*, 11, 1921–1929.

Dagg, R.A., Pickett, H.A., Neumann, A.A., Napier, C.E., Henson, J.D., Teber, E.T., Arthur, J.W., Reynolds, C.P., Murray, J., Haber, M. et al. (2017). Extensive proliferation of human cancer cells with ever-shorter telomeres. *Cell Rep.*, 19, 2544–2556.

Danforth Jr., D.N. (2016). Genomic changes in normal breast tissue in women at normal risk or at high risk for breast cancer. *Breast Cancer (Auckl)*, 10, 109–146.

Davoli, T., Denchi, E.L., de Lange, T. (2010). Persistent telomere damage induces bypass of mitosis and tetraploidy. *Cell*, 141, 81–93.

Demanelis, K., Jasmine, F., Chen, L.S., Chernoff, M., Tong, L., Delgado, D., Zhang, C., Shinkle, J., Sabarinathan, M., Lin, H. et al. (2020). Determinants of telomere length across human tissues. *Science*, 369(6509):eaaz6876.

Depinho, R.A. (2000). The age of cancer. *Nature*, 408, 248–54.

Der-Sarkissian, H., Bacchetti, S., Cazes, L., Londono-Vallejo, J.A. (2004). The shortest telomeres drive karyotype evolution in transformed cells. *Oncogene*, 23, 1221–1228.

Dilley, R.L., Verma, P., Cho, N.W., Winters, H.D., Wondisford, A.R., Greenberg, R.A. (2016). Break-induced telomere synthesis underlies alternative telomere maintenance. *Nature*, 539, 54–58.

Draskovic, I., Arnoult, N., Steiner, V., Bacchetti, S., Lomonte, P., Londono-Vallejo, A. (2009). Probing PML body function in ALT cells reveals spatiotemporal requirements for telomere recombination. *Proc. Natl. Acad. Sci. USA*, 106, 15726–15731.

Dwight, T., Flynn, A., Amarasinghe, K., Benn, D.E., Lupat, R., Li, J., Cameron, D.L., Hogg, A., Balachander, S., Candiloro, I.L.M. et al. (2018). TERT structural rearrangements in metastatic pheochromocytomas. *Endocr. Relat. Cancer*, 25, 1–9.

Falabella, M., Kolesar, J.E., Wallace, C., De Jesus, D., Sun, L., Taguchi, Y.V., Wang, C., Wang, T., Xiang, I.M., Alder, J.K. et al. (2019). G-quadruplex dynamics contribute to regulation of mitochondrial gene expression. *Sci. Rep.*, 9, 5605.

Frias, C., Pampalona, J., Genesca, A., Tusell, L. (2012). Telomere dysfunction and genome instability. *Front Biosci (Landmark Ed.)*, 17, 2181–2196.

Gisselsson, D. (2003). Chromosome instability in cancer: How, when, and why? *Adv. Cancer Res.*, 87, 1–29.

Gomez, D., Aouali, N., Renaud, A., Douarre, C., Shin-Ya, K., Tazi, J., Martinez, S., Trentesaux, C., Morjani, H., Riou, J.F. (2003). Resistance to senescence induction and telomere shortening by a G-quadruplex ligand inhibitor of telomerase. *Cancer Res.*, 63, 6149–6153.

Gonzalez-Quevedo, R., Iniesta, P., Moran, A., De Juan, C., Sanchez-Pernaute, A., Fernandez, C., Torres, A., Diaz-Rubio, E., Balibrea, J.L., Benito, M. (2002). Cooperative role of telomerase activity and p16 expression in the prognosis of non-small-cell lung cancer. *J. Clin. Oncol.*, 20, 254–262.

Gray, L.T., Puig Lombardi, E., Verga, D., Nicolas, A., Teulade-Fichou, M.P., Londono-Vallejo, A., Maizels, N. (2019). G-quadruplexes sequester free heme in living cells. *Cell Chem. Biol.*, 26, 1681–1691.e5.

Guilleret, I. and Benhattar, J. (2003). Demethylation of the human telomerase catalytic subunit (hTERT) gene promoter reduced hTERT expression and telomerase activity and shortened telomeres. *Exp. Cell Res.*, 289, 326–334.

Guittat, L., Alberti, P., Gomez, D., De Cian, A., Pennarun, G., Lemarteleur, T., Belmokhtar, C., Paterski, R., Morjani, H., Trentesaux, C. et al. (2004). Targeting human telomerase for cancer therapeutics. *Cytotechnology*, 45, 75–90.

Gunes, C. and Rudolph, K.L. (2013). The role of telomeres in stem cells and cancer. *Cell*, 152, 390–393.

Hahn, W.C. (2002). Senescence, telomere shortening and telomere maintenance. *Cancer Biol. Ther.*, 1, 398–400.

Hakin-Smith, V., Jellinek, D.A., Levy, D., Carroll, T., Teo, M., Timperley, W.R., McKay, M.J., Reddel, R.R., Royds, J.A. (2003). Alternative lengthening of telomeres and survival in patients with glioblastoma multiforme. *Lancet*, 361, 836–838.

Hampel, S.M., Pepe, A., Greulich-Bode, K.M., Malhotra, S.V., Reszka, A.P., Veith, S., Boukamp, P., Neidle, S. (2013). Mechanism of the antiproliferative activity of some naphthalene diimide G-quadruplex ligands. *Mol. Pharmacol.*, 83, 470–480.

Hanahan, D. and Weinberg, R.A. (2011). Hallmarks of cancer: The next generation. *Cell*, 144, 646–674.

Hasegawa, D., Okabe, S., Okamoto, K., Nakano, I., Shin-Ya, K., Seimiya, H. (2016). G-quadruplex ligand-induced DNA damage response coupled with telomere dysfunction and replication stress in glioma stem cells. *Biochem. Biophys. Res. Commun.*, 471, 75–81.

Heaphy, C.M., Subhawong, A.P., Hong, S.M., Goggins, M.G., Montgomery, E.A., Gabrielson, E., Netto, G.J., Epstein, J.I., Lotan, T.L., Westra, W.H. et al. (2011). Prevalence of the alternative lengthening of telomeres telomere maintenance mechanism in human cancer subtypes. *Am. J. Pathol.*, 179, 1608–1615.

Henson, J.D., Lau, L.M., Koch, S., Martin La Rotta, N., Dagg, R.A., Reddel, R.R. (2017). The C-circle assay for alternative-lengthening-of-telomeres activity. *Methods*, 114, 74–84.

Herbig, U., Jobling, W.A., Chen, B.P., Chen, D.J., Sedivy, J.M. (2004). Telomere shortening triggers senescence of human cells through a pathway involving ATM, p53, and p21(CIP1), but not p16(INK4a). *Mol. Cell*, 14, 501–513.

Herrmann, M., Pusceddu, I., Marz, W., Herrmann, W. (2018). Telomere biology and age-related diseases. *Clin. Chem. Lab. Med.*, 56, 1210–1222.

Hosseini, H., Obradovic, M.M.S., Hoffmann, M., Harper, K.L., Sosa, M.S., Werner-Klein, M., Nanduri, L.K., Werno, C., Ehrl, C., Maneck, M. et al. (2016). Early dissemination seeds metastasis in breast cancer. *Nature*, 540, 552–558.

Huang, D.S., Wang, Z., He, X.J., Diplas, B.H., Yang, R., Killela, P.J., Meng, Q., Ye, Z.Y., Wang, W., Jiang, X.T. et al. (2015). Recurrent TERT promoter mutations identified in a large-scale study of multiple tumour types are associated with increased TERT expression and telomerase activation. *Eur. J. Cancer*, 51, 969–976.

Incles, C.M., Schultes, C.M., Kelland, L.R., Neidle, S. (2003). Acquired cellular resistance to flavopiridol in a human colon carcinoma cell line involves up-regulation of the telomerase catalytic subunit and telomere elongation. Sensitivity of resistant cells to combination treatment with a telomerase inhibitor. *Mol. Pharmacol.*, 64, 1101–1108.

Jarroux, J., Foretek, D., Bertrand, C., Gabriel, M., Szachnowski, U., Saci, Z., Guo, S., Londono-Vallejo, A., Pinskaya, M., Morillon, A. (2021). HOTAIR lncRNA promotes epithelial-mesenchymal transition by redistributing LSD1 at regulatory chromatin regions. *EMBO Rep*, 22(7), e50193.

Jemal, A., Siegel, R., Ward, E., Hao, Y., Xu, J., Murray, T., Thun, M.J. (2008). Cancer statistics, 2008. *CA Cancer J. Clin.*, 58, 71–96.

Job, S., Draskovic, I., Burnichon, N., Buffet, A., Cros, J., Lepine, C., Venisse, A., Robidel, E., Verkarre, V., Meatchi, T. et al. (2019). Telomerase activation and ATRX mutations are independent risk factors for metastatic pheochromocytoma and paraganglioma. *Clin. Cancer Res.*, 25, 760–770.

Jones, S., Chen, W.D., Parmigiani, G., Diehl, F., Beerenwinkel, N., Antal, T., Traulsen, A., Nowak, M.A., Siegel, C., Velculescu, V.E. et al. (2008). Comparative lesion sequencing provides insights into tumor evolution. *Proc. Natl. Acad. Sci. USA*, 105, 4283–4288.

Kent, T., Gracias, D., Shepherd, S., Clynes, D. (2019). Alternative lengthening of telomeres in pediatric cancer: Mechanisms to therapies. *Front. Oncol.*, 9, 1518.

Khattar, E. and Tergaonkar, V. (2017). Transcriptional regulation of telomerase reverse transcriptase (TERT) by MYC. *Front. Cell Dev. Biol.*, 5, 1.

Kim, N.W., Piatyszek, M.A., Prowse, K.R., Harley, C.B., West, M.D., Ho, P.L., Coviello, G.M., Wright, W.E., Weinrich, S.L., Shay, J.W. (1994). Specific association of human telomerase activity with immortal cells and cancer. *Science*, 266, 2011–2015.

Koneru, B., Lopez, G., Farooqi, A., Conkrite, K.L., Nguyen, T.H., Macha, S.J., Modi, A., Rokita, J.L., Urias, E., Hindle, A. et al. (2020). Telomere maintenance mechanisms define clinical outcome in high-risk neuroblastoma. *Cancer Res.*, 80, 2663–2675.

Kulic, A., Plavetic, N.D., Gamulin, S., Jakic-Razumovic, J., Vrbanec, D., Sirotkovic-Skerlev, M. (2016). Telomerase activity in breast cancer patients: Association with poor prognosis and more aggressive phenotype. *Med. Oncol.*, 33, 23.

Lantuejoul, S., Soria, J.C., Morat, L., Lorimier, P., Moro-Sibilot, D., Sabatier, L., Brambilla, C., Brambilla, E. (2005). Telomere shortening and telomerase reverse transcriptase expression in preinvasive bronchial lesions. *Clin. Cancer Res.*, 11, 2074–2082.

Lee, Y.K., Park, N.H., Lee, H. (2012). Prognostic value of alternative lengthening of telomeres-associated biomarkers in uterine sarcoma and uterine carcinosarcoma. *Int. J. Gynecol. Cancer*, 22, 434–441.

Lee, J.C., Jeng, Y.M., Liau, J.Y., Tsai, J.H., Hsu, H.H., Yang, C.Y. (2015). Alternative lengthening of telomeres and loss of ATRX are frequent events in pleomorphic and dedifferentiated liposarcomas. *Mod. Pathol.*, 28, 1064–1073.

Lee, D.D., Leao, R., Komosa, M., Gallo, M., Zhang, C.H., Lipman, T., Remke, M., Heidari, A., Nunes, N.M., Apolonio, J.D. et al. (2019). DNA hypermethylation within TERT promoter upregulates TERT expression in cancer. *J. Clin. Invest.*, 129, 223–229.

Lee, D.D., Komosa, M., Nunes, N.M., Tabori, U. (2020). DNA methylation of the TERT promoter and its impact on human cancer. *Curr. Opin. Genet. Dev.*, 60, 17–24.

Londono-Vallejo, J.A. (2008). Telomere instability and cancer. *Biochimie*, 90, 73–82.

Lopez-Otin, C., Blasco, M.A., Partridge, L., Serrano, M., Kroemer, G. (2013). The hallmarks of aging. *Cell*, 153, 1194–1217.

Maciejowski, J., Li, Y., Bosco, N., Campbell, P.J., de Lange, T. (2015). Chromothripsis and kataegis induced by telomere crisis. *Cell*, 163, 1641–1654.

Martens, U.M., Zijlmans, J.M., Poon, S.S., Dragowska, W., Yui, J., Chavez, E.A., Ward, R.K., Lansdorp, P M. (1998). Short telomeres on human chromosome. *Nat. Genet.*, 18, 76–80.

Matano, M., Date, S., Shimokawa, M., Takano, A., Fujii, M., Ohta, Y., Watanabe, T., Kanai, T., Sato, T. (2015). Modeling colorectal cancer using CRISPR-Cas9-mediated engineering of human intestinal organoids. *Nat. Med.*, 21, 256–262.

Matsuo, T., Shimose, S., Kubo, T., Fujimori, J., Yasunaga, Y., Ochi, M. (2010). Alternative lengthening of telomeres as a prognostic factor in malignant fibrous histiocytomas of bone. *Anticancer Res.*, 30, 4959–4962.

Merle, P., Evrard, B., Petitjean, A., Lehn, J.M., Teulade-Fichou, M.P., Chautard, E., De Cian, A., Guittat, L., Tran, P.L., Mergny, J.L. et al. (2011). Telomere targeting with a new G4 ligand enhances radiation-induced killing of human glioblastoma cells. *Mol. Cancer Ther.*, 10, 1784–1795.

Müller, P.A. and Vousden, K.H. (2013). Mutations in cancer. *Nat. Cell Biol.*, 15, 2–8.

Muraki, K., Nyhan, K., Han, L., Murnane, J.P. (2012). Mechanisms of telomere loss and their consequences for chromosome instability. *Front. Oncol.*, 2, 135.

Murnane, J.P., Sabatier, L., Marder, B.A., Morgan, W.F. (1994). Telomere dynamics in an immortal human cell line. *EMBO J.*, 13, 4953–4962.

Nault, J.C., Mallet, M., Pilati, C., Calderaro, J., Bioulac-Sage, P., Laurent, C., Laurent, A., Cherqui, D., Balabaud, C., Zucman-Rossi, J. (2013). High frequency of telomerase reverse-transcriptase promoter somatic mutations in hepatocellular carcinoma and preneoplastic lesions. *Nat. Commun.*, 4, 2218.

Nowak, M.A., Komarova, N.L., Sengupta, A., Jallepalli, P.V., Shih IE, M., Vogelstein, B., Lengauer, C. (2002). The role of chromosomal instability in tumor initiation. *Proc. Natl. Acad. Sci. USA*, 99, 16226–16231.

Nowak, T., Januszkiewicz, D., Zawada, M., Pernak, M., Lewandowski, K., Rembowska, J., Nowicka, K., Mankowski, P., Nowak, J. (2006). Amplification of hTERT and hTERC genes in leukemic cells with high expression and activity of telomerase. *Oncol. Rep.*, 16, 301–305.

Ortiz-Montero, P., Londono-Vallejo, A., Vernot, J.P. (2017). Senescence-associated IL-6 and IL-8 cytokines induce a self- and cross-reinforced senescence/inflammatory milieu strengthening tumorigenic capabilities in the MCF-7 breast cancer cell line. *Cell Commun. Signal.*, 15, 17.

Ozden, S., Tiber, P.M., Ozgen, Z., Ozyurt, H., Serakinci, N., Orun, O. (2014). Expression of TRF2 and its prognostic relevance in advanced stage cervical cancer patients. *Biol. Res.*, 47, 61.

Panero, J., Stanganelli, C., Arbelbide, J., Fantl, D.B., Kohan, D., Garcia Rivello, H., Rabinovich, G.A., Slavutsky, I. (2014). Expression profile of shelterin components in plasma cell disorders. Clinical significance of POT1 overexpression. *Blood Cells Mol. Dis.*, 52, 134–139.

Pastushenko, I. and Blanpain, C. (2019). EMT transition states during tumor progression and metastasis. *Trends Cell Biol.*, 29, 212–226.

Peifer, M., Hertwig, F., Roels, F., Dreidax, D., Gartlgruber, M., Menon, R., Kramer, A., Roncaioli, J.L., Sand, F., Heuckmann, J.M. et al. (2015). Telomerase activation by genomic rearrangements in high-risk neuroblastoma. *Nature*, 526, 700–704.

Potrony, M., Badenas, C., Aguilera, P., Puig-Butille, J.A., Carrera, C., Malvehy, J., Puig, S. (2015). Update in genetic susceptibility in melanoma. *Ann. Transl. Med.*, 3, 210.

Potts, P.R. and Yu, H. (2007). The SMC5/6 complex maintains telomere length in ALT cancer cells through SUMOylation of telomere-binding proteins. *Nat. Struct. Mol. Biol.*, 14, 581–590.

Queisser, A., Heeg, S., Thaler, M., Von Werder, A., Opitz, O.G. (2013). Inhibition of telomerase induces alternative lengthening of telomeres during human esophageal carcinogenesis. *Cancer Genet.*, 206, 374–386.

Rajagopalan, H. and Lengauer, C. (2004). CIN-ful cancers. *Cancer Chemother. Pharmacol.*, 54(Suppl 1), S65–S68.

Raynaud, C.M., Jang, S.J., Nuciforo, P., Lantuejoul, S., Brambilla, E., Mounier, N., Olaussen, K.A., Andre, F., Morat, L., Sabatier, L. et al. (2008). Telomere shortening is correlated with the DNA damage response and telomeric protein down-regulation in colorectal preneoplastic lesions. *Ann. Oncol.*, 19, 1875–1881.

Reddel, R.R. (2000). The role of senescence and immortalization in carcinogenesis. *Carcinogenesis*, 21, 477–484.

Reddel, R.R. (2003). Alternative lengthening of telomeres, telomerase, and cancer. *Cancer Lett.*, 194, 155–162.

Rode, L., Nordestgaard, B.G., Bojesen, S.E. (2015). Peripheral blood leukocyte telomere length and mortality among 64,637 individuals from the general population. *J. Natl. Cancer Inst.*, 107, djv074.

Roger, L., Jones, R.E., Heppel, N.H., Williams, G.T., Sampson, J.R., Baird, D.M. (2013). Extensive telomere erosion in the initiation of colorectal adenomas and its association with chromosomal instability. *J. Natl. Cancer Inst.*, 105, 1202–1211.

Roumelioti, F.M., Sotiriou, S.K., Katsini, V., Chiourea, M., Halazonetis, T.D., Gagos, S. (2016). Alternative lengthening of human telomeres is a conservative DNA replication process with features of break-induced replication. *EMBO Rep.*, 17, 1731–1737.

Sanz-Casla, M.T., Vidaurreta, M., SANCHEZ-Rueda, D., Maestro, M.L., Arroyo, M., Cerdan, F.J. (2005). Telomerase activity as a prognostic factor in colorectal cancer. *Onkologie*, 28, 553–557.

Shay, J.W. and Wright, W.E. (1996). The reactivation of telomerase activity in cancer progression. *Trends Genet.*, 12, 129–131.

Shay, J.W. and Wright, W.E. (2000). Hayflick, his limit, and cellular ageing. *Nat. Rev. Mol. Cell Biol.*, 1, 72–76.

Shay, J.W., Werbin, H., Wright, W.E. (1997). Telomerase assays in the diagnosis and prognosis of cancer. *Ciba Found. Symp.*, 211, 148–155, discussion 155–159.

Shih, I.M., Zhou, W., Goodman, S.N., Lengauer, C., Kinzler, K.W., Vogelstein, B. (2001). Evidence that genetic instability occurs at an early stage of colorectal tumorigenesis. *Cancer Res.*, 61, 818–822.

Speedy, H.E., Kinnersley, B., Chubb, D., Broderick, P., Law, P.J., Litchfield, K., Jayne, S., Dyer, M.J.S., Dearden, C., Follows, G.A. et al. (2016). Germ line mutations in shelterin complex genes are associated with familial chronic lymphocytic leukemia. *Blood*, 128, 2319–2326.

Stern, J.L., Theodorescu, D., Vogelstein, B., Papadopoulos, N., Cech, T.R. (2015). Mutation of the TERT promoter, switch to active chromatin, and monoallelic TERT expression in multiple cancers. *Genes Dev.*, 29, 2219–2224.

Strahl, C. and Blackburn, E.H. (1996). Effects of reverse transcriptase inhibitors on telomere length and telomerase activity in two immortalized human cell lines. *Mol. Cell Biol.*, 16, 53–65.

Sung, W.K., Zheng, H., Li, S., Chen, R., Liu, X., Li, Y., Lee, N.P., Lee, W.H., Ariyaratne, P.N., Tennakoon, C. et al. (2012). Genome-wide survey of recurrent HBV integration in hepatocellular carcinoma. *Nat. Genet.*, 44, 765–769.

Suraweera, N., Mouradov, D., Li, S., Jorissen, R.N., Hampson, D., Ghosh, A., Sengupta, N., Thaha, M., Ahmed, S., Kirwan, M. et al. (2016). Relative telomere lengths in tumor and normal mucosa are related to disease progression and chromosome instability profiles in colorectal cancer. *Oncotarget*, 7, 36474–36488.

Swiggers, S.J., Kuijpers, M.A., De Cort, M.J., Beverloo, H.B., Zijlmans, J.M. (2006). Critically short telomeres in acute myeloid leukemia with loss or gain of parts of chromosomes. *Genes Chromosomes Cancer*, 45, 247–256.

Sze, K.M., Ho, D.W., Chiu, Y.T., Tsui, Y.M., Chan, L.K., Lee, J.M., Chok, K.S., Chan, A.C., Tang, C.N., Tang, V.W. et al. (2021). Hepatitis B virus-telomerase reverse transcriptase promoter integration harnesses Host ELF4, resulting in telomerase reverse transcriptase gene transcription in hepatocellular carcinoma. *Hepatology*, 73, 23–40.

Taylor, B.S., Schultz, N., Hieronymus, H., Gopalan, A., Xiao, Y., Carver, B.S., Arora, V.K., Kaushik, P., Cerami, E., Reva, B. et al. (2010). Integrative genomic profiling of human prostate cancer. *Cancer Cell*, 18, 11–22.

Tijhuis, A.E., Johnson, S.C., Mcclelland, S.E. (2019). The emerging links between chromosomal instability (CIN), metastasis, inflammation and tumour immunity. *Mol. Cytogenet.*, 12, 17.

Tomoda, R., Seto, M., Tsumuki, H., Iida, K., Yamazaki, T., Sonoda, J., Matsumine, A., Uchida, A. (2002). Telomerase activity and human telomerase reverse transcriptase mRNA expression are correlated with clinical aggressiveness in soft tissue tumors. *Cancer*, 95, 1127–1133.

Ulaner, G.A., Huang, H.Y., Otero, J., Zhao, Z., Ben-Porat, L., Satagopan, J.M., Gorlick, R., Meyers, P., Healey, J.H., Huvos, A.G. et al. (2003). Absence of a telomere maintenance mechanism as a favorable prognostic factor in patients with osteosarcoma. *Cancer Res.*, 63, 1759–1763.

Venturini, L., Motta, R., Gronchi, A., Daidone, M., Zaffaroni, N. (2010). Prognostic relevance of ALT-associated markers in liposarcoma: A comparative analysis. *BMC Cancer*, 10, 254.

Vinagre, J., Almeida, A., Populo, H., Batista, R., Lyra, J., Pinto, V., Coelho, R., Celestino, R., Prazeres, H., Lima, L. et al. (2013). Frequency of TERT promoter mutations in human cancers. *Nat. Commun.*, 4, 2185.

Watanabe, T., Kobunai, T., Yamamoto, Y., Matsuda, K., Ishihara, S., Nozawa, K., Yamada, H., Hayama, T., Inoue, E., Tamura, J. et al. (2012). Chromosomal instability (CIN) phenotype, CIN high or CIN low, predicts survival for colorectal cancer. *J. Clin. Oncol.*, 30, 2256–2264.

Wentzensen, I.M., Mirabello, L., Pfeiffer, R.M., Savage, S.A. (2011). The association of telomere length and cancer: A meta-analysis. *Cancer Epidemiol. Biomarkers Prev.*, 20, 1238–1250.

Willeit, P., Willeit, J., Mayr, A., Weger, S., Oberhollenzer, F., Brandstatter, A., Kronenberg, F., Kiechl, S. (2010). Telomere length and risk of incident cancer and cancer mortality. *JAMA*, 304, 69–75.

Wright, W.E., Piatyszek, M.A., Rainey, W.E., Byrd, W., Shay, J.W. (1996). Telomerase activity in human germline and embryonic tissues and cells. *Dev. Genet.*, 18, 173–179.

Wu, Y., Poulos, R.C., Reddel, R.R. (2020). Role of POT1 in human cancer. *Cancers (Basel)*, 12(10), 2739.

Yang, Z., Takai, K.K., Lovejoy, C.A., de Lange, T. (2020). Break-induced replication promotes fragile telomere formation. *Genes Dev.*, 34, 1392–1405.

Yeager, T.R., Neumann, A.A., Englezou, A., Huschtscha, L.I., Noble, J.R., Reddel, R.R. (1999). Telomerase-negative immortalized human cells contain a novel type of promyelocytic leukemia (PML) body. *Cancer Res.*, 59, 4175–4179.

Zhang, J.M., Yadav, T., Ouyang, J., Lan, L., Zou, L. (2019). Alternative lengthening of telomeres through two distinct break-induced replication pathways. *Cell Rep.*, 26, 955–968.e3.

Zhang, H., Zhao, R., Tones, J., Liu, M., Dilley, R.L., Chenoweth, D.M., Greenberg, R.A., Lampson, M.A. (2020). Nuclear body phase separation drives telomere clustering in ALT cancer cells. *Mol. Biol. Cell*, 31, 2048–2056.

Zheng, Y.L., Hu, N., Sun, Q., Wang, C., Taylor, P.R. (2009). Telomere attrition in cancer cells and telomere length in tumor stroma cells predict chromosome instability in esophageal squamous cell carcinoma: A genome-wide analysis. *Cancer Res.*, 69, 1604–1614.

Zhu, C.Q., Cutz, J.C., Liu, N., Lau, D., Shepherd, F.A., Squire, J.A., Tsao, M.S. (2006). Amplification of telomerase (hTERT) gene is a poor prognostic marker in non-small-cell lung cancer. *Br. J. Cancer*, 94, 1452–1459.

12

Physiopathology of Telomeres

Caroline KANNENGIESSER[1] and Patrick REVY[2]

[1]APHP Department of Genetics, Hôpital Bichat, Paris Cité University, France

[2]Imagine Institute, INSERM, Paris Cité University, France

12.1. Introduction

Telomeres are essential for chromosome protection and genome stability. Elizabeth H. Blackburn, Carol W. Greider and Jack W. Szostak were awarded the Nobel Prize in Medicine in 2009 for the description of the nature of telomeres and the identification of the enzyme telomerase.

Two types of diseases associated with telomere abnormalities can be distinguished: constitutional telomere biology disorders, also called telomeropathies, due to the presence of germline mutations affecting the protection and/or length of telomeres and acquired diseases such as cancer which in 90% of cases are associated with telomerase reactivation leading to a constant maintenance of the length of telomeres, a phenomenon that allows cellular immortalization specific to cancerous cells. Several mechanisms, sometimes additive in a tumor, can contribute to the reactivation of telomerase: presence of somatic mutations in the *TERT* promoter or structural rearrangements (insertion of retroviral sequences to the *TERT* promoter or amplification of *TERT* or *TERC*). Furthermore in about 10% of cancers, telomere length maintenance is guaranteed by a telomerase-independent mechanism called ALT (alternative lengthening of telomeres) (Cesare and Reddel 2010; Yuan et al. 2019).

This chapter will deal exclusively with telomeropathies, a group of Mendelian diseases caused by pathogenic variations in genes encoding factors involved in the protection and/or maintenance of telomeres. In these telomeropathies, the clinical expression and severity of symptoms are very variable. Historically, the first clinically identified telomeropathies were dyskeratosis congenita (DC) and Hoyeraal–Hreidarsson (HH) syndrome. Biologically, patients with telomeropathies most often have abnormally shortened telomeres, explaining other names for telomeropathies as "syndromes of telomeres shortening" or "short telomere syndrome" (STS) (Armanios 2009; Armanios and Blackburn 2012). However, as discussed below, it appears that this terminology may be inappropriate since some telomeropathies are not associated with a telomere length abnormality. It seems to us that "telomere biology disorders" is more appropriate (Savage and Bertuch 2010; Savage 2014).

These diseases can be considered to form a continuum between the most severe forms of pediatric onset represented by HH syndrome and Revesz syndrome with adult forms consisting of pulmonary fibrosis (PF), bone marrow failure or liver disease. Between these extremes, other diseases coexist including DC and Coats plus (CP) syndrome. In this chapter, we will report the genes associated with these diseases. We will mainly describe the clinical characteristics of the different diseases and/or organ damage associated with telomere defects. In addition, we will briefly present genetic testing and counseling, as well as therapeutic alternatives proposed or currently under development.

12.2. Telomere length in blood

As mentioned above, most telomeropathies are associated with accelerated telomere shortening. Measurement of telomere length in blood cells has therefore emerged as a very effective test for the diagnosis of telomeropathies.

12.2.1. *Measurement of telomere length*

12.2.1.1. *Flow-FISH*

Flow-FISH is the reference technique for assessing telomere length (Alder et al. 2018) (Table 12.1). This technique is performed on blood cells where in situ hybridization is performed with a fluorescent probe which interacts specifically with telomeres. The fluorescence intensity, determined with a flow cytometer, is directly correlated to the length of telomeres. This technique requires fluorescence data to be obtained in a large "control" population (100 individuals). The results are then expressed by comparing the fluorescence intensity of an individual with those obtained in these "healthy" controls (Alder et al. 2018).

12.2.1.2. *Techniques other than flow-FISH*

There are several other methods of measuring telomere length (listed in Table 12.1), which may have other advantages and disadvantages in comparison to Flow-FISH. However, these techniques, performed either from DNA or from cells, require special expertise. They are therefore not routine biological analyses offered in laboratories and are only used to answer specific questions (for more details, see Chapter 2).

Method	Principle	Advantages	Disadvantages	Reference technique
Flow-FISH	Determination of fluorescence in interphase cells by flow cytometry (telomere-specific fluorescent probe)	Reproducible and accurate	– Need for living cells (blood samples) – Equipment (cytometer) – Cell culture (line 1301)	YES +++
TRF (telomere restriction fragment)	Southern blot method based on the hybridization of radioactive probes recognizing telomeric DNA (after digestion of genomic DNA with specific restriction enzymes)	– From DNA (can be tested on samples from deceased individuals) – Technique widely used in early DC publications	– Radio-marked probe (specialized installation) – Cumbersome and time consuming technique (enzymatic restriction, hybridization, etc.) – Not automated (small series) – Large amount of DNA (1–2 μg)	YES ++
Q-PCR	Quantification of the number of telomeric repeats by a PCR amplification technique	– From DNA –Small amount of DNA – Relatively fast – Large series (automatable)	– Lack of robustness and reproducibility – Need for duplicates or triplicates	No

Method	Principle	Advantages	Disadvantages	Reference technique
Q-FISH	Determination of fluorescence intensity by telomere microscopy after hybridization of telomere-specific fluorescent probes in metaphase blocked cells	– Accurate – Sensitive – Ability to determine telomere length for each chromosome	– Small series – Need for live cells – Time consuming	No
STeLA (single telomere length analysis)	Method based on a specific PCR after primer ligation. Detection by Southern blot	– Low amount of DNA – Specific quantification – Little equipment	– Radio-labeled probe (specialized installation) – Heavy and time consuming technique (enzymatic restriction, hybridization, etc.) – Small series – Limited to certain chromosomes	No

Table 12.1. *Comparison of different techniques for measuring telomere length*

12.2.2. *Telomere length in patients with telomeropathy*

Numerous studies have revealed that in the majority of patients, the common feature of the different forms of telomeropathies is the presence of abnormally short telomeres in the nucleated cells of the blood. It is therefore generally accepted that the main cause of the clinical signs observed in patients is the accelerated reduction in telomere length leading to a decrease in the proliferative capacity of cells and premature aging of organs. As telomere attrition is essentially associated with cell division, a defect in telomere maintenance will mainly affect tissues in which a high rate of proliferation and/or renewal is present. This particularly concerns the hematopoietic system and the mucocutaneous tissues.

However, recent work (Alder et al. 2018) has made it possible to reconsider the dogma that patients with telomeropathies must necessarily have very short telomeres and to redefine the thresholds according to age and disease. For example, children with severe diseases (hematological, such as bone marrow failure and/or neurological disease) have, in the vast majority of cases, telomeres below the first percentile. They thus belong to the 1% of the population of their age with the shortest telomeres. To interpret telomere length results, it is necessary to take into account the number of generations carrying the mutation (first or Nth generation) in the family of the proband because of the transmission of short telomeres (see section 4.4.3). Among patients with PF (often the first generations) with a *TERT* or *TERC* mutation, 85% have telomeres below the 10th percentile. Some may have telomeres between the 10th and 25th percentiles.

Finally, it is difficult to establish for all telomeropathies a correlation between the severity and early onset of symptoms and the extent of telomere shortening measured in the blood, especially since some asymptomatic relatives belonging to telomeropathy families with identified mutations have abnormally shortened telomeres without clinical manifestation (see section 12.5.2).

12.3. Different genes associated with telomere defects

To date, germline mutations in 16 genes, encoding factors involved in the stability and/or regulation of telomere length, have been associated with constitutional telomere pathologies: *TERT*, *TERC* and *DKC1* (encoding Dyskerin); *NOP10*, *NHP2* and WRAP53 (encoding TCAB1), *PARN*, *RTEL1* and *ACD* (encoding TPP1); *TINF2* (encoding TIN2); *NAF1*, *ZCHCC8*, *SHQ1*, *POT1*, *CTC1* and *STN1* (Table 12.2 and Figure 12.1). Depending on the nature of the genes, and the associated disease, all types of genetic transmission are encountered (X-linked, autosomal recessive, autosomal dominant). It is important to note that bi-allelic mutations of a gene most often cause DC or HH syndrome before adolescence, whereas a heterozygous mutation of the same gene will be responsible for the development of PF, progressive bone marrow failure or other complications in adulthood.

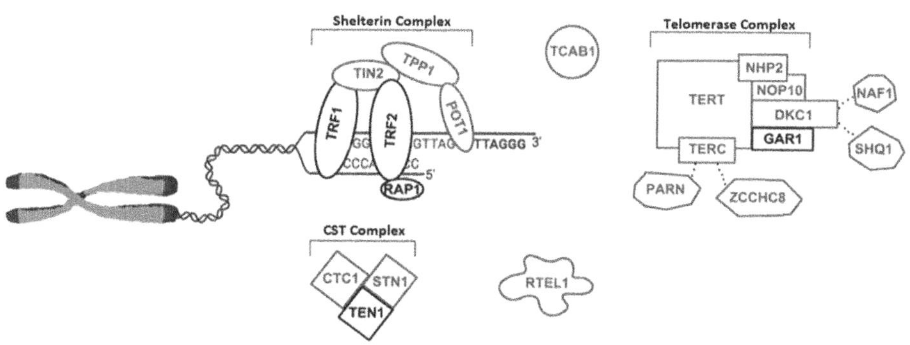

Figure 12.1. *Factors involved in telomeropathy. For a color version of this figure, see www.iste.co.uk/saintome/telomeres.zip*

International guidelines suggest that the word "mutation" should be replaced by the term "pathogenic variations" and that variations should be interpreted according to a five-class system. However, the word "mutation" is widely used and understandable to most readers. We have therefore decided to keep this term and use it in this chapter.

We will distinguish two groups of telomeropathy:

– Telomeropathy with shortened telomere length.

For several genes, mutations can be transmitted in a recessive (bi-allelic mutations) or dominant (mono-allelic mutation) mode.

– Coats plus telomeropathy, autosomal recessive transmission and not necessarily associated with a shortened telomere length.

Schematically, the factors involved in telomeropathies are associated with different complexes or pathways: the telomerase complex, the shelterin complex, the CST complex, and telomere stability (Figure 12.1). These four categories are briefly described.

12.3.1. *Telomerase complex*

TERT, TERC, DKC1, NAF1, NOP10, NHP2, TCAB1, SHQ1 and PARN factors are found within the telomerase complex and/or participate in TERC maturation and/or telomerase recruitment to telomeres. TERT represents the catalytic subunit of the telomerase complex and has reverse transcription activity. In combination with the hTR/TERC RNA template and several cofactors (DKC1, NOP10, NHP2, NAF1, SHQ1), the processive activity of TERT allows for the addition of telomeric sequences to telomeres and thus participates in the maintenance of telomere length.

For several factors involved in telomeropathies, there is an alteration in TERC function, stability and/or maturation. For example, the factors PARN (Moon et al. 2015; Stuart et al. 2015) and ZCCHC8 (Gable et al. 2019) are not directly involved in telomerase activity but are required for the production of a mature and functional form of TERC.

The TCAB1 factor is involved in the recruitment of the telomerase complex to telomeres and stimulates its telomere elongation activity.

12.3.2. *Shelterin complex*

Within the shelterin complex, defects in TIN2 (encoded by TINF2), POT1, TPP1 (encoded by ACD) have been associated with telomeropathies. Although not part of the telomerase complex, TIN2, POT1 and TPP1 factors participate in telomerase recruitment, stabilization and activity at telomeres (Pike et al. 2019; Laprade et al. 2020).

12.3.3. *CST complex (CTC1/STN1/TEN1)*

The CST complex formed by CTC1, TEN1 and STN1 appears to participate in telomere replication and the formation of the 3' extension that allows telomerase to perform its reverse transcriptase activity. However, the link between CTC1 and STN1 mutations, associated pathologies and telomere length is not well understood to date (Stewart et al. 2018).

12.3.4. *Telomere stability factor*

RTEL1 is a DNA helicase that participates in multiple functions required for genome stability. In particular, RTEL1 is involved in DNA replication in regions with difficult to replicate structures including telomeres. Defects in RTEL1 cause high genetic instability associated with losses of telomeric sequences causing significant telomere shortening (Vannier et al. 2012; Le Guen et al. 2013; Sarek et al. 2015). In addition, it was recently shown that RTEL1 was also involved in the generation of the 3' telomere overhang allowing telomerase to regulate telomere length (Awad et al. 2020).

In Table 12.2, the name of the genes is in italics, and then the products of this gene are listed. HH: Hoyeraal–Hreidarsson syndrome, DC: dyskeratosis congenita, CP: Coats plus, RS: Revesz syndrome, PF: pulmonary fibrosis, KD: kidney disease, LD: liver disease. Modes of transmission are also listed. AD: autosomal dominant, AR: autosomal recessive, XL: X-linked.

In addition to the above-mentioned genes involved in telomeropathies, other works have proposed the involvement of new genes responsible for these diseases such as MDM4 and NPM1 (Nachmani et al. 2019; Toufektchan et al. 2020). However, in these studies, the clinical and/or genetic descriptions are insufficient or unconvincing and will require further work to formally demonstrate their implications in human pathologies. Finally, a splicing anomaly in the gene encoding the Apollo factor, involved in DNA repair and telomere protection, was identified in a patient with HH. The protein resulting from this splicing defect was shown to exert a dominant negative effect on telomere protection. However, the molecular origin of the splicing defect could not be established and consequently the causal link between the splicing anomaly of the Apollo gene and the disease could not be formally demonstrated (Touzot et al. 2010).

Gene/protein	Disease (transmission)	References
Telomeropathy with shortened telomere length		
TERT/TERT	DC (AD and AR)	(Vulliamy et al. 2005; Marrone et al. 2007b; Du et al. 2009; Carillo et al. 2011; Collopy et al. 2015)
	HH/(AR)	(Vulliamy et al. 2005; Marrone et al. 2007b; Gramatges et al. 2013)
	PF	(Armanios et al. 2007; Tsakiri et al. 2007; Parry et al. 2011; Borie et al. 2016; Newton et al. 2016; Judge et al. 2017; Maryoung et al. 2017)
	LD (AD)	(Calado et al. 2009; Kapuria et al. 2019)
TERC/hTR or TERC	DC (AD)	(Vulliamy et al. 2004; Abdallah et al. 2009; Du et al. 2009; Jongmans et al. 2012; Collopy et al. 2015)
	PF (AD)	(Armanios et al. 2007; Tsakiri et al. 2007; Alder et al. 2008, 2011; Dai et al. 2015; Borie et al. 2016)
	HH (AD and AR)	(Vulliamy et al. 2011)
	LD (AD)	(Kapuria et al. 2019)
TINF2/TIN2	HH (AD)	(Walne et al. 2008; Touzot et al. 2012)
	DC (AD)	(Savage et al. 2008; Walne et al. 2008; Vulliamy et al. 2011; Vulliamy et al. 2012)
	RS (AD)	(Savage et al. 2008; Sasa et al. 2012; Gupta et al. 2017; Sakwit et al. 2019)
	PF (AD)	(Yamaguchi et al. 2010; Alder et al. 2015)
	LD (AD)	(Kapuria et al. 2019)
ACD/TPP1	HH (AR)	(Kocak et al. 2014)
	DC (AD)	(Guo et al. 2014)
	PF (AD)	(Hoffman et al. 2019)

Gene/protein	Disease (transmission)	References
RTEL1/RTEL1	DC (AR)	(Ballew et al. 2013a; Walne et al. 2013; Speckmann et al. 2017)
	HH (AR)	(Ballew et al. 2013a, 2013b; Deng et al. 2013; Le Guen et al. 2013; Walne et al. 2013; Jullien et al. 2016; Moriya et al. 2016; Touzot et al. 2016)
	PF (AD)	(Cogan et al. 2015; Juge et al. 2017; Kannengiesser et al. 2015; Stuart et al. 2015; Borie et al. 2019)
PARN/PARN	HH (AR)	(Burris et al. 2016; Benyelles et al. 2019a; Dodson et al. 2019)
	PF (AD and AR)	(Dhanraj et al. 2015; Stuart et al. 2015; Xing and Garcia 2016; Juge et al. 2017; Kropski et al. 2017; Verduyn et al. 2017; Alder et al. 2018; Zhang et al. 2019)
	DC (AR)	(Dhanraj et al. 2015; Moon et al. 2015; Tummala et al. 2015)
	KD (AD)	(Lata et al. 2018)
NOP10/NOP10	DC (AR)	(Walne et al. 2007)
	PF (AD)	(Kannengiesser et al. 2020)
NHP2/NHP2	DC (AR)	(Vulliamy et al. 2008)
	HH (AR)	(Benyelles et al. 2020)
	PF (AD)	(Benyelles et al. 2020)
DKC1/DKC1	DC (XL)	(Heiss et al. 1998; Knight et al. 1999b, 2001; Vulliamy et al. 2001, 2006; Cossu et al. 2002; Wong et al. 2004; Kanegane et al. 2005; Vulliamy et al. 2006)
	HH (XL)	(Knight et al. 1999a; Cossu et al. 2002; Vulliamy et al. 2011; Vogiatzi et al. 2013)
	PF (XL)	(Hisata et al. 2013; Kropski et al. 2014)
NAF1/NAF1	PF (AD)	(Stanley et al. 2016)
ZCCHC8/ZCCHC8	PF (AD)	(Gable et al. 2019)
WRAP53/TCAB1	DC (AR)	(Zhong et al. 2011)
SHQ1/ SHQ1	Neurodevelopmental disorder	(Bizarro and Meier 2017)
Plus syndrome with autosomal recessive transmission and not necessarily associated with a shortened telomere length		
CTC1/CTC1	CP (AR)	(Anderson et al. 2012; Polvi et al. 2012; Bisserbe et al. 2015; Netravathi et al. 2015; Lin et al. 2017)
	DC (AR)	(Keller et al. 2012; Walne et al. 2012)
POT1/POT1	CP (AR)	(Takai et al. 2016)
STN1/STN1	CP (AR)	(Simon et al. 2016)

Table 12.2. *List of genes involved in telomerpathy*

12.4. Different diseases associated with telomere defects

12.4.1. *Dyskeratosis congenita (DC)*

DC is a syndromic disease initially described as a combination of a triad of mucocutaneous manifestations: nail dysplasia (small, brittle, thin nails with ridges and pits), abnormal (hypo- or hyper-) reticular pigmentation of the skin, especially of the chest and neck, and oral leukoplakia (Table 12.3). However, the presentations of DC are very heterogeneous and apart from the classic triad of DC (which is however not present in all DC patients), individuals with DC may develop progressive bone marrow failure. In addition to the above-mentioned complications, individuals with DC have a predisposition to develop solid tumors (mainly of the head and neck type) and leukemias (also section 12.4.8) (Alter et al. 2009; Schratz et al. 2020), PF and progressive organ disease. Patients with DC may also have dental abnormalities (caries, periodontal disease, taurodontism), signs of aging including early hair graying (canitis) or premature hair loss (alopecia) and retinopathy.

Type of cells or tissues	Clinical manifestations in men with telomeropathy	References
Skin	DC triad (oral leukoplakia, nail dystrophy, pigmentation abnormalities)	(Marrone et al. 2007a; Vulliamy et al. 2008; Du et al. 2009; Carillo et al. 2011; Collopy et al. 2015; Benyelles et al. 2019b)
	Early hair graying	(Diaz de Leon et al. 2011; Borie et al. 2018)
Bone marrow and Blood	Macrocytosis, thrombocytopenia	(Diaz de Leon et al. 2011; Borie et al. 2016; Newton et al. 2016)
	Bone marrow hypoplasia, aplastic anemia, pancytopenia	(Yamaguchi et al. 2003; Touzot et al. 2012, 2016; Le Guen et al. 2013; Yamaguchi et al. 2005)
	Myelodysplasia	
	Acute myeloid leukemia	
Immune system	Opportunistic infections	
	Immunodeficiency of B, T and NK cells	(Glousker et al. 2015; Le Guen et al. 2013; Touzot et al. 2012, 2016; Cossu et al. 2002; Jyonouchi et al. 2011; Wagner et al. 2018)
Gastro-intestinal	Enterocolitis	(Touzot et al. 2012; Alder et al. 2018)

Type of cells or tissues	Clinical manifestations in men with telomeropathy	References
Lung	Pulmonary fibrosis Early onset and severe emphysema	(Armanios et al. 2007; Tsakiri et al. 2007; Parry et al. 2011; Borie et al. 2016; Newton et al. 2016; Juge et al. 2017; Maryoung et al. 2017)
Liver	Non-alcoholic steatohepatitis cirrhosis, cryptogenic hepatic fibrosis, hepato-pulmonary syndrome Regenerative nodular hyperplasia Portal hypertension without cirrhosis	(Calado et al. 2009, 2011; Gorgy et al. 2015; Kapuria et al. 2019)
CNS neurological	Cerebellar hypoplasia Microcephaly Mental retardation Neurodevelopmental or psychomotor disorder Depression	(Bhala et al. 2019; Touzot et al. 2012; Le Guen et al. 2013; Benyelles et al. 2019b, 2020)
Bones	Osteopenia Osteoporosis Non-vascular necrosis	
Eyes	Coats plus syndrome Revesz syndrome Retinitis pigmentosa	(Takai et al. 2016)
Kidney	Kidney disease	(Lata et al. 2018)

Table 12.3. *List of clinical manifestations in telomeropathies. Adapted from Armanios and Blackburn (2012)*

Bone marrow failure can progress to bone marrow aplasia and pancytopenia, myelodysplastic syndrome (MDS) or acute myeloid leukemia. These situations require the consideration of a hematopoietic stem cell transplantation (bone marrow transplantation). The results of the first bone marrow transplantations of DC patients with conventional conditioning were mixed. Survival was poor and many patients developed complications, reflecting the high sensitivity of these patients to immunosuppressive therapies. Alkylating agents and radiation exposure were removed and a promising conditioning regimen was proposed combining fludarabine and alemtuzumab (Dalle and Peffault de Latour 2016; Agarwal 2018). A 2016 review of the literature reviews bone marrow transplantation in DC (Barbaro and Vedi 2016). The survival of 109 patients described in the literature is poor with a survival of 57% at 5 years and 23% at 10 years, despite attempts to modify graft conditioning.

In practice, at present, it is necessary for these patients with bone marrow aplasia of constitutional origin to discuss in a multidisciplinary manner the possibility of bone marrow transplantation in this indication and to evaluate the benefit-risk balance beyond the adaptation of the conditioning. Unfortunately, bone marrow transplantation in telomeropathies does not correct other extra-medullary disorders.

12.4.2. *Hoyeraal–Hreidarsson (HH) syndrome*

HH syndrome is considered as one of the rarest and most severe clinical forms of DC. Patients with HH represent less than 100 cases described in the literature. The main clinical features observed in HH syndrome are intrauterine growth retardation, cerebellar hypoplasia and/or microcephaly, progressive bone marrow failure and immune deficiency (Hoyeraal et al. 1970; Hreidarsson et al. 1988; Aalfs and Hennekam 1995; Berthet et al. 1995; Revy et al. 2000; Cossu et al. 2002; Glousker et al. 2015). Immune deficiency is often manifested by a virtual absence of B lymphocytes and NK lymphocytes from birth (Revy et al. 2000; Jyonouchi et al. 2011; Touzot et al. 2012; Le Guen et al. 2013; Jullien et al. 2016; Touzot et al. 2016; Benyelles et al. 2019a).

The mucocutaneous triad seen in DC is not necessarily found in HH patients, although it may appear gradually over time. Most HH patients die within the first decade of life due to repeated infections secondary to the severe immune deficiency. In contrast to DC, the development of cancer has not, to our knowledge, been reported in patients with HH. This is probably due to the early age of death in children with HH, which does not allow the development of cancers classically observed in DC patients. The severe phenotype of HH has been correlated with the drastic reduction in telomere length (Alter et al. 2007, 2012; Alder et al. 2018). To date, mutations in seven genes have been associated with HH syndrome (*TINF2, ACD, DKC1, TERT, RTEL1, PARN* and *NHP2*).

However, interestingly, most of these factors have both telomeric and extra-telomeric functions, suggesting that the severity of symptoms observed in this disease may be a combination of a defect in telomere maintenance accompanied by abnormalities in other biological processes (such as DNA replication, DNA damage repair and ribosomal RNA biogenesis). This hypothesis is supported by the analysis of mouse models deficient for these factors. Indeed, mice deficient (KO) for factors involved only in the regulation of telomere length (such as TERT or TERC) are viable and show signs of premature aging only after several generations, when the length of the telomeres starts to be abnormally short. On the other hand, mice deficient in RTEL1, PARN and DKC1 are not viable, demonstrating, at least in mice, that their function extends beyond the regulation of telomere length. It has been reported that DKC1, PARN and NHP2, in addition to their function of maturation and stabilization of the telomerase complex, are involved in the biogenesis of ribosomal RNAs, which is necessary for the production of the ribosomes which guarantee protein translation (Mochizuki et al. 2004; Benyelles et al. 2019b, 2020).

12.4.3. *The Revesz syndrome*

Revesz syndrome is a very rare and severe variant of DC characterized by the combination of bilateral exudative retinopathy (characterized by vascular abnormalities around the retina) with intracranial calcifications and DC symptoms including early bone marrow failure, nail dystrophy and growth retardation (Revesz et al. 1992; Walne et al. 2008; Sasa et al. 2012). To date, to our knowledge, only mutations in *TINF2* (encoding TIN2) have been associated with Revesz syndrome (Savage et al. 2008; McElnea et al. 2013; Gupta et al. 2017; Sakwit et al. 2019). Similar to HH syndrome, patients with Revesz syndrome have extremely short telomeres compared to patients of the same age with DC (Alter et al. 2007, 2012).

12.4.4. *Coats plus syndrome*

Coats plus syndrome is a rare and severe disease characterized by occlusive retinal vasculopathy associated primarily with intrauterine growth retardation, developmental delay and multiple visceral involvement including cerebral calcifications and cysts, leukoencephalopathy and recurrent gastrointestinal bleeding. Bleeding, infection and liver disease (portal hypertension) are the main causes of death in patients between the ages of 2 and 30 years. Marrow involvement is sometimes observed. Mutations in the telomere factors *CTC1*, *POT1* and *STN1* have been associated with Coats plus syndrome.

In recent years, the spectrum of telomeropathy has been broadened and it is now recognized that the diagnosis can be made in adulthood. The triad of mucocutaneous manifestations is rarely present in these so-called “cryptic” forms, but patients may frequently present with early hair graying and various other disorders: bone marrow failure, PF, cutaneous disorders and hepatic disorders, associated or not, and in an apparently sporadic or familial context. For this type of clinical manifestations, the best described patients are those with *TERT* mutations (the most frequently mutated gene).

12.4.5. *Bone marrow failure*

Bone marrow failure presents as marrow-poor bi- or pancytopenia, bone marrow aplasia, myelodysplasia or leukemia.

Patients develop severe pancytopenia or MDS after several years.

The first mutations of genes involved in telomeropathies (TERTand TERC) were described in patients with aplastic anemia, myelodysplasia or acute leukemia (Yamaguchi et al. 2003, 2005).

Anemia is present in 13% of patients in a series of 134 *TERT* mutation carriers from 20 independent families (Diaz de Leon et al. 2010). In another paper on the clinico-biological comparison of 20 *TERT* mutation carriers and 20 non-carriers from 11 independent families, an increase in mean corpuscular volume (MCV) of red blood cells, a decrease in platelet count and a decrease in red blood cell count were significantly observed (Diaz de Leon et al. 2011).

Macrocytosis (MCV) >100 μm^3 and thrombocytopenia (<150 G/L) are often present in patients affected by PF.

In a cohort of French patients, 12 patients among 40 *TERT* or *TERC* mutation carriers (i.e. 30%) had abnormalities of the blood count: thrombocytopenia (n = 6; 15%), macrocytosis (n = 3; 7.5%) or both (n = 3; 7.5%). Bone marrow failure was objectified: myelodysplasia (n = 5; 12.5%), bone marrow aplasia (n = 2; 5%), myelofibrosis (n = 1; 2.5%) and acute myeloid leukemia (n = 1; 2.5%) (Borie et al. 2016).

In the 32 patients with PF with *RTEL1* mutation, macrocytosis was found in 15.6% and thrombocytopenia in 12.5% of patients (Borie et al. 2019).

Thrombocytopenia is also found in the recent cohort of Kapuria describing liver involvement. However, in this case, it is not easy to distinguish thrombocytopenia as a result of liver involvement or bone marrow involvement (Kapuria et al. 2019).

An immune deficiency (B, T and NK cell immunodeficiency) may be present. In addition, it has been reported that STS may be associated with T-cell immunodeficiency demonstrated in humans and mice as well (Wagner et al. 2018).

12.4.6. *Lung disease*

Monogenic PF (familial, associated with other diseases, early age of onset) is telomeropathies in the same way as DC, whose clinical expression reveals itself earlier at the hematological and/or cutaneous level.

Patients with pulmonary involvement are the most numerous among patients with telomeropathies (Alder et al. 2018).

The most common lung involvement is diffuse interstitial lung disease (ILD) and presents as PF. The probability that a patient with a *TERT* mutation will develop PF by age 60 is 60% (Diaz de Leon et al. 2011).

The pleuroparenchymal fibroelastosis (PPFE) subtype appears to be more frequently associated with pathogenic variations than other IPF subtypes (Newton et al. 2016).

The largest series of mutation-carrier patients with lung disease is that of Newton et al. (2016) with 115 patients with mutations in the *TERT* (n = 75), *TERC* (n = 7), *RTEL1* (n = 14) and *PARN* (n = 19) genes.

In most cohorts of patients with lung disease, the age of onset of symptoms is earlier for individuals carrying mutations in telomere-related genes (TRG) (about 10 years earlier on average compared to the age of onset in the general population (65 years)) (Kannengiesser et al. 2015; Borie et al. 2016; Newton et al. 2016; Borie et al. 2019).

The age of disease onset is significantly earlier in patients with *TERC* mutations compared to those with *PARN* mutations (Newton et al. 2016). Furthermore, it has been observed that the progression of lung disease caused by telomeropathy is particularly rapid (median period of less than 3 years from diagnosis to death or transplantation; Newton et al. 2016).

In retrospective series, an unfavorable environment is described with exposure to aero contaminants in 40–96% of cases (Diaz de Leon et al. 2010; Borie et al. 2016; Newton et al. 2016). Thus, it appears that telomeropathies are associated with a predisposition to develop lung damage, especially when individuals are exposed to harmful volatile agents (dust, toxic gases).

An alternative treatment for patients with PF is lung transplantation, which can be considered before age 65. Several studies, performed on small numbers, have shown the risk of severe hematological (especially thrombocytopenia) and renal complications after lung transplantation in patients with telomeropathy (Silhan et al. 2014; Borie et al. 2015; Tokman et al. 2015). In practice, care should be taken with IMUREL (azathioprine) immunosuppression in these patients with predisposition to blood disease.

Mutational status is assessed during the pre-transplantation workup. A patient with a mutation in TRG may be excluded from lung transplantation if extra-pulmonary involvement (such as myelodysplasia) may compromise the successful completion of the transplantation.

12.4.7. *Liver disease*

The first descriptions of patients with a telomerase complex mutation (*TERT*; *TERC*) with liver disease were made by the team of Rodrigo Calado concerning 69 patients from five independent families (Calado et al. 2009).

Damage is very variable: elevation of liver enzymes with a variable degree of necrosis, inflammation, fibrosis and histological liver regeneration, cirrhosis, cryptogenic, post viral or alcoholic hepatitis. The presence of a TRG mutation increases the risk of developing cirrhosis in case of hepatitis C virus infection and/or alcohol use.

Intra-familial comparison of 40 patients from 11 independent families, that is, 20 pairs of *TERT* mutation carriers and non-carriers, showed the presence of liver cytolysis in 5–8% of *TERT* mutation carrier patients (Diaz de Leon et al. 2011).

Interestingly, *TERT* and *TERC* mutation testing in a cohort of 134 patients with liver cirrhosis was positive in 10 patients (7.5%). All had cirrhosis considered secondary, due to hepatitis C (n = 4), alcohol consumption (n = 3), Wilson's disease (n = 1), non-alcoholic steatohepatitis (n = 1) or primary biliary cirrhosis (n = 1) (Calado et al. 2009). Thus, the gene–environment interaction is also strong for liver disease expression.

In a large study of 521 patients with cirrhosis secondary to chronic liver disease and 600 non-cirrhotic controls, *TERT* and *TERC* mutation testing was performed.

The number of mutation carriers was significantly higher in the group of patients with cirrhosis compared to non-cirrhotic controls (1.7% versus 0.3%) (Hartmann et al. 2011).

Mary Armanios' team sought to understand the cause of dyspnea (difficult breathing) in patients with telomeropathy without lung parenchymal involvement (Gorgy et al. 2015). In a retrospective cohort of 42 dyspneic patients with STS, nine (21%) had a hepato-pulmonary syndrome, that is, liver disease associated with intra-pulmonary vascular dilatations causing hypoxemia. Genetic analysis revealed a pathogenic variation in six patients (66%): *TERT* (n = 4), *RTEL1* (n = 1) or *DKC1* (n = 1). The most common histo-pathologic abnormality was regenerative nodular hyperplasia (RNH). The retained diagnosis was portal hypertension without cirrhosis for four of the six patients for whom a liver biopsy was available (Gorgy et al. 2015).

A clinical case description combined with a literature review provided an update on liver involvement (Patnaik et al. 2018).

A recent article describes liver damage in 30% (n = 41) of 121 patients studied with telomeropathy (*TERT, TERC, TINF2*) (Kapuria et al. 2019), although there are some reservations about the interpretation of some variants retained as pathogenic.

Collectively, this work highlights that liver disease is not uncommon in telomeropathies.

12.4.8. *Other clinical impairments*

A recent article reports the *neurological impairments* present in a cohort of 44 patients with telomeropathy including 31 DC, 12 HH and 1 RS (Bhala et al. 2019). Note that 25 of the patients (57%) had at least one structural abnormality, most commonly cerebellar hypoplasia (39%), 21 (48%) had a neurodevelopmental or psychomotor disorder and 12 (27%) had a psychiatric diagnosis including depression (Bhala et al. 2019). Patients with autosomal recessive or X-linked disease had more neurological abnormalities than patients with dominantly inherited disease (Bhala et al. 2019).

The recent description of two patients with heterozygous *PARN* mutation in a cohort of adult patients with unexplained chronic kidney disease may raise questions about the frequency of *renal impairment* in telomeropathies (Lata et al. 2018).

Of the 180 individuals evaluated at the John Hopkins Center in the United States, 12.8% had *cancer*. Solid tumors are rare (2.8%) (Schratz et al. 2020). Most cases represent young boys with *DKC1* mutations (4/5 or 80%) who developed squamous cell carcinoma of the tongue (n = 2), anal (n = 2), oral cavity (n = 1) or rectal (n = 1). The tumors were resectable and generally of good prognosis. The risk of cancer is mainly related to the hematological risk, most often MDS followed by AML (75% of cancers) (Schratz et al. 2020).

Gastrointestinal impairment is poorly evaluated, yet in extreme forms such as Coats plus or DKC1-related forms, this impairment is severe and prominent.

Skeletal injuries such as osteoporosis (5%), osteopenia (10%) and non-vascular necrosis have been reported in DC. The evaluation of these injuries in adult patients with late onset is lacking.

12.5. Genetic analysis and counseling

12.5.1. *Genetic analysis of telomeropathies*

In compliance with regulations (Bioethics law), a constitutional genetic analysis is offered in case of suspected Mendelian disease after the patient signed an informed consent. A next-generation sequencing (NGS) panel allows targeting a growing number of genes (including *TERT, TERC, DKC1, RTEL1, PARN, TINF2, NOP10, NHP2, ACD, NAF1, WRAP53, SHQ1, ZCCHC8, POT1, CTC1, STN1*). Whole-exome sequencing can also be implemented in some laboratories.

On the telomerase database site, in the "disease" tab, most of the TRG mutations identified in patients are listed (see: http://telomerase.asu.edu/diseases.html). Although very useful, this site is not regularly updated.

12.5.2. *Genetics of telomeropathies*

A constitutional pathogenic mutation is identified in a large number of cases in the mono-allelic or bi-allelic state. This explains 60–70% of pediatric patients and 30% of familial forms of PF (adult).

The assessment of penetrance (risk of developing the disease at a given age) has been made possible by the identification of a large number of adult cases, including those with a lung disease. Penetrance is incomplete and increases with age since 50% of *TERT, TERC, RTEL1, PARN, DKC1, TINF2* and *NAF1* mutation carriers in the 0–20 years category are symptomatic, for 67% between 20 and 40 years, for 80% between 40 and 60 years and 88% between 60 and 80 years (Alder et al. 2018).

The gene–environment interaction is important for lung and liver diseases expression. The exogenous agents promoting the development of bone marrow failure are less clearly identified (infectious agents such as viruses).

12.5.3. *Telomeropathy genetic counseling in practice, management of rare diseases in France*

It is possible to discuss these cases in multidisciplinary consultation meetings (RCP) within the various reference centers, supervised by the networks: the RCP of the reference center for aplastic anemia, within the MaRiH French "filère" (Professor Régis Peffault de Latour, Dr. Flore Sicre de Fontbrune, Dr. Thierry Leblanc, Hôpital Saint Louis) and the Pneumo-Genetic MDD (multidisciplinary discussion) at Bichat (reference center for rare pulmonary diseases, Professor Bruno Crestani, Dr. Raphael Borie, RESPIFIL filère). Hematologists (Dr. Flore Sicre de Fontbrune) and hepatologists from the reference center for vascular diseases of the liver (Dr. Aurélie Plessier), a genetic counselor and a psychologist are part of the quorum.

The identification of a pathogenic mutation in an affected individual requires specific management, particularly in the context of an organ transplantation (bone marrow or lung). For bone marrow transplantation, the diagnosis of telomeropathy will lead to the implementation of conditioning drugs adapted to monogenic bone marrow failure and will most often contraindicate intra-familial bone marrow transplantations (to be discussed during MDD after determining, among other things, the mutation status, the length of the telomeres and extra medullary diseases). For lung transplantation, phenotype/genotype correlations have demonstrated hematological toxicity of some immunosuppressive drugs after lung transplantation in patients with a mutation in the *TERT* or *TERC* genes (Silhan et al. 2014; Borie et al. 2015). In current practice – in pre-lung transplantation assessment, a genetic test is proposed – adaptation of conditioning according to the mutation status and thorough hematological evaluation if mutation is identified or telomeropathy is suspected.

Pre-symptomatic screening may be offered to relatives. In asymptomatic relatives, an interview with a psychologist, as well as monitoring of organs affected in STS is proposed.

Pre-natal screening may be discussed in cases of particular severity in a given family.

12.5.4. *Epigenetic heritability*

There is an epigenetic heritability linked to the transmission of short telomeres, independently of the mutation. Indeed, parents with short telomeres can transmit short telomeres to their offspring without necessarily transmitting the germline mutation. In a series of 134 individuals carrying a *TERT* mutation from 20 independent families, it has been shown that telomere length decreases with age, that a progressive decrease in telomere length is observed with transmission to successive generations and that members of these families non-carrier of the mutation have shorter telomeres than expected, demonstrating the so-called "epigenetic" transmission of short telomeres, independently of the familial mutation (Diaz de Leon et al. 2010)

Thus, within families carrying mutations, non-carriers may have shortened telomeres. This epigenetic heritability explains the observation of a genetic anticipation in mutated families (younger age of onset and worsening of symptoms over the generations) and may probably contribute to the occurrence of rare cases of phenocopies. Thus, in a series of families with DC, the existence of phenocopy (existence of DC symptoms without carrying the germline mutation implied in familial telomeropathy) has been demonstrated in some individuals (Collopy et al. 2015). For example, in a family with a heterozygous mutation in *TERC*, a 10-year-old sister of the proband, who did not carry the familial mutation, had cerebellar hypoplasia, microcephaly and a telomere length at the first percentile. In a family with a heterozygous *TERT* mutation, an 18-year-old brother of the proband, who was not a carrier of the familial mutation, had leukoplakia and a telomere length below the first percentile.

Similarly, phenocopy has been reported in a *TERT*-related PF family (Alder et al. 2011a). An affected PF relative did not carry the familial *TERT* mutation but had shorter than expected telomeres and had high tobacco exposure (>50 pack years) (Alder et al. 2011a).

12.6. Treatment of diseases associated with telomere defects

There is currently no "personalized" treatment targeting the constitutional defect in these patients. Theoretically, telomerase activity and/or telomere length should be restored. As patients have a multi-organ defect, the gene therapy approach seems to be difficult to implement at this time.

Therapeutic alternatives are those available for each form of organ damage (organ transplantations: lung, bone marrow, liver and specific drugs: anti-fibrotic agents, etc.).

Results of a clinical trial testing the efficacy of androgens (danazol 800 mg/d) in a series of 27 patients with telomeropathies with moderate hematological involvement appeared promising (Townsley et al. 2016). Out of 24 evaluable patients, a hematological response was obtained for 79% at 3 months and for 83% at 24 months. An increase in telomere length was reported (175 base pairs (bp) at 6 months, 360 bp at 12 months). However, another study questioned the increase in telomere length obtained by the not very robust quantitative PCR technique (Khincha et al. 2018). No increase in telomere length could be demonstrated in an independent series of patients for whom telomere length was measured by Flow-FISH (Khincha et al. 2018). French clinicians have therefore set up a national hospital-based clinical research program (PHRC) (ANDROTELO; reference center for aplastic anemia) focusing on the treatment with androgens of affected patients and the objective is to recruit 40 patients with BM or lung disease (Reference Center for Rare Lung Diseases, Hôpital Bichat, France).

12.7. Compensatory somatic mutations in the *TERT* promoter

Somatic mutations located in the *TERT* promoter have been identified in several cancers (melanoma, liver cancer, brain tumors, etc.). These mutations mainly target three positions so far (–124, –146 and –57 upstream of the initiation codon at position +1) and create new binding sites for transcription factors. Therefore, these mutations induce TERT expression and cause telomerase activation that facilitates cell proliferation and survival (Heidenreich et al. 2014; Chiba et al. 2015; Huang et al. 2015; Heidenreich and Kumar 2017; Yuan et al. 2019).

Surprisingly, in patients with telomeropathy and no cancer pathology, somatic mutations in the *TERT* promoter in a mosaic pattern have been shown in blood cells. This observation suggests that in the context of telomeropathy, somatic *TERT* promoter-activating mutations may counteract the constitutional defect in the telomerase pathway (Revy et al. 2019). Indeed, a first study reported the presence of a somatic mutation in the *TERT* promoter in patients with PF and carrying germline mutations in a TRG (*TERT* and *PARN*) (Maryoung et al. 2017). Furthermore, in

EBV immortalized lymphoblastoid lineages, these mutations confer an advantage for clonal expansion (positive selection) (Maryoung et al. 2017). In this study, only the most frequent *TERT* promoter mutations (–124, –146) were analyzed, and the technique used (Sanger sequencing) did not allow sensitive detection of these mutations, and probably underestimated (5%) the frequency of patients carrying somatic *TERT* promoter mutations in their blood cells. A second study, using more sensitive techniques (next-generation sequencing), revealed that somatic mutations in the *TERT* promoter were also detected in individuals with hematological presentation and TRG mutation (TERT, TERC, PARN) (Gutierrez-Rodrigues et al. 2018), although the frequency of patients with mosaicism appeared to be lower than in patients with PF, presumably because they are older in the latter case. Finally, a third study detected a somatic mutation of the *TERT* promoter in blood cells in an individual with PF and hematological abnormalities caused by a germline mutation of *NHP2* (Benyelles et al. 2020). Current data in the literature do not indicate whether these somatic mutations are associated with a particular clinical translation, beneficial or deleterious.

12.8. Missing heredity

A number of cases of telomeropathy presenting as DC, HH syndromes or familial PF remain unexplained. A number of monogenic (one major causative gene) forms of these telomeropathies remain to be elucidated. Decreased telomere length is not an exclusive marker of telomeropathies and may reflect bone marrow failure outside the context of genetic telomeropathy.

Several complementary hypotheses can be put forward to explain the difficulty in elucidating the genetic basis of these as yet unexplained telomeropathies: oligogenism (contribution of several germline genetic variants with a strong to moderate effect), polygenism (contribution of several constitutional genetic variants with a weak effect) and the role of the environment and epigenetics (modification of DNA in non-coding sequences).

Interestingly, in genetic association studies of idiopathic PF, non-rare variants are identified in telomere maintenance genes as associated with this disease. The clinical implication of these genetic associations remains to be elucidated (*TERT*, rs2736100; *TERC*: rs6793295; *STN1/OBFC1*: rs11191865), suggesting, however, that they may contribute to telomere length (Kropski and Loyd 2015).

12.9. Conclusion

The clinical expression of telomeropathies is variable. Gene–environment interaction is important. Genetic counseling is difficult because of the anticipation and transmission of short telomeres.

It is estimated that the genetic cause of these diseases has been identified in about 30–40% of cases. It is therefore very likely that other genes involved in telomere pathologies will be characterized in the future, and it is possible to export a list of more than 2,000 genes from a German website whose objective is to list the genes directly and indirectly involved in telomere regulation mechanisms in humans.

As described above, telomere pathologies are most commonly associated with abnormal telomere length. A subtype of DC and HH has been described that could be caused by telomere instabilities leading to increased cell apoptosis without affecting telomere length (Touzot et al. 2010, 2012). However, to date, no genetic cause has been associated with this type of atypical pathology. It is conceivable that in the coming years new telomeropathies could be identified in which the telomere defect would be qualitative and not quantitative.

Two recent papers show that inhibition of PAPD5 factor by chemical compounds, increasing the amount of TERC could be a possible strategy to improve the blood phenotype of patients with DC. This promising work paves the way for possible clinical trials to test the efficacy and tolerability of PAPD5 inhibitors in the numerous patients with telomeropathy and lung diseases (Fok et al. 2019; Nagpal et al. 2020).

12.10. References

Aalfs, C.M. and Hennekam, R.C. (1995). Differences between the Hoyeraal-Hreidarsson syndrome and an autosomal recessive congenital intrauterine infection-like syndrome. *Am. J. Med. Genet.*, 58, 385.

Abdallah, P., Luciano, P., Runge, K.W., Lisby, M., Géli, V., Gilson, E., Teixeira, M.T. (2009). A two-step model for senescence triggered by a single critically short telomere. *Nat. Cell Biol.*, 11, 988–993.

Agarwal, S. (2018). Evaluation and management of hematopoietic failure in dyskeratosis congenita. *Hematology/Oncology Clinics of North America*, 32, 669–685.

Alder, J.K., Chen, J.J., Lancaster, L., Danoff, S., Su, S.C., Cogan, J.D., Vulto, I., Xie, M., Qi, X., Tuder, R.M. et al. (2008). Short telomeres are a risk factor for idiopathic pulmonary fibrosis. *Proceedings of the National Academy of Sciences of the United States of America*, 105, 13051–13056.

Alder, J.K., Cogan, J.D., Brown, A.F., Anderson, C.J., Lawson, W.E., Lansdorp, P.M., Phillips III, J.A., Loyd, J.E., Chen, J.J., Armanios, M. (2011a). Ancestral mutation in telomerase causes defects in repeat addition processivity and manifests as familial pulmonary fibrosis. *PLoS Genet.*, 7, e1001352.

Alder, J.K., Guo, N., Kembou, F., Parry, E.M., Anderson, C.J., Gorgy, A.I., Walsh, M.F., Sussan, T., Biswal, S., Mitzner, W. et al. (2011b). Telomere length is a determinant of emphysema susceptibility. *Am. J. Respir. Crit. Care Med.*, 184, 904–912.

Alder, J.K., Stanley, S.E., Wagner, C.L., Hamilton, M., Hanumanthu, V.S., Armanios, M. (2015). Exome sequencing identifies mutant TINF2 in a family with pulmonary fibrosis. *Chest*, 147, 1361–1368.

Alder, J.K., Hanumanthu, V.S., Strong, M.A., DeZern, A.E., Stanley, S.E., Takemoto, C.M., Danilova, L., Applegate, C.D., Bolton, S.G., Mohr, D.W. et al. (2018). Diagnostic utility of telomere length testing in a hospital-based setting. *Proceedings of the National Academy of Sciences of the United States of America*, 115, E2358–E2365.

Alter, B.P., Baerlocher, G.M., Savage, S.A., Chanock, S.J., Weksler, B.B., Willner, J.P., Peters, J.A., Giri, N., Lansdorp, P.M. (2007). Very short telomere length by flow fluorescence in situ hybridization identifies patients with dyskeratosis congenita. *Blood*, 110, 1439–1447.

Alter, B.P., Giri, N., Savage, S.A., Rosenberg, P.S. (2009). Cancer in dyskeratosis congenita. *Blood*, 113, 6549–6557.

Alter, B.P., Rosenberg, P.S., Giri, N., Baerlocher, G.M., Lansdorp, P.M., Savage, S.A. (2012). Telomere length is associated with disease severity and declines with age in dyskeratosis congenita. *Haematologica*, 97, 353–359.

Anderson, B.H., Kasher, P.R., Mayer, J., Szynkiewicz, M., Jenkinson, E.M., Bhaskar, S.S., Urquhart, J.E., Daly, S.B., Dickerson, J.E., O'Sullivan, J. et al. (2012). Mutations in CTC1, encoding conserved telomere maintenance component 1, cause Coats plus. *Nature Genetics*, 44, 338–342.

Armanios, M. (2009). Syndromes of telomere shortening. *Annu. Rev. Genomics Hum. Genet.*, 10, 45–61.

Armanios, M. and Blackburn, E.H. (2012). The telomere syndromes. *Nature Reviews Genetics*, 13, 693–704.

Armanios, M.Y., Chen, J.J., Cogan, J.D., Alder, J.K., Ingersoll, R.G., Markin, C., Lawson, W.E., Xie, M., Vulto, I., Phillips III, J.A. et al. (2007). Telomerase mutations in families with idiopathic pulmonary fibrosis. *The New England Journal of Medicine*, 356, 1317–1326.

Awad, A., Glousker, G., Lamm, N., Tawil, S., Hourvitz, N., Smoom, R., Revy, P., Tzfati, Y. (2020). Full length RTEL1 is required for the elongation of the single-stranded telomeric overhang by telomerase. *Nucleic Acids Research*, 48, 7239–7251.

Ballew, B.J., Joseph, V., De, S., Sarek, G., Vannier, J.B., Stracker, T., Schrader, K.A., Small, T.N., O'Reilly, R., Manschreck, C. et al. (2013a). A recessive founder mutation in regulator of telomere elongation helicase 1, RTEL1, underlies severe immunodeficiency and features of Hoyeraal Hreidarsson syndrome. *PLoS Genet.*, 9, e1003695.

Ballew, B.J., Yeager, M., Jacobs, K., Giri, N., Boland, J., Burdett, L., Alter, B.P., Savage, S.A. (2013b). Germline mutations of regulator of telomere elongation helicase 1, RTEL1, in dyskeratosis congenita. *Hum. Genet.*, 132, 473–480.

Barbaro, P. and Vedi, A. (2016). Survival after hematopoietic stem cell transplant in patients with dyskeratosis congenita: Systematic review of the literature. *Biology of Blood and Marrow Transplantation: Journal of the American Society for Blood and Marrow Transplantation*, 22, 1152–1158.

Benyelles, M., Episkopou, H., O'Donohue, M.F., Kermasson, L., Frange, P., Poulain, F., Burcu Belen, F., Polat, M., Bole-Feysot, C., Langa-Vives, F. et al. (2019a). Impaired telomere integrity and rRNA biogenesis in PARN-deficient patients and knock-out models. *EMBO Molecular Medicine*, 11, e10201.

Benyelles, M., Episkopou, H., O'Donohue, M.F., Kermasson, L., Frange, P., Poulain, F., Burcu Belen, F., Polat, M., Bole-Feysot, C., Langa-Vives, F. et al. (2019b). Impaired telomere integrity and rRNA biogenesis in PARN-deficient patients and knock-out models. *EMBO Molecular Medicine*, 11, e10201.

Benyelles, M., O'Donohue, M.F., Kermasson, L., Lainey, E., Borie, R., Lagresle-Peyrou, C., Nunes, H., Cazelles, C., Fourrage, C., Ollivier, E. et al. (2020). NHP2 deficiency impairs rRNA biogenesis and causes pulmonary fibrosis and Hoyeraal-Hreidarsson syndrome. *Hum. Mol. Genet.*, 29, 907.

Berthet, F., Tuchschmid, P., Boltshauser, E., Seger, R.A. (1995). The Hoyeraal-Hreidarsson syndrome: Don't forget the associated immunodeficiency. *Eur. J. Pediatr.*, 154, 998.

Bhala, S., Best, A.F., Giri, N., Alter, B.P., Pao, M., Gropman, A., Baker, E.H., Savage, S.A. (2019). CNS manifestations in patients with telomere biology disorders. *Neurology Genetics*, 5, 370.

Bisserbe, A., Tertian, G., Buffet, C., Turhan, A., Lambotte, O., Nasser, G., Alvin, P., Tardieu, M., Riant, F., Bergametti, F. et al. (2015). Cerebro-retinal microangiopathy with calcifications and cysts due to recessive mutations in the CTC1 gene. *Revue neurologique*, 171, 445–449.

Bizarro, J. and Meier, U.T. (2017). Inherited SHQ1 mutations impair interaction with NAP57/dyskerin, a major target in dyskeratosis congenita. *Molecular Genetics & Genomic Medicine*, 5, 805–808.

Borie, R., Kannengiesser, C., Hirschi, S., Le Pavec, J., Mal, H., Bergot, E., Jouneau, S., Naccache, J.M., Revy, P., Boutboul, D. et al. (2015). Severe hematologic complications after lung transplantation in patients with telomerase complex mutations. *The Journal of Heart and Lung Transplantation: The Official Publication of the International Society for Heart Transplantation*, 34, 538–546.

Borie, R., Tabeze, L., Thabut, G., Nunes, H., Cottin, V., Marchand-Adam, S., Prevot, G., Tazi, A., Cadranel, J., Mal, H. et al. (2016). Prevalence and characteristics of TERT and TERC mutations in suspected genetic pulmonary fibrosis. *The European Respiratory Journal*, 48, 1721–1731.

Borie, R., Bouvry, D., Cottin, V., Gauvain, C., Cazes, A., Debray, M.P., Cadranel, J., Dieude, P., Degot, T., Dominique, S. et al. (2018). Regulator of telomere length 1 (RTEL1) mutations are associated with heterogeneous pulmonary and extra-pulmonary phenotypes. *The European Respiratory Journal*, 53, 1800508.

Burris, A.M., Ballew, B.J., Kentosh, J.B., Turner, C.E., Norton, S.A., Laboratory, N.D.C.G.R., Group, N.D.C.S.W., Giri, N., Alter, B.P., Nellan, A. et al. (2016). Hoyeraal-Hreidarsson syndrome due to PARN mutations: Fourteen years of follow-up. *Pediatr. Neurol.*, 56, 62–68.e61.

Calado, R.T., Regal, J.A., Kleiner, D.E., Schrump, D.S., Peterson, N.R., Pons, V., Chanock, S.J., Lansdorp, P.M., Young, N.S. (2009). A spectrum of severe familial liver disorders associate with telomerase mutations. *PLoS One*, 4, e7926.

Calado, R.T., Brudno, J., Mehta, P., Kovacs, J.J., Wu, C., Zago, M.A., Chanock, S.J., Boyer, T.D., Young, N.S. (2011). Constitutional telomerase mutations are genetic risk factors for cirrhosis. *Hepatology*, 53, 1600–1607.

Carillo, S., Henry, L., Lippert, E., Girodon, F., Guiraud, I., Richard, C., Dubois Galopin, F., Cleyrat, C., Jourdan, E., Kralovics, R. et al. (2011). Nested high-resolution melting curve analysis: A highly sensitive, reliable, and simple method for detection of JAK2 exon 12 mutations – Clinical relevance in the monitoring of polycythemia. *J. Mol. Diagn.*, 13, 263–270.

Cesare, A.J. and Reddel, R.R. (2010). Alternative lengthening of telomeres: Models, mechanisms and implications. *Nature Reviews Genetics*, 11, 319–330.

Chiba, K., Johnson, J.Z., Vogan, J.M., Wagner, T., Boyle, J.M., Hockemeyer, D. (2015). Cancer-associated TERT promoter mutations abrogate telomerase silencing. *Elife*, 4.

Cogan, J.D., Kropski, J.A., Zhao, M., Mitchell, D.B., Rives, L., Markin, C., Garnett, E.T., Montgomery, K.H., Mason, W.R., McKean, D.F. et al. (2015). Rare variants in RTEL1 are associated with familial interstitial pneumonia. *Am. J. Respir. Crit. Care Med.*, 191, 646.

Collopy, L.C., Walne, A.J., Cardoso, S., de la Fuente, J., Mohamed, M., Toriello, H., Tamary, H., Ling, A.J., Lloyd, T., Kassam, R. et al. (2015). Triallelic and epigenetic-like inheritance in human disorders of telomerase. *Blood*, 126, 176–184.

Cossu, F., Vulliamy, T.J., Marrone, A., Badiali, M., Cao, A., Dokal, I. (2002). A novel DKC1 mutation, severe combined immunodeficiency (T+B-NK- SCID) and bone marrow transplantation in an infant with Hoyeraal-Hreidarsson syndrome. *British Journal of Haematology*, 119, 765–768.

Dai, J., Cai, H., Zhuang, Y., Wu, Y., Min, H., Li, J., Shi, Y., Gao, Q., Yi, L. (2015). Telomerase gene mutations and telomere length shortening in patients with idiopathic pulmonary fibrosis in a Chinese population. *Respirology*, 20, 122–128.

Dalle, J.H. and Peffault de Latour, R. (2016). Allogeneic hematopoietic stem cell transplantation for inherited bone marrow failure syndromes. *International Journal of Hematology*, 103, 373–379.

Deng, Z., Glousker, G., Molczan, A., Fox, A.J., Lamm, N., Dheekollu, J., Weizman, O.E., Schertzer, M., Wang, Z., Vladimirova, O. et al. (2013). Inherited mutations in the helicase RTEL1 cause telomere dysfunction and Hoyeraal-Hreidarsson syndrome. *Proceedings of the National Academy of Sciences of the United States of America*, 110, E3408–E3416.

Dhanraj, S., Gunja, S.M., Deveau, A.P., Nissbeck, M., Boonyawat, B., Coombs, A.J., Renieri, A., Mucciolo, M., Marozza, A., Buoni, S. et al. (2015). Bone marrow failure and developmental delay caused by mutations in poly(A)-specific ribonuclease (PARN). *J. Med. Genet.*, 52, 738.

Diaz de Leon, A., Cronkhite, J.T., Katzenstein, A.L., Godwin, J.D., Raghu, G., Glazer, C.S., Rosenblatt, R.L., Girod, C.E., Garrity, E.R., Xing, C. et al. (2010). Telomere lengths, pulmonary fibrosis and telomerase (TERT) mutations. *PLoS One*, 5, e10680.

Diaz de Leon, A., Cronkhite, J.T., Yilmaz, C., Brewington, C., Wang, R., Xing, C., Hsia, C.C.W., Garcia, C.K. (2011). Subclinical lung disease, macrocytosis, and premature graying in kindreds with telomerase (TERT) mutations. *Chest*, 140, 753–763.

Dodson, L.M., Baldan, A., Nissbeck, M., Gunja, S.M.R., Bonnen, P.E., Aubert, G., Birchansky, S., Virtanen, A., Bertuch, A.A. (2019). From incomplete penetrance with normal telomere length to severe disease and telomere shortening in a family with monoallelic and biallelic PARN pathogenic variants. *Human Mutation*, 40, 2414–2429.

Du, H.Y., Pumbo, E., Ivanovich, J., An, P., Maziarz, R.T., Reiss, U.M., Chirnomas, D., Shimamura, A., Vlachos, A., Lipton, J.M. et al. (2009). TERC and TERT gene mutations in patients with bone marrow failure and the significance of telomere length measurements. *Blood*, 113, 309–316.

Fok, W.C., Shukla, S., Vessoni, A.T., Brenner, K.A., Parker, R., Sturgeon, C.M., Batista, L.F.Z. (2019). Posttranscriptional modulation of TERC by PAPD5 inhibition rescues hematopoietic development in dyskeratosis congenita. *Blood*, 133, 1308–1312.

Gable, D.L., Gaysinskaya, V., Atik, C.C., Talbot Jr., C.C., Kang, B., Stanley, S.E., Pugh, E.W., Amat-Codina, N., Schenk, K.M., Arcasoy, M.O. et al. (2019). ZCCHC8, the nuclear exosome targeting component, is mutated in familial pulmonary fibrosis and is required for telomerase RNA maturation. *Genes & Development*, 33, 1381–1396.

Glousker, G., Touzot, F., Revy, P., Tzfati, Y., Savage, S.A. (2015). Unraveling the pathogenesis of Hoyeraal-Hreidarsson syndrome, a complex telomere biology disorder. *British Journal of Haematology*, 170, 457.

Gorgy, A.I., Jonassaint, N.L., Stanley, S.E., Koteish, A., DeZern, A.E., Walter, J.E., Sopha, S.C., Hamilton, J.P., Hoover-Fong, J., Chen, A.R. et al. (2015). Hepatopulmonary syndrome is a frequent cause of dyspnea in the short telomere disorders. *Chest*, 148, 1019–1026.

Gramatges, M.M., Qi, X., Sasa, G.S., Chen, J.J., Bertuch, A.A. (2013). A homozygous telomerase T-motif variant resulting in markedly reduced repeat addition processivity in siblings with Hoyeraal Hreidarsson syndrome. *Blood*, 121, 3586–3593.

Guo, Y., Kartawinata, M., Li, J., Pickett, H.A., Teo, J., Kilo, T., Barbaro, P.M., Keating, B., Chen, Y., Tian, L. et al. (2014). Inherited bone marrow failure associated with germline mutation of ACD, the gene encoding telomere protein TPP1. *Blood*, 124, 2767–2774.

Gupta, M.P., Talcott, K.E., Kim, D.Y., Agarwal, S., Mukai, S. (2017). Retinal findings and a novel TINF2 mutation in Revesz syndrome: Clinical and molecular correlations with pediatric retinal vasculopathies. *Ophthalmic Genetics*, 38, 51–60.

Gutierrez-Rodrigues, F., Donaires, F.S., Pinto, A., Vicente, A., Dillon, L.W., Cle, D.V., Santana, B.A., Pirooznia, M., Ibanez, M., Townsley, D.M. et al. (2018). Pathogenic TERT promoter variants in telomere diseases. *Genet. Med.*, 21, 1594.

Hartmann, D., Srivastava, U., Thaler, M., Kleinhans, K.N., N'Kontchou, G., Scheffold, A., Bauer, K., Kratzer, R.F., Kloos, N., Katz, S.F. et al. (2011). Telomerase gene mutations are associated with cirrhosis formation. *Hepatology*, 53, 1608–1617.

Heidenreich, B. and Kumar, R. (2017). TERT promoter mutations in telomere biology. *Mutat. Res.*, 771, 15–31.

Heidenreich, B., Rachakonda, P.S., Hemminki, K., Kumar, R. (2014). TERT promoter mutations in cancer development. *Curr. Opin. Genet. Dev.*, 24, 30–37.

Heiss, N.S., Knight, S.W., Vulliamy, T.J., Klauck, S.M., Wiemann, S., Mason, P.J., Poustka, A., Dokal, I. (1998). X-linked dyskeratosis congenita is caused by mutations in a highly conserved gene with putative nucleolar functions. *Nature Genetics*, 19, 32–38.

Hisata, S., Sakaguchi, H., Kanegane, H., Hidaka, T., Shiihara, J., Ichinose, M., Kojima, S., Nukiwa, T., Ebina, M. (2013). A novel missense mutation of DKC1 in dyskeratosis congenita with pulmonary fibrosis. *Sarcoidosis, Vasculitis, and Diffuse Lung Diseases: Official Journal of WASOG*, 30, 221–225.

Hoffman, T.W., van der Vis, J.J., van der Smagt, J.J., Massink, M.P.G., Grutters, J.C., van Moorsel, C.H.M. (2019). Pulmonary fibrosis linked to variants in the ACD gene, encoding the telomere protein TPP1. *The European Respiratory Journal*, 54, 1900809.

Hoyeraal, H.M., Lamvik, J., Moe, P.J. (1970). Congenital hypoplastic thrombocytopenia and cerebral malformations in two brothers. *Acta Paediatr. Scand.*, 59, 185–191.

Hreidarsson, S., Kristjansson, K., Johannesson, G., Johannsson, J.H. (1988). A syndrome of progressive pancytopenia with microcephaly, cerebellar hypoplasia and growth failure. *Acta Paediatr. Scand.*, 77, 773–775.

Huang, D.S., Wang, Z., He, X.J., Diplas, B.H., Yang, R., Killela, P.J., Meng, Q., Ye, Z.Y., Wang, W., Jiang, X.T. et al. (2015). Recurrent TERT promoter mutations identified in a large-scale study of multiple tumour types are associated with increased TERT expression and telomerase activation. *Eur. J. Cancer*, 51, 969–976.

Jongmans, M.C., Verwiel, E.T., Heijdra, Y., Vulliamy, T.J., Kamping, E.J., Hehir-Kwa, J.Y., Bongers, E.M., Pfundt, R., van Emst, L., van Leeuwen, F.N. et al. (2012). Revertant somatic mosaicism by mitotic recombination in dyskeratosis congenita. *Am. J. Hum. Genet.*, 90, 426–433.

Juge, P.A., Borie, R., Kannengiesser, C., Gazal, S., Revy, P., Wemeau-Stervinou, L., Debray, M.P., Ottaviani, S., Marchand-Adam, S., Nathan, N. et al. (2017). Shared genetic predisposition in rheumatoid arthritis-interstitial lung disease and familial pulmonary fibrosis. *The European Respiratory Journal*, 49, 1602314.

Jullien, L., Kannengiesser, C., Kermasson, L., Cormier-Daire, V., Leblanc, T., Soulier, J., Londono-Vallejo, A., de Villartay, J.P., Callebaut, I., Revy, P. (2016). Mutations of the RTEL1 helicase in a Hoyeraal-Hreidarsson syndrome patient highlight the importance of the ARCH domain. *Hum. Mutat.*, 37, 469–472.

Jyonouchi, S., Forbes, L., Ruchelli, E., Sullivan, K.E. (2011). Dyskeratosis congenita: A combined immunodeficiency with broad clinical spectrum – A single-center pediatric experience. *Pediatr. Allergy Immunol.*, 22, 313–319.

Kanegane, H., Kasahara, Y., Okamura, J., Hongo, T., Tanaka, R., Nomura, K., Kojima, S., Miyawaki, T. (2005). Identification of DKC1 gene mutations in Japanese patients with X-linked dyskeratosis congenita. *British Journal of Haematology*, 129, 432–434.

Kannengiesser, C., Borie, R., Menard, C., Reocreux, M., Nitschke, P., Gazal, S., Mal, H., Taille, C., Cadranel, J., Nunes, H. et al. (2015). Heterozygous RTEL1 mutations are associated with familial pulmonary fibrosis. *The European Respiratory Journal*, 46, 474–485.

Kannengiesser, C., Manali, E.D., Revy, P., Callebaut, I., Ba, I., Borgel, A., Oudin, C., Haritou, A., Kolilekas, L., Malagari, K. et al. (2020). First heterozygous NOP10 mutation in familial pulmonary fibrosis. *The European Respiratory Journal*, 55, 1902465.

Kapuria, D., Ben-Yakov, G., Ortolano, R., Cho, M.H., Kalchiem-Dekel, O., Takyar, V., Lingala, S., Gara, N., Tana, M., Kim, Y.J. et al. (2019). The spectrum of hepatic involvement in patients with telomere disease. *Hepatology*, 69, 2579–2585.

Keller, R.B., Gagne, K.E., Usmani, G.N., Asdourian, G.K., Williams, D.A., Hofmann, I., Agarwal, S. (2012). CTC1 mutations in a patient with dyskeratosis congenita. *Pediatric Blood & Cancer*, 59, 311–314.

Khincha, P.P., Bertuch, A.A., Gadalla, S.M., Giri, N., Alter, B.P., Savage, S.A. (2018). Similar telomere attrition rates in androgen-treated and untreated patients with dyskeratosis congenita. *Blood Advances*, 2, 1243–1249.

Knight, S.W., Heiss, N.S., Vulliamy, T.J., Aalfs, C.M., McMahon, C., Richmond, P., Jones, A., Hennekam, R.C., Poustka, A., Mason, P.J. et al. (1999a). Unexplained aplastic anaemia, immunodeficiency, and cerebellar hypoplasia (Hoyeraal-Hreidarsson syndrome) due to mutations in the dyskeratosis congenita gene, DKC1. *British Journal of Haematology*, 107, 335–339.

Knight, S.W., Heiss, N.S., Vulliamy, T.J., Greschner, S., Stavrides, G., Pai, G.S., Lestringant, G., Varma, N., Mason, P.J., Dokal, I. et al. (1999b). X-linked dyskeratosis congenita is predominantly caused by missense mutations in the DKC1 gene. *Am. J. Hum. Genet.*, 65, 50–58.

Knight, S.W., Vulliamy, T.J., Morgan, B., Devriendt, K., Mason, P.J., Dokal, I. (2001). Identification of novel DKC1 mutations in patients with dyskeratosis congenita: Implications for pathophysiology and diagnosis. *Hum. Genet.*, 108, 299–303.

Kocak, H., Ballew, B.J., Bisht, K., Eggebeen, R., Hicks, B.D., Suman, S., O'Neil, A., Giri, N., Maillard, I., Alter, B.P. et al. (2014). Hoyeraal-Hreidarsson syndrome caused by a germline mutation in the TEL patch of the telomere protein TPP1. *Genes Dev.*, 28, 2090–2102.

Kropski, J.A. and Loyd, J.E. (2015). Telomeres revisited: RTEL1 variants in pulmonary fibrosis. *The European Respiratory Journal*, 46, 312–314.

Kropski, J.A., Mitchell, D.B., Markin, C., Polosukhin, V.V., Choi, L., Johnson, J.E., Lawson, W.E., Phillips III, J.A., Cogan, J.D., Blackwell, T.S. et al. (2014). A novel dyskerin (DKC1) mutation is associated with familial interstitial pneumonia. *Chest*, 146, e1–e7.

Kropski, J.A., Reiss, S., Markin, C., Brown, K.K., Schwartz, D.A., Schwarz, M.I., Loyd, J.E., Phillips III, J.A., Blackwell, T.S., Cogan, J.D. (2017). Rare genetic variants in PARN are associated with pulmonary fibrosis in families. *Am. J. Respir. Crit. Care Med.*, 196, 1481.

Laprade, H., Querido, E., Smith, M.J., Guerit, D., Crimmins, H., Conomos, D., Pourret, E., Chartrand, P., Sfeir, A. (2020). Single-molecule imaging of telomerase RNA reveals a recruitment-retention model for telomere elongation. *Molecular Cell*, 79, 115–126.e116.

Lata, S., Marasa, M., Li, Y., Fasel, D.A., Groopman, E., Jobanputra, V., Rasouly, H., Mitrotti, A., Westland, R., Verbitsky, M. et al. (2018). Whole-exome sequencing in adults with chronic kidney disease: A pilot study. *Annals of Internal Medicine*, 168, 100–109.

Le Guen, T., Jullien, L., Touzot, F., Schertzer, M., Gaillard, L., Perderiset, M., Carpentier, W., Nitschke, P., Picard, C., Couillault, G. et al. (2013). Human RTEL1 deficiency causes Hoyeraal-Hreidarsson syndrome with short telomeres and genome instability. *Hum. Mol. Genet.*, 22, 3239–3249.

Lin, H., Gong, L., Zhan, S., Wang, Y., Liu, A. (2017). Novel biallelic missense mutations in CTC1 gene identified in a Chinese family with Coats plus syndrome. *Journal of the Neurological Sciences*, 382, 142–145.

Marrone, A., Sokhal, P., Walne, A.J., Beswick, R., Kirwan, M., Killick, S., Williams, M., Marsh, J., Vulliamy, T.J., Dokal, I. (2007a). Functional characterization of novel telomerase RNA (TERC) mutations in patients with diverse clinical and pathological presentations. *Haematologica*, 92, 1013–1020.

Marrone, A., Walne, A.J., Tamary, H., Masunari, Y., Kirwan, M., Beswick, R., Vulliamy, T.J., Dokal, I. (2007b). Telomerase reverse-transcriptase homozygous mutations in autosomal recessive dyskeratosis congenita and Hoyeraal-Hreidarsson syndrome. *Blood*, 110, 4198–4205.

Maryoung, L., Yue, Y., Young, A., Newton, C.A., Barba, C., van Oers, N.S., Wang, R.C., Garcia, C.K. (2017). Somatic mutations in telomerase promoter counterbalance germline loss-of-function mutations. *J. Clin. Invest.*, 127, 982–986.

McElnea, E.M., van der Spek, N., Smith, O., Fitzsimon, S., Patel, C.K., O'Marcaigh, A. (2013). Revesz syndrome masquerading as bilateral cicatricial retinopathy of prematurity. *Journal of AAPOS: The Official Publication of the American Association for Pediatric Ophthalmology and Strabismus*, 17, 634–636.

Mochizuki, Y., He, J., Kulkarni, S., Bessler, M., Mason, P.J. (2004). Mouse dyskerin mutations affect accumulation of telomerase RNA and small nucleolar RNA, telomerase activity, and ribosomal RNA processing. *Proceedings of the National Academy of Sciences of the United States of America*, 101, 10756–10761.

Moon, D.H., Segal, M., Boyraz, B., Guinan, E., Hofmann, I., Cahan, P., Tai, A.K., Agarwal, S. (2015). Poly(A)-specific ribonuclease (PARN) mediates 3'-end maturation of the telomerase RNA component. *Nature Genetics*, 47, 1482–1488.

Moriya, K., Niizuma, H., Rikiishi, T., Yamaguchi, H., Sasahara, Y., Kure, S. (2016). Novel compound heterozygous RTEL1 gene mutations in a patient with Hoyeraal-Hreidarsson syndrome. *Pediatric Blood & Cancer*, 63, 1683–1684.

Nachmani, D., Bothmer, A.H., Grisendi, S., Mele, A., Bothmer, D., Lee, J.D., Monteleone, E., Cheng, K., Zhang, Y., Bester, A.C. et al. (2019). Germline NPM1 mutations lead to altered rRNA 2'-O-methylation and cause dyskeratosis congenita. *Nature Genetics*, 51, 1518–1529.

Nagpal, N., Wang, J., Zeng, J., Lo, E., Moon, D.H., Luk, K., Braun, R.O., Burroughs, L.M., Keel, S.B., Reilly, C. et al. (2020). Small-molecule PAPD5 inhibitors restore telomerase activity in patient stem cells. *Cell Stem Cell*, 26, 896–909.e898.

Netravathi, M., Kumari, R., Kapoor, S., Dakle, P., Dwivedi, M.K., Roy, S.D., Pandey, P., Saini, J., Ramakrishna, A., Navalli, D. et al. (2015). Whole exome sequencing in an Indian family links Coats plus syndrome and dextrocardia with a homozygous novel CTC1 and a rare HES7 variation. *BMC Medical Genetics*, 16, 5.

Newton, C.A., Batra, K., Torrealba, J., Kozlitina, J., Glazer, C.S., Aravena, C., Meyer, K., Raghu, G., Collard, H.R., Garcia, C.K. (2016). Telomere-related lung fibrosis is diagnostically heterogeneous but uniformly progressive. *Eur. Respir. J.*, 48, 1710–1720.

Parry, E.M., Alder, J.K., Qi, X., Chen, J.J., Armanios, M. (2011). Syndrome complex of bone marrow failure and pulmonary fibrosis predicts germline defects in telomerase. *Blood*, 117, 5607–5611.

Patnaik, M.M., Kamath, P.S., Simonetto, D.A. (2018). Hepatic manifestations of telomere biology disorders. *J. Hepatol.*, 69, 736–743.

Pike, A.M., Strong, M.A., Ouyang, J.P.T., Greider, C.W. (2019). TIN2 functions with TPP1/POT1 to stimulate telomerase processivity. *Molecular and Cellular Biology*, 39, e00593-18.

Polvi, A., Linnankivi, T., Kivela, T., Herva, R., Keating, J.P., Makitie, O., Pareyson, D., Vainionpaa, L., Lahtinen, J., Hovatta, I. et al. (2012). Mutations in CTC1, encoding the CTS telomere maintenance complex component 1, cause cerebroretinal microangiopathy with calcifications and cysts. *Am. J. Hum. Genet.*, 90, 540–549.

Revy, P., Busslinger, M., Tashiro, K., Arenzana, F., Pillet, P., Fischer, A., Durandy, A. (2000). A syndrome involving intrauterine growth retardation, microcephaly, cerebellar hypoplasia, B lymphocyte deficiency, and progressive pancytopenia. *Pediatrics*, 105, E39.

Revy, P., Kannengiesser, C., Fischer, A. (2019). Somatic genetic rescue in Mendelian haematopoietic diseases. *Nature Reviews Genetics*, 20, 582–598.

Sakwit, A., Rojanaporn, D., Mekjaruskul, P., Suriyajakryuththana, W., Sasanakul, W., Sirachainan, N. (2019). Novel mutation of the TINF2 gene resulting in severe phenotypic Revesz syndrome. *Pediatric Blood & Cancer*, 66, e27557.

Sarek, G., Vannier, J.B., Panier, S., Petrini, J.H., Boulton, S.J. (2015). TRF2 recruits RTEL1 to telomeres in S phase to promote T-loop unwinding. *Molecular Cell.*, 61, 788.

Sasa, G.S., Ribes-Zamora, A., Nelson, N.D., Bertuch, A.A. (2012). Three novel truncating TINF2 mutations causing severe dyskeratosis congenita in early childhood. *Clinical Genetics*, 81, 470–478.

Savage, S.A. (2014). Human telomeres and telomere biology disorders. *Prog. Mol. Biol. Transl. Sci.*, 125, 41–66.

Savage, S.A. and Bertuch, A.A. (2010). The genetics and clinical manifestations of telomere biology disorders. *Genet. Med.*, 12, 753–764.

Savage, S.A., Giri, N., Baerlocher, G.M., Orr, N., Lansdorp, P.M., Alter, B.P. (2008). TINF2, a component of the shelterin telomere protection complex, is mutated in dyskeratosis congenita. *Am. J. Hum. Genet.*, 82, 501–509.

Schratz, K.E., Haley, L., Danoff, S.K., Blackford, A.L., DeZern, A.E., Gocke, C.D., Duffield, A.S., Armanios, M. (2020). Cancer spectrum and outcomes in the Mendelian short telomere syndromes. *Blood*, 135, 1946–1956.

Silhan, L.L., Shah, P.D., Chambers, D.C., Snyder, L.D., Riise, G.C., Wagner, C.L., Hellstrom-Lindberg, E., Orens, J.B., Mewton, J.F., Danoff, S.K. et al. (2014). Lung transplantation in telomerase mutation carriers with pulmonary fibrosis. *The European Respiratory Journal*, 44, 178–187.

Simon, A.J., Lev, A., Zhang, Y., Weiss, B., Rylova, A., Eyal, E., Kol, N., Barel, O., Cesarkas, K., Soudack, M. et al. (2016). Mutations in STN1 cause Coats plus syndrome and are associated with genomic and telomere defects. *The Journal of Experimental Medicine*, 213, 1429–1440.

Speckmann, C., Sahoo, S.S., Rizzi, M., Hirabayashi, S., Karow, A., Serwas, N.K., Hoemberg, M., Damatova, N., Schindler, D., Vannier, J.B. et al. (2017). Clinical and molecular heterogeneity of RTEL1 deficiency. *Front. Immunol.*, 8, 449.

Stanley, S.E., Gable, D.L., Wagner, C.L., Carlile, T.M., Hanumanthu, V.S., Podlevsky, J.D., Khalil, S.E., DeZern, A.E., Rojas-Duran, M.F., Applegate, C.D. et al. (2016). Loss-of-function mutations in the RNA biogenesis factor NAF1 predispose to pulmonary fibrosis-emphysema. *Sci. Transl. Med.*, 8, 351ra107.

Stewart, J.A., Wang, Y., Ackerson, S.M., Schuck, P.L. (2018). Emerging roles of CST in maintaining genome stability and human disease. *Frontiers in Bioscience*, 23, 1564–1586.

Stuart, B.D., Choi, J., Zaidi, S., Xing, C., Holohan, B., Chen, R., Choi, M., Dharwadkar, P., Torres, F., Girod, C.E. et al. (2015). Exome sequencing links mutations in PARN and RTEL1 with familial pulmonary fibrosis and telomere shortening. *Nature Genetics*, 47, 512–517.

Takai, H., Jenkinson, E., Kabir, S., Babul-Hirji, R., Najm-Tehrani, N., Chitayat, D.A., Crow, Y.J., de Lange, T. (2016). A POT1 mutation implicates defective telomere end fill-in and telomere truncations in Coats plus. *Genes Dev.*, 30, 812–826.

Tokman, S., Singer, J.P., Devine, M.S., Westall, G.P., Aubert, J.D., Tamm, M., Snell, G.I., Lee, J.S., Goldberg, H.J., Kukreja, J. et al. (2015). Clinical outcomes of lung transplant recipients with telomerase mutations. *The Journal of Heart And Lung Transplantation: The Official Publication of the International Society for Heart Transplantation*, 34, 1318–1324.

Toufektchan, E., Lejour, V., Durand, R., Giri, N., Draskovic, I., Bardot, B., Laplante, P., Jaber, S., Alter, B.P., Londono-Vallejo, J.A. et al. (2020). Germline mutation of MDM4, a major p53 regulator, in a familial syndrome of defective telomere maintenance. *Science Advances*, 6, eaay3511.

Touzot, F., Callebaut, I., Soulier, J., Gaillard, L., Azerrad, C., Durandy, A., Fischer, A., de Villartay, J.P., Revy, P. (2010). Function of Apollo (SNM1B) at telomere highlighted by a splice variant identified in a patient with Hoyeraal-Hreidarsson syndrome. *Proceedings of the National Academy of Sciences of the United States of America*, 107, 10097–10102.

Touzot, F., Gaillard, L., Vasquez, N., Le Guen, T., Bertrand, Y., Bourhis, J., Leblanc, T., Fischer, A., Soulier, J., de Villartay, J.P. et al. (2012). Heterogeneous telomere defects in patients with severe forms of dyskeratosis congenita. *The Journal of Allergy and Clinical Immunology*, 129, 473–482, e471-473.

Touzot, F., Kermasson, L., Jullien, L., Moshous, D., Ménard, C., Ikincioğullari, A., Doğu, F., Sari, S., Giacobbi-Milet, V., Etzioni, A. et al. (2016). Extended clinical and genetic spectrum associated with biallelic RTEL1 mutations. *Blood Advances*, 1, 36.

Townsley, D.M., Dumitriu, B., Liu, D., Biancotto, A., Weinstein, B., Chen, C., Hardy, N., Mihalek, A.D., Lingala, S., Kim, Y.J. et al. (2016). Danazol treatment for telomere diseases. *The New England Journal of Medicine*, 374, 1922–1931.

Tsakiri, K.D., Cronkhite, J.T., Kuan, P.J., Xing, C., Raghu, G., Weissler, J.C., Rosenblatt, R.L., Shay, J.W., Garcia, C.K. (2007). Adult-onset pulmonary fibrosis caused by mutations in telomerase. *Proceedings of the National Academy of Sciences of the United States of America*, 104, 7552–7557.

Tummala, H., Walne, A.J., Collopy, L., Cardoso, S., de la Fuente, J., Lawson, S., Powell, J., Cooper, N., Foster, A., Mohammed, S. et al. (2015). Poly(A)-specific ribonuclease deficiency impacts telomere biology and causes dyskeratosis congenita. *J. Clin. Invest.*, 125, 2151–2160.

Vannier, J.B., Pavicic-Kaltenbrunner, V., Petalcorin, M.I., Ding, H., Boulton, S.J. (2012). RTEL1 dismantles T loops and counteracts telomeric G4-DNA to maintain telomere integrity. *Cell*, 149, 795–806.

Verduyn, M., Rigaud, M., Dromer, C. (2017). A rare familial form of idiopathic pulmonary fibrosis with Poly(A)-specific ribonuclease (PARN) mutation. *Revue de pneumologie clinique*, 73, 272–275.

Vogiatzi, P., Perdigones, N., Mason, P.J., Wilson, D.B., Bessler, M. (2013). A family with Hoyeraal-Hreidarsson syndrome and four variants in two genes of the telomerase core complex. *Pediatric Blood & Cancer*, 60, E4–E6.

Vulliamy, T.J., Knight, S.W., Mason, P.J., Dokal, I. (2001). Very short telomeres in the peripheral blood of patients with X-linked and autosomal dyskeratosis congenita. *Blood Cells Mol. Dis.*, 27, 353–357.

Vulliamy, T.J., Marrone, A., Szydlo, R., Walne, A.J., Mason, P.J., Dokal, I. (2004). Disease anticipation is associated with progressive telomere shortening in families with dyskeratosis congenita due to mutations in TERC. *Nature Genetics*, 36, 447–449.

Vulliamy, T.J., Walne, A.J., Baskaradas, A., Mason, P.J., Marrone, A., Dokal, I. (2005). Mutations in the reverse transcriptase component of telomerase (TERT) in patients with bone marrow failure. *Blood Cells Mol. Dis.*, 34, 257–263.

Vulliamy, T.J., Marrone, A., Knight, S.W., Walne, A.J., Mason, P.J., Dokal, I. (2006). Mutations in dyskeratosis congenita: Their impact on telomere length and the diversity of clinical presentation. *Blood*, 107, 2680–2685.

Vulliamy, T.J., Beswick, R., Kirwan, M., Marrone, A., Digweed, M., Walne, A.J., Dokal, I. (2008). Mutations in the telomerase component NHP2 cause the premature ageing syndrome dyskeratosis congenita. *Proceedings of the National Academy of Sciences of the United States of America*, 105, 8073–8078.

Vulliamy, T.J., Kirwan, M.J., Beswick, R., Hossain, U., Baqai, C., Ratcliffe, A., Marsh, J., Walne, A.J., Dokal, I. (2011). Differences in disease severity but similar telomere lengths in genetic subgroups of patients with telomerase and shelterin mutations. *PLoS One*, 6, e24383.

Vulliamy, T.J., Beswick, R., Kirwan, M.J., Hossain, U., Walne, A.J., Dokal, I. (2012). Telomere length measurement can distinguish pathogenic from non-pathogenic variants in the shelterin component, TIN2. *Clin. Genet.*, 81, 76–81.

Wagner, C.L., Hanumanthu, V.S., Talbot Jr., C.C., Abraham, R.S., Hamm, D., Gable, D.L., Kanakry, C.G., Applegate, C.D., Siliciano, J., Jackson, J.B. et al. (2018). Short telomere syndromes cause a primary T cell immunodeficiency. *J. Clin. Invest.*, 128, 5222–5234.

Walne, A.J., Vulliamy, T.J., Marrone, A., Beswick, R., Kirwan, M., Masunari, Y., Al-Qurashi, F.H., Aljurf, M., Dokal, I. (2007). Genetic heterogeneity in autosomal recessive dyskeratosis congenita with one subtype due to mutations in the telomerase-associated protein NOP10. *Hum. Mol. Genet.*, 16, 1619–1629.

Walne, A.J., Vulliamy, T.J., Beswick, R., Kirwan, M., Dokal, I. (2008). TINF2 mutations result in very short telomeres: Analysis of a large cohort of patients with dyskeratosis congenita and related bone marrow failure syndromes. *Blood*, 112, 3594–3600.

Walne, A.J., Bhagat, T., Kirwan, M., Gitaux, C., Desguerre, I., Leonard, N., Nogales, E., Vulliamy, T.J., Dokal, I. (2012). Mutations in the telomere capping complex in bone marrow failure and related syndromes. *Haematologica*, 98, 334.

Walne, A.J., Vulliamy, T.J., Kirwan, M., Plagnol, V., Dokal, I. (2013). Constitutional mutations in RTEL1 cause severe dyskeratosis congenita. *Am. J. Hum. Genet.*, 92, 448–453.

Wong, J.M., Kyasa, M.J., Hutchins, L., Collins, K. (2004). Telomerase RNA deficiency in peripheral blood mononuclear cells in X-linked dyskeratosis congenita. *Hum. Gen.*, 115, 448–455.

Xing, C. and Garcia, C.K. (2016). Epigenetic inheritance of telomere length obscures identification of causative PARN locus. *J. Med. Genet.*, 53, 356–358.

Yamaguchi, H., Baerlocher, G.M., Lansdorp, P.M., Chanock, S.J., Nunez, O., Sloand, E., Young, N.S. (2003). Mutations of the human telomerase RNA gene (TERC) in aplastic anemia and myelodysplastic syndrome. *Blood*, 102, 916–918.

Yamaguchi, H., Calado, R.T., Ly, H., Kajigaya, S., Baerlocher, G.M., Chanock, S.J., Lansdorp, P.M., Young, N.S. (2005). Mutations in TERT, the gene for telomerase reverse transcriptase, in aplastic anemia. *The New England Journal of Medicine*, 352, 1413–1424.

Yamaguchi, H., Inokuchi, K., Takeuchi, J., Tamai, H., Mitamura, Y., Kosaka, F., Ly, H., Dan, K. (2010). Identification of TINF2 gene mutations in adult Japanese patients with acquired bone marrow failure syndromes. *British Journal of Haematology*, 150, 725–727.

Yuan, X., Larsson, C., Xu, D. (2019). Mechanisms underlying the activation of TERT transcription and telomerase activity in human cancer: Old actors and new players. *Oncogene*, 38, 6172–6183.

Zhang, D., Zhou, Z., Abu-Hijleh, M., Batra, K., Xing, C., Garcia, C.K. (2019). Homozygous rare PARN missense mutation in familial pulmonary fibrosis. *American Journal of Respiratory and Critical Care Medicine*, 199, 797–799.

Zhong, F., Savage, S.A., Shkreli, M., Giri, N., Jessop, L., Myers, T., Chen, R., Alter, B.P., Artandi, S.E. (2011). Disruption of telomerase trafficking by TCAB1 mutation causes dyskeratosis congenita. *Genes Dev.*, 25, 11–16.

13

Telomeres Targeting Agents

Sophie BOMBARD[1], Patrick MAILLIET[2] and Marie-Paule TEULADE-FICHOU[1]

[1] *Institut Curie, Paris-Saclay University, Orsay, France*

[2] *Structure and Instability of Genomes, National Museum of Natural History, CNRS, INSERM, Paris, France*

13.1. Introduction

Telomeres are nucleoproteic entities with specific DNA structures at the ends of chromosomes that have many functions, mostly devoted to maintaining the genetic stability. *Interalia* telomeres protect chromosomes from end-to-end fusion and recombination, maintain chromosome integrity and play a role in chromosome positioning in the nucleus. In humans, the telomeric DNA is typically composed of a double-stranded domain (length ca. 5–15 kb) terminated by a 3′-single-stranded overhang protruding from the double stranded part (length ca. 50–600 nucleotides). In humans, as well as in most eukaryotic organisms, the telomeric DNA harbors a tandem G-repeat sequence d(5′-TTAGGG)n, a characteristic feature that confers a particularly high propensity to form secondary structures in particular G-quadruplexes (G4).

Indeed, the single-stranded overhang of human telomeres (TTAGGG)n has been the first example of quadruplex forming sequences structurally characterized by

Telomeres,
coordinated by Carole SAINTOMÉ. © ISTE Ltd 2023.

NMR (Wang and Patel 1993). Therefore, this DNA structure (abbreviated G4-htelo) rapidly became the focus of intensive consideration and the paradigm of G4 forming sequences (see Chapter 1). Consequently, a privileged strategy emerged rapidly by the late 1990s that was based on pharmacological targeting of G4-htelo. Initially, this strategy was intended to provide a means for inhibiting the elongation activity of the telomerase through a mechanism of sequestration of its DNA substrate. Thus, the first generations of G4-htelo interactive agents were expected to behave as indirect telomerase inhibitors and were systematically evaluated for their capacity to inhibit telomerase activity. Subsequently, it appeared that perturbing the telomeric structure by G4-binding agents was inducing rapid apoptosis of cancer cells in a manner independent of telomerase inhibition (independently of the time lag required after telomerase inhibition). In support of this notion, G4 ligands appeared to be valuable tools for targeting cancer cells possessing the ALT (alternative lengthening of telomere) phenotype, which corresponds to a pathway of telomere maintenance independent of telomerase, and may represent up to 80% of the cell population in certain cancers (e.g. gliomas and sarcomas). Altogether this led to the emergence of telomere targeting agents (TTA) as potential anticancer agents possessing a site-specific action, which may make them better targeted drugs with lower toxicity and mutagenicity as compared to the traditional anticancer chemotherapy drugs (for review, see Neidle (2002, 2010, 2016); Ohnmacht and Neidle (2014)). Currently, the simultaneous targeting of telomeres via G4 ligands and inhibition of telomerase are considered valuable synergistic strategies to induce cancer cell death (De Cian et al. 2008a; Lee et al. 2018).

Twenty years after the first G4 gold rush, several hundreds of G4 ligands targeting telomeres have been synthesized and evaluated *in vitro* and in cells, sometimes with mitigating results. This review is not intended to provide a comprehensive list of all of the compounds studied so far, but proposes a survey of the seminal works that have paved the way and led to the emergence of leading compounds, which are still currently used as G4-benchmarks.

13.2. Chemistry and *in vitro* properties of G4 ligands targeting telomeres

Seminal work: Neidle et al. (1997) reported for the first time that the quadruplex form of telomeric DNA can be targeted by aromatic compounds derived from the anthraquinone family. Following this report, several groups started to target G4-htelo with various classes of compounds mostly derived from classical duplex and triplex DNA intercalators, which display extended aromatic scaffolds (amidoanthraquinones, amido/aminoacridines (BRACO-19), condensed acridines

(MMQ, RHPS4) and naphthalene diimide (NDI)) (Figure 13.1) (Monchaud and Teulade-Fichou 2008). Subsequently, a variety of chemical classes have been designed for targeting G4-htelo. For the sake of clarity, we will follow the classification by chemical scaffold (condensed heteroaromatics, macrocycles, bisquinoliniums derivatives).

13.2.1. *Condensed heteroaromatics: the first generations*

Most of these compounds have been engineered with the addition of two or three cationic chains for targeting G4 grooves with the aim of adding selectivity for G4 with regard to duplex DNA (Figure 13.1). Although X-ray and NMR structures of G4-htelo were reported early in the 1990s, it was another 5–10 years before well-resolved structures of G4-ligands bound to G4-structures appeared (Gavathiotis et al. 2003; Campbell et al. 2008; De Cian et al. 2008a). These seminal works provided evidence that the main binding mode of G4 ligands is based on π-π stacking on the external (top and bottom) G-quartets, thereby indicating that extended aromatic surfaces are a strong determinant of G4 recognition, and providing guidelines for affinity and selectivity improvement.

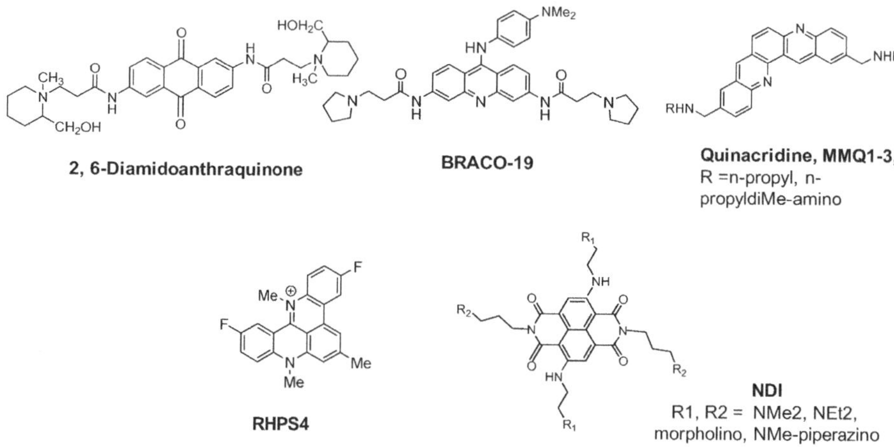

Figure 13.1. *Structures of condensed heteroaromatics*

Design of BRACO-19: a comprehensive structure–activity relationship (SAR) study within a series of di- and tri-substituted acridines demonstrated that the anilino substituent in position 9 (i.e. parallel to the ring nitrogen) was playing a crucial role to optimize the G4 selectivity with regard to duplex DNA, thereby validating the trisubstituted acridine (BRACO-19) as a privileged chemotype for G4 binding (Read et al. 2001; Campbell et al. 2008). The measurement of affinity constants by the surface plasmon resonance (SPR) method evidenced a strong binding interaction of BRACO-19 with G4-htelo ($Ka = 1.6 \times 10^7$ M^{-1}), a value that is 60× higher than that obtained with a control duplex hairpin (CGAATTGC stem sequence). Molecular modeling showed that each pendant side arm is lying in a groove, thus providing strong support to a specific interaction with G4-htelo (White et al. 2007; Campbell et al. 2008). A crystal structure reported in 2008 showed BRACO-19 stacked at the interface of two quadruplex units (each unit is a bimolecular G4-htelo of sequence: TAGGGTTAGGGT) (Figure 13.2(A)). This model indicates that the acridine core is overlapping two guanines over the four constituting G-quartets and confirmed the importance of the three side-chains for promoting simultaneous anchorage into the G4-grooves (Cuenca et al. 2008).

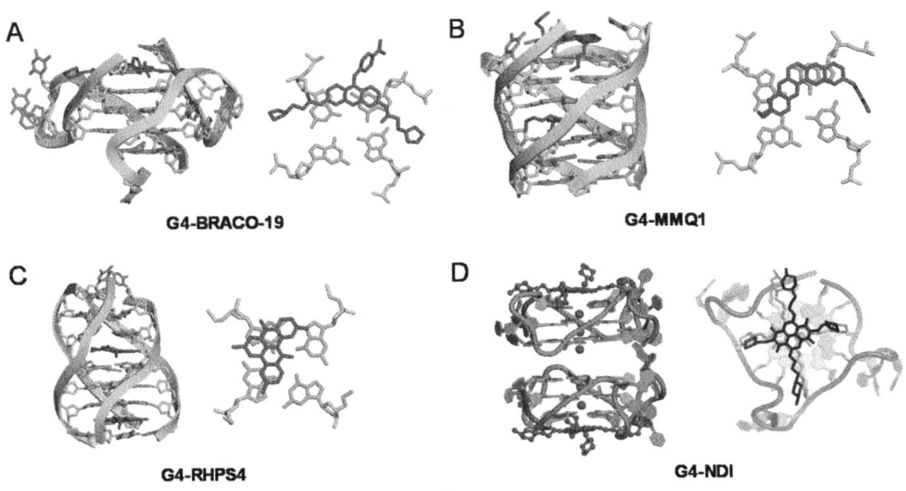

Figure 13.2. *Side and top view of G4-ligand/G4-htelo complexes determined by X-ray or NMR showing 1:1 binding (A, PDB 3CE5) and 2:1 binding with symmetrical tetrastranded G4s (B, C PDB 2JWQ, 1NZM). 1:1 binding observed with a dimeric G4-htelo (D, PDB 3UYH); color code: G: light blue, T: orange, A: green; phosphodiester backbone in gray, ligand in pink (ABCD) and dark blue (D). For a color version of this figure, see www.iste.co.uk/saintome/telomeres.zip*

Design of amino-quinacridines (MMQ): quinacridines are condensed aromatics with a pentacyclic scaffold and can be viewed as extended acridines or dibenzophenanthrolines. Initially developed for targeting triplex DNA (Teulade-Fichou et al. 2001), the crescent-shape quinacridine motif was known for its capacity to overlap with guanine pairs (Baudoin et al. 1999). Substitution by two amino side chains of the meta series MMQ (monomeric meta-quinacridine) induced a strong and selective binding of G4-htelo, as initially shown using the FRET-melting assay (Fluorescence Resonance Energy Transfer) (Mergny et al. 2001) (ΔTm from +12.5°C to +19.7°C, Table 13.1). This was further confirmed by binding constant measurements using fluorimetric titration, since MMQ are fluorescent motifs strongly quenched upon association with G4, which gave *Ka* values in the 10^8 M^{-1} range (Table 13.1). Subsequently, an NMR study of MMQ1 interacting with a tetramolecular quadruplex 5′-d(TTAGGGT)$_4$ showed the extensive π-stacking of the quinacridine core on G-quartets with overlapping of three guanine rings (Figure 13.2(B)). This finding demonstrated the strong interaction between the MMQ series and G4-htelo, which is consistent with the high affinity constants, and highlighted the advantage of the pentacyclic crescent-shape quinacridine scaffold to optimize the aromatic/aromatic overlap. Interestingly, this NMR model also indicated that the two cationic side chains of MMQ1 fit snugly into the hydrophobic grooves, which is again in full agreement with what was previously observed for the trisubstituted acridines series. However, the weaknesses of this series lie in a relatively moderate G4/duplex selectivity (two- to four-fold) when compared to the other G4 ligand classes (Table 13.1) (Teulade-Fichou et al. 2003).

Design of RHPS4: RHPS4 belongs to a pentacyclic acridinium salt series that was initially developed as an analogue of the natural product Necatarone (Cheng et al. 2008). In view of its structural similarity with the first G4-bioactive acridines, this series has been selected for potential telomerase inhibition and G4-telomere binding properties. Initial work (Gowan et al. 2001) showed that RHPS4 is indeed a strong telomerase inhibitor *in vitro* and a potent telomeric G4 binding agent, both activities being correlated. A very high affinity has been measured by SPR together with a G4/Duplex selectivity up to 35 (and up to 60 for two amino acetyl NHAc analogues); this ensemble of favorable parameters has promoted RHPS4 as the prototype of TTA.

Interestingly, in a 2001 study, the structure of RHPS4 bound to the parallel stranded tetramolecular quadruplex(TTAGGG)$_4$ was resolved by NMR showing ligand stacking above and below the quadruplex core (Figure 13.2(C)). This study provided a paradigm for the G4–ligand interaction mode. As tetramolecular G4 is a structure model of relatively poor biological relevance, several follow up studies using the native sequence forming an intramolecular G4 were conducted and determined that RHPS4 was binding with high efficiency to the intramolecular G4-htelo, as measured by CD, SPR and ESI-MS (Cheng et al. 2008; Rizzo et al. 2014).

Design of NDIs: subsequently polysubstituted NDIs have emerged as promising G4 ligands owing to the large aromatic surface (NDI core), as well as to its chemical accessibility and easy functionalization. These characteristics allowed the production of a large number of derivatives. In particular, the possibility of grafting four cationic chains (Figure 13.1) afforded extremely potent G4-htelo binders with an impressive stabilization effect and affinity (ΔTm up to 35°C and *K*a value lying in the 10^7 M^{-1} range, Table 13.1). This compound series also displayed exquisite selectivity for G4 versus duplex DNA *in vitro* (Cuenca et al. 2008; Hampel et al. 2010; Micco et al. 2013). The exceptional G4-recognition properties of tetrasusbstituted NDI were assumed to result from extensive electrostatic interactions of the four chains within the four grooves; however, X-ray structures of the tetramorpholino or tetramethylpiperazine derivatives bound to a dimeric G4-htelo instead demonstrated the occurrence of multiple electrostatic/H-bonded contacts with phosphates groups (Figure 13.2(D)) (Collie et al. 2012; Micco et al. 2013).

13.2.2. *Macrocyclic polyheteroaryles*

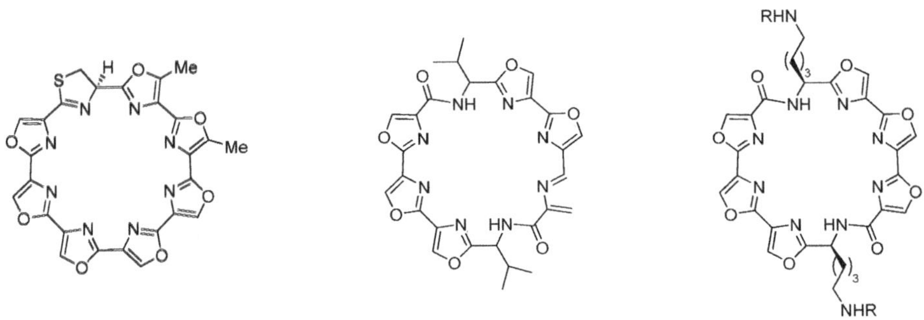

Telomestatin **HXDV series** **6ODT series**

Figure 13.3. *Structures of telomestatin and two series of hexaoxazoles macrocyles*

Design of telomestatin and derivatives (Figure 13.3): telomestatin is a natural compound isolated from *Streptomyces anulatus* 3533-SV4 and displays a macrocyclic scaffold comprised of seven oxazole rings and one dihydro-thiazole(thiazoline) ring. It was initially selected as a hit through a screening test aimed at the discovery of indirect telomerase inhibitors acting by sequestration of the 3′-end telomeric substrate in an unavailable folded state. Indeed, the first report on telomestatin showed that this compound was a more potent telomerase inhibitor than any of the previously reported G-quadruplex interactive compounds, at least *in* *vitro* (Shin-Ya et al. 2001; Kim et al. 2002). Soon after, this activity was correlated to G4 binding and in particular G4-htelo recognition. Telomestatin exhibits an exquisite selectivity for G4 structures and in particular G4-htelo when compared to duplex DNA. In fact, little to no association of telomestatin was observed with duplex DNA (see Table 13.1), which can be attributed both to the steric hindrance of the macrocyclic skeleton and its neutral feature. These outstanding properties combined with a potent telomerase inhibitory activity brought keen interest for this compound, thereby stimulating numerous studies on its biological effects in the years following its discovery (see below). On the structural level, the binding parameters have never been determined, but the stabilization of G4-htelo ($\Delta Tm = +17°C$, see Table 13.1) indicated a strong binding strength, in particular considering that there is no electrostatic contribution to the interaction. The latter thus appeared exclusively based on hydrophobic and Van der Waals forces, although some studies have suggested that telomestatin may behave like a crown ether coordinating to bulk K^+ cation, which could contribute to its high affinity (Agrawal et al. 2008; Rosu et al. 2010). There is no structural data concerning telomestatin interaction with G4, but a number of molecular modeling analyses indicated 1:1 or 2:1 stacking on G4-htelo (Figure 13.4(A)), while evidencing a concave shape of telomestatin challenging the dogma that rigid planar systems are prerequisite for strong π-stacking on G-quartets.

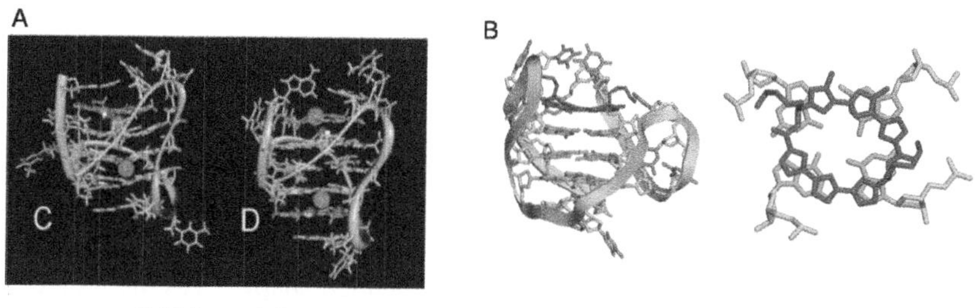

Figure 13.4. *Binding mode of telomestatin and analogues on telomeric quadruplexes. (A) Telomestatin bound to the hybrid form of G4-htelo (Tel 22 sequence: A(GGGTTA)3GGG) model generated by molecular dynamics (from PDB 2HY9). (B) Structure of the macrocycle 6ODT bound to G4-htelo (HT sequence: 5'-TT(GGGTTA)3GGGA) model generated on NMR constraints (PBD: 2MB3). For a color version of this figure, see www.iste.co.uk/saintome/telomeres.zip*

The development of telomestatin has nonetheless been strongly limited by its poor chemical availability due to laborious synthesis and its limited stability in biological media. Therefore, a number of telomestatin mimics have been developed, most being polyoxazoles macrocycles exemplified by the HDXV and 6-ODT series

(Figure 13.3) designed by Pilch et al. (2007) and Nagasawa et al. (2008), respectively. Strikingly these compounds retain most of the binding characteristics of the telomestatin with regard to G4-htelo (high affinity with $Ka = 7 \times 10^7$ M^{-1}, $\Delta Tm = +18.8°C$, see Table 13.1) together with an exquisite G4/Duplex selectivity, as no significant stabilization of duplex DNA is induced by 6ODT ($\Delta Tm = +0.3°C$), thus revealing poor to no binding (Maleki et al. 2017).

Finally, an NMR structure of 6OTD bound to G4-htelo (3+1 sequence) was reported by Phan et al. (2014). The data showed a very large overlap of the quartet with ligand interaction involving the four guanine residues (Figure 13.4(B)). Remarkably, this structure is fully consistent with the models provided for the interaction of G4-htelo with telomestatin.

13.2.3. *Bisquinolinium and bisquinoline compounds with an aromatic dicarboxamide core (triazines, 360A, PhenDC3, PDS)*

Design of bisquinolinium derivatives: note that 360A (also called pyridine dicarboxamide core (PDC)) is a second-generation compound resulting from optimization of the triazine series initially discovered by Mailliet et al. (2002) (Figure 13.5). Since triazines showed moderate affinity (see Table 13.1) and poor metabolic/chemical stability, it was decided to replace the N-C bonds by amide bonds and the central triazine by a pyridine unit for the purpose of synthesis. These modifications thus afforded the series with a PDC to which 360A belongs.

Very rapidly, 360A appeared as an excellent G4-ligand, combining high affinity and high selectivity for G4-htelo (Table 13.1). These remarkable properties were rationalized in terms of preorganization of the molecule that adopts a V-shape upon internal H-bonding (De Cian et al. 2007b). This bended shape gives the molecule the capacity to overlap 3 G, as schematized in Figure 13.6. However, the existence of equilibrium between the V-shape and the linear conformation is not excluded, as suggested by calculations on PDC analogues reported recently (Xie et al. 2018). On these grounds, it was decided to further optimize this chemotype with the replacement of the pyridine core by a more rigid extended core like phenanthroline, which afforded the PhenDC3 compound (Figure 13.5). Simple geometric considerations (De Cian et al. 2007a) indicated that PhenDC3 exhibits a U-shape with a perfect size matching with that of a quartet in the two dimensions (lateral and diagonal) (Figures 13.6 (A) and (B)). Systematic comparison of the two series confirmed this prediction, and PhenDC3 was shown to retain the high selectivity of 360A, while displaying a significantly increased binding affinity for G4-htelo. The higher binding of PhenDC3 when compared to 360A has since been verified on a large panel of G4 structures and varies from slight ($\Delta\Delta Tm$ (PhenDC3-360A) > +1–5°C) to significant

(ΔΔTm (PhenDC3-360A) > 10°C) as a function of the G4 structural features (topology, loop length) (Saha et al. 2020).

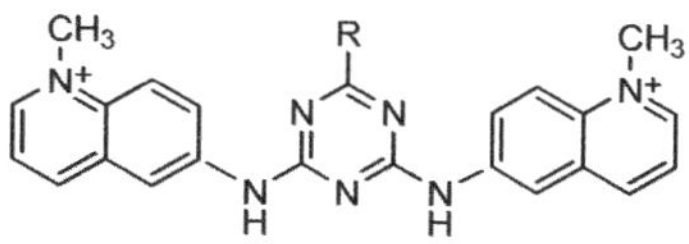

Triazine series
12459 (R=NH2)
115405 (R=SCH3)

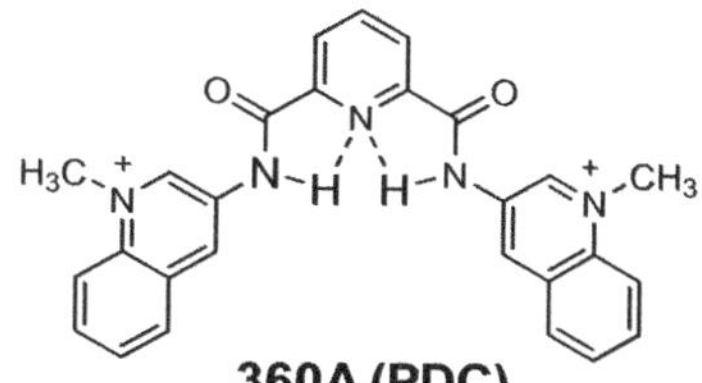

360A (PDC)

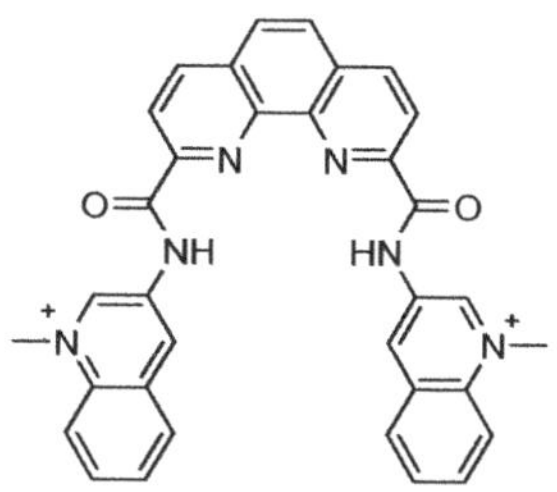

PhenDC3

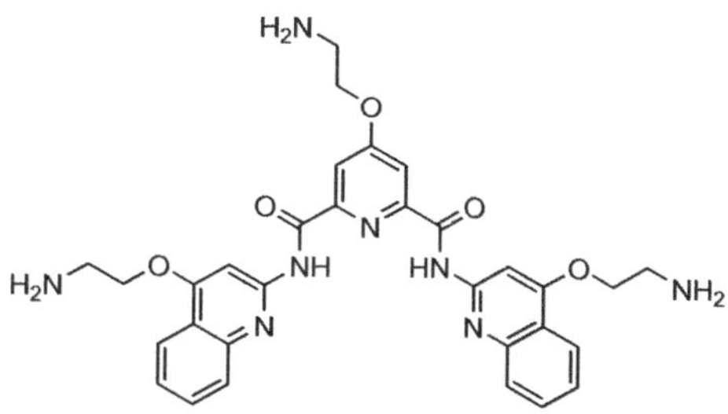

Pyridostatin (PDS)

Figure 13.5. *Structures of bisquinolinium compounds and pyridostatin*

Design of PDS: in 2008, Balasubramanian and Rodriguez reported a new scaffold designed for G4 recognition, termed pyridostatin (PDS, Figure 13.5). Obviously, the authors capitalized on the molecular guidelines mentioned above for the trisubstituted acridines and the 360A series. Actually, PDS combines the pyridodicarboxamide (PDC) core (for preorganization) with three cationic amino terminated side-chains distributed on the central pyridine and two quinoline units (for groove anchorage). Evaluation by the classical binding FRET-melting assay revealed that PDS is an excellent G4-htelo binder, as shown from the large induced stabilization (ΔTm = +28–30°C, Table 13.1), thereby rivaling the performances of the parent compound 360A, as well as that of its phenanthroline analogue PhenDC3. This was verified by the determination of affinity constant of PDS for G4-hetlo that showed a very high value lying in between that of the two bisquinolinium compounds (Bonnat et al. 2017) (Table 13.1). The following affinity ranking PhenDC3 > PDS > 360A can thus be assessed and was further confirmed by other methods, in particular, ESI-MS (Marchand et al. 2015). Remarkably, the triscationic charge of PDS is not deleterious to G4/duplex selectivity in consistency with the trend observed with trisubstituted acridines. Nonetheless, the degree of preorganization of PDC and PDS can be variable, especially in water, as shown by two studies (Rocca et al. 2017; Xie et al. 2018). This could explain the superiority of the PhenDC3 chemotype that provides a larger overlap (4G vs. 3G) and displays a higher rigidity/preorganization due to the curved shape of the phenanthroline motif. The structure of PhenDC3 bound to the oncogene myc G4 forming sequence (Figure 13.6(A)) was solved in 2014 and demonstrated the perfect match of this scaffold with the aromatic area of a G4 quartet. Interestingly, 6ODT and PhenDC3 are the only G4 interactive compounds to have shown this capacity, so far, making them privileged scaffolds and lead series for G4 recognition. Surprisingly, there is currently no structure available on the interaction of 360A or PDS with G4-htelo or other G4 originating from different sequences. However, recently an NMR structure showing intercalation of 360A between quartets of bases (A2G2 and C2G2) in a particular quadruplex-like structure (VK2) was reported, evidencing for the first time the V-shape adopted by 360A upon π-π stacking (Kotar et al. 2020).

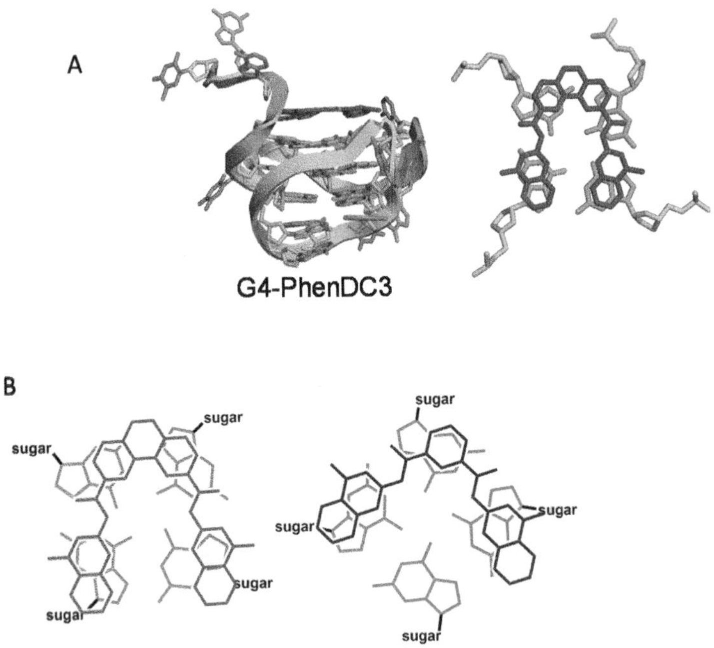

Figure 13.6. *(A) Structure of PhenDC3 (pink) bound to the Myc quadruplex (Pu24T sequence: 5'-TGAG$_3$TG$_2$TAG$_3$TG4AAG$_2$) model generated on NMR constraints (PDB: 2MGN). (B) Overlap of PhenDC3 (pink) with a G-quartet (blue) from NMR, (Chung 2014) and the same with 360A (purple) manually generated. For a color version of this figure, see www.iste.co.uk/saintome/telomeres.zip*

Ligand	ΔTm (°C)	*Ka* M^{-1}	G4/duplex = *Ka* G4 /*Ka* duplex
BRACO-19	+27.5	$1.6 \times 10^{7*}$	30–60
MMQ1	+12.5	$2.0 \times 10^{5*}$–$1.6 \times 10^{7**}$	nd&
MMQ3	+19.7	10^{9**}	nd&
RHPS4	+8.0–+20.0#	$1.1 \times 10^{7*}$	32.3
Telomestatin	+17	nd	nd&&
6ODT	+18.8	$7 \times 10^{7***}$	nd&&
Triazine 12459	+8.0	nd	nd&
Triazine 115405	+20.0	nd	nd&
360A	+25.0–30.0##	$1.0 \times 10^{8***}$	60–70
PhenDC3	+30.0–35.0##	1.0×10^{8}–$4 \times 10^{8****}$	>100
PDS	+28–30.0 ##	8.0×10^{7}–1.0×10^{8}	nd&
NDI (tetraetN)	+35.2°C	$3.1 \times 10^{7*}$	nd&

Table 13.1. *ΔTm (°C) = variation in melting temperature induced by the ligand in the conditions of the FRET-melting assay using the doubly labeled sequence F21T derived from G4-htelo, same conditions for all ligands, [DNA] = 0.2 μM, [ligand] = 1 μM, buffer lithium cacodylate 90 mM, KCl 10 mM. #The low value was reported in Cheng and Modi (2008), and the high value is taken from the PhD thesis of Lefebvre (2017) (unpublished data). This discrepancy can be explained by interference of RHPS4 fluorescence with the FRET signal, as mentioned in the literature (De Cian et al. 2007b). Ka = Equilibrium constant determined by *SPR, **fluorimetric titration, ***single molecule experiment. ****Determined by SPR, ITC, ESI-MS (when several studies or methods were applied, a range is given as a Ka value, and is strongly dependent on the method used and the experimental conditions). &Qualitative determination of preference for G4 in conditions of competition with excess of duplex by various methods (equilibrium dialysis and FRET-melting). &&Poor to no binding to duplex DNA in the measurements conditions (Abraham Punnoose et al. 2017; Bonnat et al. 2017)*

13.2.4. *Alternative hybrid scaffolds derived from the series above*

The majority of studies dedicated to designing G4 ligands used (and are still using) the monomeric G4-htelo motif as a target. However, presumably, multiple G4-motifs are likely to exist simultaneously within the telomeric repeated domains in view of the length of the single stranded overhang. This possibility is supported by a number of studies that evidenced the formation of multiple G4 units in telomeric, or more largely in tandem G4-repeat containing sequences (Amrane et al.

2012; Bugaut and Alberti 2015). For this reason, several studies explored alternative scaffolds more apt to target at least two neighboring G4 motifs. Logically, these are based on multimers of 360A, 6ODT, PDS or NDI (Figures 13.7(A–D)).

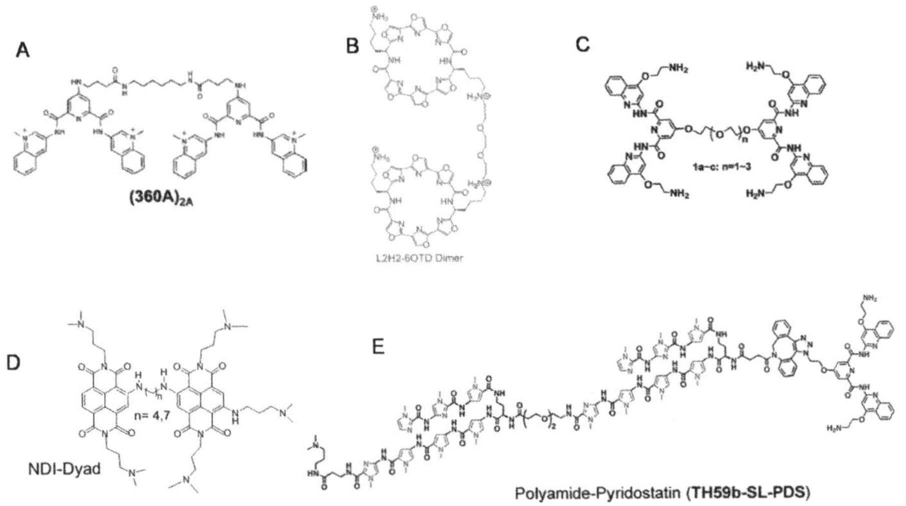

Figure 13.7. *Structures of (A) 360A dimer (Saintome et al. 2018; Hwang et al. 2019). (B) 6ODT dimer (Abraham Punnoose et al. 2017). (C) PDS dimer (Ma et al. 2020). (D) NDI-dimer (Doria et al. 2019). (E) PDS-hairpin hybrid (Mandal et al. 2019). For a color version of this figure, see www.iste.co.uk/saintome/telomeres.zip*

Finally, considering that telomeric G4 may form at the junction between the duplex domain and the single stranded overhang, this structural peculiarity was chosen as a target to build hybrid conjugates mixing a G4 motif (PDS or 360A) and a minor groove binding motif (polyamide hairpins), with the aim of obtaining highly selective agents combining sequence and structure recognition motifs (Figure 13.7(E)) (Lejault 2017; Mandal et al. 2019). Most of these dimeric forms and conjugates exhibit enhanced binding properties, both in terms of affinity and selectivity, which validated the hybrid design. However, their cellular behaviors in cells have been poorly documented so far and issues with solubility and cellular uptake may be expected.

13.2.5. *Telomeric localization of G4 ligands through imaging*

Telomeric effects could result from indirect action of G4-ligands elsewhere in the genome or from off-target interaction with proteins and eventually from partial trapping in subcellular compartments (e.g. mitochondria). Therefore, it appeared necessary to develop methods for direct visualization of G4 ligand localization at telomeres. Two methods were carried out to visualize the localization of 360A and PhenDC3 at the level of metaphasic chromosomes. The first study (case of 360A, Figure 13.5) was based on autoradiography using the tritium-labeled derivative ^{3}H-360A (Figure 13.8(A)). The autoradiograph showed that the ligand was preferentially bound at terminal regions of chromosomes (with minor distribution at interstitial regions) (Granotier et al. 2005). The second study (case of PhenDC3, Figure 13.5) was based on NanoSIMS (nanoscale secondary ion mass spectrometry) imaging and conducted with a derivative of PhenDC3 tagged with a bromine atom (Br-PhenDC3, Figure 13.8(B)), as this heavy element can be detected easily by mass spectrometry microscopy. The analysis showed that the ligand is present at 100% of all chromosome ends and in non-telomeric domains, which is variable as a function of the chromosome (Verga et al. 2017). This result is fully consistent with the data obtained with 360A.

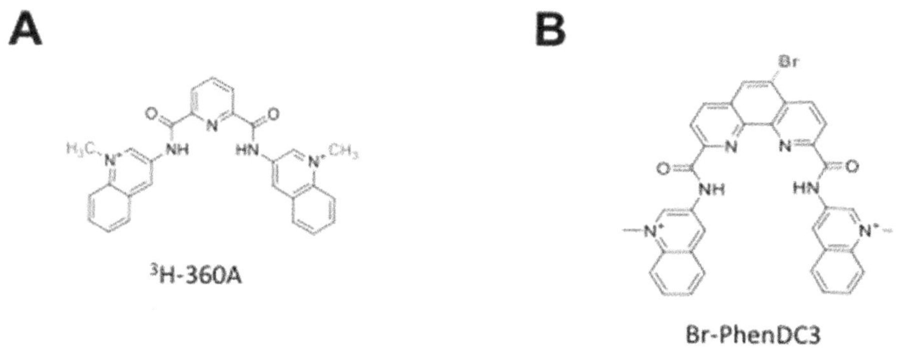

Figure 13.8. *(A) ^{3}H-360A (^{3}H are in red) used on metaphasisc chromosomes, incubation in cells for 48 h (ligand 0.3 μM, T98G cells, glioblastoma) (Granotier et al. 2005). (B) Br-PhenDC3 used on metaphasic chromosomes, incubation in cells for 24 h (ligand 0.3 μM, HEK 293 cells, human epithelial kidney) (Verga et al. 2017). For a color version of this figure, see www.iste.co.uk/saintome/telomeres.zip*

13.3. Biological properties of G4 ligands targeting telomeres

The first G4 ligand generations have been systematically evaluated for their telomerase inhibitory properties, but it rapidly turned out that the short time scale of the cellular responses observed were not consistent with the long-term consequences of telomerase inhibition. It was thus rapidly concluded that the cellular telomeric phenotypes of G4 ligands were the result of a dual mechanism of action based on telomerase inhibition and direct action at telomeres. Hence, over the past two decades, a large body of knowledge has been accumulated on the telomeric effects triggered by G4 ligands, and in particular, those described above have been instrumental to gaining understanding and establishing common trends shared by different classes of ligands. Although there may be significant differences from one ligand to another, we chose to outline the common typical markers of telomeric effects observed in cells upon ligand treatment. For specific details of each ligand, we invite the reader to refer to original works, as well as the numerous reviews and books published on telomere-targeted quadruplex ligands and their therapeutic applications (De Cian et al. 2008b; Neidle 2010, 2012; Xu 2011).

The broad spectrum of telomeric effects of G4 ligands in cells includes the following trends:

– *Rapid apoptosis of cancer cells that is independent of telomere length and telomerase activity*: this rapid growth inhibitory effect has been observed on telomerase positive or negative cancer cells lines, in particular the ALT cells that maintain their telomere length by an alternative mechanism. Depending on the ligand, this antitumor activity is apparent within 24 h to a few days and occurs in both short and long telomere cell lines.

– *Selective antiproliferative action in cancer cells*: the selectivity for cancer cells versus normal cells is highly variable from one ligand to another and has not been systematically explored. The case of 360A is worth mentioning as this compound has limited effects on proliferation of normal cells (primary fibroblasts or peripheral blood lymphocytes), even at elevated concentration (>10–20 μM), and does not induce apoptosis or telomere aberrations (see below).

– *Induction of telomeres shortening and/or single stranded overhang degradation*: telomere shortening has been observed for several ligands (triazines, BRACO-19, RHPS4, telomestatin) after long-term treatment using the Q-FISH technique. Initially attributed to telomerase inhibition, this effect appeared to be independent of telomerase in some cases. In addition, 360A that is a potent inhibitor of telomerase *in vitro* did not induce telomere shortening, but caused rapid degradation of the single stranded telomeric overhang.

– *Induction of telomere deprotection/uncapping*: most ligands induced rapid capping alteration, which was evidenced by displacement of telomeric binding proteins such as TRF1/TRF2 and hPOT1. Displacement of telomerase (hTERT (human telomerase reverse transcriptase)) was also reported in some cases (BRACO-19). It is now widely admitted that the disruption of the shelterin complex integrity by G4 drugs is responsible for the disassembly of the T-loop structure and for a number of macroscopic aberrations observed on metaphasic chromosomes, such as anaphase bridges, telomere fusion and sister telomere losses.

– *DNA damage located at telomeres (and other sites)*: many studies indicated that G4 ligands activate an extensive DNA damage response, which is likely to result from the exposure of telomeric DNA through uncapping. The localization at telomeres was assessed using colocalization of DDR markers (γH2AX and 53B1 foci, named telomere dysfunction induced foci (TIF)) and PNA telomeric signals (Telo-FISH). In several cases (360A, BRACO19, telomestatin), the dependency of induced telomeric DNA damages with regard to ATM (ataxia-telangiectasia mutated) or ATR (ataxia telangiectasia mutated and Rad3-related) repair pathways have been shown.

– *Telomerase inhibition correlated to replicative senescence*: all compounds of the panel described above are potent inhibitors of telomerase *in vitro* with IC_{50} (concentration inhibiting 50% of activity), ranging from low μM down to nM (e.g. telomestatin and PhenDC3). For a number of these (triazines, BRACO-19, 360A, RHPS4, telomestatin), the reduction of telomerase activity in cancers cells has been reported in correlation with telomere erosion and induction of replicative senescence.

The main references related to the cellular activity of G4 ligands described above are as follows: BRACO-19: (Harrison et al. 2003; Burger et al. 2005; Moore et al. 2006); RHPS4: (Salvati et al. 2007; Rizzo et al. 2009, 2014; Berardinelli et al. 2015); 360A: (Pennarun et al. 2005, 2008, 2010); triazines: (Riou et al. 2002); telomestatin (Shin-Ya et al. 2001; Kimet al. 2002; Gomez et al. 2004; Tahara et al. 2006); PDS (Rodriguez et al. 2008; Müller et al. 2012).

Also, the NDI compounds, both monomers and dimers, demonstrated highly potent antiproliferative activity at nanomolar concentrations specific to cancer versus normal cells and were shown to induce, among others, telomeric DNA damage (Doria et al. 2019; Platella et al. 2020). However, very few data are available on the cellular effects of the macrocyclic series 6ODT. The telomeric effects of PhenDC3 have been poorly explored, with the exception of a study that demonstrated the strong impact of the ligand on the origins of replication at telomeres (Drosopoulos et al. 2015). On the other hand, PhenDC3 has been extensively used as a mechanistic tool for probing G4 in yeast and mammalian cells, and is currently considered the best G4 probe both in terms of affinity and selectivity. Its therapeutic potential as G4-RNA ligand for treating cancers derived from the Epstein–Barr virus has been recently reported (Lista et al. 2017; Reznichenko et al. 2019).

Besides their cellular effects, G4 ligands (BRACO19, RHPS4 and telomestatin) have been evaluated *in vivo* on xenograft models (cancerous cells graphted in animals to allow tumour model growth), either as single agent or in combination with cytotoxic drugs or repair inhibitors (PARP inhibitor) used in the clinical treatment of cancers. The resulting anti-tumor responses observed in animal models can be directly attributable to DNA damage. RHPS4 appeared particularly effective against human tumor xenografts in combination with taxol, irinotecan and a Parp-1 inhibitor. Synergistic effects have been reported for BRACO-19 and taxol. On the other hand, antagonist effects with alkylating agents that target G-rich sequences (cisplatin, temozolomide) have also been observed with BRACO-19 (for reviews, see Neidle (2010, 2012)). Interestingly, combination with radiation has been explored in only two cases, namely the platinum complex Pt-ctpy (see below) and RHPS4. Convincing data have been obtained in cells and on xenografted mice (Berardinelli et al. 2015; Merle et al. 2015), thereby confirming that G4 ligands could be valuable drug prototypes for treating radioresistant cancers.

However, toxicity and poor drug likeliness features have further limited the development of these first-generation G4 ligands, which nonetheless remain useful therapeutic tools and have contributed to provide the proof of concept that targeting telomeres throughout shelterin inhibition is an effective strategy for treating cancer. Many studies have been developed on glioblastoma, but numerous works tend to indicate that cancer with particular genetic deficiencies (BRCA deficiency) could be highly sensitive to G4 drug treatment.

13.4. Metal complexes targeting telomeres

Another promising recent approach to target G4 from telomeres (and others loci) is based on the use of metal complexes that display a broad range of structural and electronic properties. In fact, the metal center (mostly transition metal cation) has been used as a specific locus that can organize aromatic organic ligands in various geometries (e.g. octahedral, square based pyramidal and square planar) and orientations to optimize G4 binding, allowing the easy generation of small libraries. In addition, the electronic withdrawing properties of metal centers can reduce the electron density on the organic ligand surrounding the metal, which is expected to induce stronger interaction with the external G-quartet(s). Also, the cationic center c promotes the association with G4 by substitution of the alkali metal cation (K^+, Na^+) that stabilizes G4. The first examples of metallic G4 ligands were reported in the year 2000 and are mainly square planar transition metal complexes coordinated to porphyrins, salphen, phenantroline and terpyridine. All these complexes highlight the importance of π-π stacking in binding to G-quartets and the crucial role of the metallic cation. In addition, one more advantage of metal complexes is that some (e.g. Pt^{II}) have the ability to directly coordinate nucleobases, which strongly reinforce the interaction and may allow irreversible trapping of G4 by establishing stable adducts. In the same line, alternative hybrid scaffolds linking a known G4 ligand (360A, MMQ) to a triamino Pt^{II} center bearing a labile chloride ion (Cl) (see below) have been conceived to enable single site coordination to G4-nucleobases. A number of reviews have focused on the development of metal complexes able to bind G4 structures (Georgiades et al. 2010; Cao et al. 2017; Vilar 2018). So far, only a few complexes have been evaluated for their biological activities and a selection of these are reported herein. G4 ligands metal complexes with octahedral geometries and supramolecular metallo assemblies, mostly studied *in vitro*, will not be discussed here (Zheng et al. 2016).

Metal-porphyrin complexes (Figure 13.9(A)): the earliest G4 metal complexes reported are metalloporphyrins. These series were derived from the widely studied tetra meso-methylpyridinium porphyrin TMPyP4 that was first shown to stabilize G4-htelo by end-stacking and inhibit telomerase (Shi et al. 2001). The metal complexes of TMPyP4 (using various metals) were then shown to be more potent both for binding G4-htelo and inhibiting telomerase, thereby highlighting the critical role of the metal center in engaging further electrostatic interactions with G4. Due to poor selectivity G4/duplex of these first series, several modifications on the side arms were investigated, leading to noticeable improvement of G4-binding affinity and selectivity over duplex. Remarkably, the Mn^{III} porphyrin **1** was shown to exhibit a 1000-fold increased selectivity for G4 over duplex (Dixon et al. 2007), while the Ni^{II} porphyrin **2** was reported to induce deprotection of telomeres (Sabater et al. 2015).

Metal–salphen complexes (Figure 13.9(B)): this series has been widely studied by the group of Vilar. The shiff base provides a good basis to explore the relevance of metal geometry for G4 binding. Indeed, square planar salphens with Ni^{II}, Cu^{II} and Pt^{II} centers display a higher affinity for G4-htelo than their square based pyramidal V^{II} or distorted Zi^{II} analogues, with various selectivities over duplex DNA ($Ka = 10^3$ to 10^7 M^{-1}) (Campbell et al. 2012). Two X-ray crystal structures of Ni^{II} and Cu^{II} salphen complexes bound to G4-htelo revealed the end-stacking binding mode on the G-quartet with the metal positioned in line with the K^+ ion channel of the G4 (Campbell et al. 2012). These structures confirmed the advantages of metal complexes over organic G4-ligands. The nature and the charge of the substituents on the salphen core were shown to modify the G4-affinity and selectivity of the complexes (Rakers et al. 2018). This series was then extended to dimeric metal–salphen complexes in order to target multiple telomeric G4 repeats more efficiently, as mentioned previously (see section 13.2.4 on alternative scaffolds). As expected, these complexes have higher affinity for dimeric G4-htelo repeat than monomeric G4-htelo (Zhou et al. 2017). Few complexes have been evaluated for their biological activity. For example, the monomeric Ni^{II}-salphen complex **3** was shown to inhibit telomerase ($IC_{50} = 9$ μM) and display cytotoxicity against various cancerous cell lines, whereas its impact on telomeres has not been explored. On the other hand, the telomerase inhibition capacity of **4** that is 10 times more potent than **3** *in vitro* could be correlated to telomerase inhibition in cellulo, as this complex only induced cytotoxicity after long-term treatments (Mancini et al. 2016).

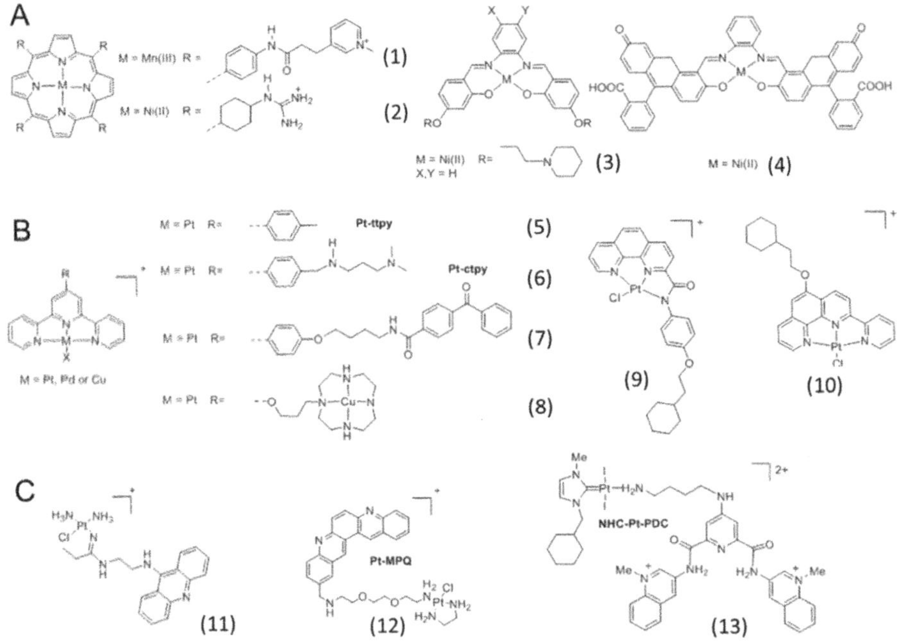

Figure 13.9. *Representative examples of metal complexes developed for targeting G4 structures. (A) Metal–porphyrin, (B) metal–salphen complexes, (C) metal–terpyridine, (D) metal–phenanthroline complexes, and (E) hybrid scaffolds containing a Pt-complex linked to a G4 ligand*

Metal terpyridine complexes (Figure 13.9(C)): the series was pioneered by the groups of Teulade Fichou and Bombard and subsequently developed by Vilar et al. The design is based on the coordination of a transition metal cation to a terpyridine (tpy-derivatives) or an extended tolylterpyridine (ttpy-derivatives). On the whole, the metal geometry, the number of aromatic rings, as well as the number and position of the substituents were shown to dictate the affinity of the complexes for G4-htelo (Bertrand et al. 2007b; Suntharalingam et al. 2009). Actually, the most potent G4-htelo binders are Pt-ttpy complex (5) obviously due to the square planar geometry (Pt^{II}) or the presence of a square planar face Cu^{II}, which is consistent with the poor binding of the Zn^{II-} analogue exhibiting a trigonal bipyramidal geometry.

Interestingly the direct metal coordination of Pt-ttpy (**5**) to the adenines of the loops was shown for G4-htelo, suggesting that a high specificity may be achieved in this way (Bertrand et al. 2009). This unique dual quartet/loop interaction evidenced for the first time with G4-htelo was then observed with the G4 myc. Subsequently, an NMR study of Pt-ttpy (**5**) interacting with a variant of G4-myc comforted the quartet end-stacking mode previously established for G4 ligands, while evidencing the simultaneous occurrence of metal/base coordination (Trajkovski et al. 2015).

Metal-terpyridine have also been functionalized with photoactivable groups (Pt-ttpy-Bn, 7) to reinforce anchoring to G4 structures by photo-crosslinking (Morel et al. 2016), and with an azamacrocyclic unit coordinated to Zn^{II} or Cu^{II} (**8**) assumed to display interactions with the G4 loops and grooves (Stafford et al. 2015). While the cytotoxicity of the photoactivable metal terpyridin (Pt-ttpy-Bn, **7**) increased upon light activation, the bi-metallic complexes (**8**) were not studied in cells, and they retained a high affinity for G4 ($Ka = 10^7$ to 10^8 M^{-1}). Pt-ttpy (**5**) and Pt-ctpy (**6**) have been widely studied for their cellular activity and were shown to target telomeres in cells, together with other loci in genomic DNA. In particular, recent investigations revealed that Pt-ttpy quantitatively binds to telomeres by coordination (Saker et al. 2018), induces telomeric DNA damage following chromosome mis-segregation (Lee et al. 2018) and causes telomere deprotection (Bombard et al. unpublished results). Finally, the dimethylamino derivative Pt-ctpy was shown to be an effective radiosensitizer in multiple glioblastoma and non-small cell lung cancer cell lines, furthermore able to delay tumor growth over a long period (Merle et al. 2015). This radiosensitizing activity was correlated to an increase of telomeric damages post-irradiation. Moreover, Pt-ttpy was also recently shown to be an efficient radiosensitizer by a telomere-independent mechanism (Bombard et al. unpublished data). Their radiosensitizing effect highlights the potential therapeutic terpyridine-G4 complexes to be used at subtoxic concentration in combination with ionizing radiations.

Metal-phenanthroline complexes (Figure 13.9(D)): another class of metal complexes exhibiting high affinity for G4-htelo corresponds to the phenanthroline derivatives. The first derivative of this series (**9**) displayed telomerase inhibition activity (Reed et al. 2007). The optimized derivative (**10**) was shown to be a potent inhibitor of telomerase *in vitro* and decrease telomere length as a function of time in non-small lung cancer cell lines, A549 cell lines, thereby supporting telomerase inhibition in cells (Mancini et al. 2016).

Hybrid G4-Pt complexes (Figure 13.9(E)): for Pt^{II} complexes, a labile chloride (Cl^-) can ensure platination to DNA, as demonstrated for Pt-ttpy (**5**). Another strategy to anchor the ligand irreversibly to their G4 target has been developed, which is based on combining within the same molecular architecture a classical G4 ligand to afford specific π-stacking and a platinum moiety for coordination to nucleobases. Therefore, this new class of hybrid conjugates allows interactions with G4 through a double non-covalent/covalent binding. The proof of concept *in vitro* was held by the groups of Bierbach and Teulade-Fichou by combining an acridine (Rao and Bierbach 2007; Rao et al. 2011) (11) or a quinacridine (MPQ) (Bertrand et al. 2007) (Pt-MPQ, 12) respectively, with an ethylene diamine Pt complex using a linker suitable for spanning the two external G-quartets. Then a stronger G4 ligand, 360A (PDC) was combined with NHC-Pt (N-heterocycle-carbene) complex **13** with linker chains of various lengths. These hybrid complexes were able to stabilize G4-htelo *in vitro*, in addition to coordinating an adenine of the loop or a guanine from one external G-tetrad. From a biological point of view, Pt-MPQ **(12)** (Charif et al. 2017) and the NHC-Pt-PDC series **13** (Betzer et al. 2016) were shown to target telomeres by inducing telomeric DNA damages (telomere loss and telomere fusion) and/or deprotecting them (loss of TRF2 foci), with a significantly higher efficiency than the G4 ligand alone. These studies suggest that the association of a G4 ligand to a platinum complex direct the complex to the G4 structure and improve the telomere disruption by the G4 ligand. However, it still remains to demonstrate if the direct coordination of these hybrid complexes to telomeres is a primary factor of their cellular phenotype.

In conclusion, metal complexes may display a strong binding activity towards G4-htelo *in vitro*, which has been related to telomerase inhibition ability *in vitro* and/or to telomere disruption in cellulo. It is important to highlight that these complexes can also stabilize other biologically relevant G4 localized elsewhere in the genome, as in oncogene promoters. Moreover, the downregulation of some identified genes have been shown and can contribute, in addition to telomere targeting, to the cytotoxic activity of the complexes. Therefore, the determination of their exact cellular target remains challenging, but is crucial for the development of more potent and more selective metal complexes directed to telomeres.

13.5. Compounds targeting the proteins of the human shelterin complex

G4 ligands and metal complexes, disclosed in the previous paragraphs, target the telomeric DNA. In tumor cells, among other effects, such compounds induce a disruption of the telomere-protective shelterin complex, followed by growth inhibition.

Human shelterin is the protein component of telomeres, composed of six proteins: TRF1 (telomeric repeat-binding factor 1), TRF2 (telomeric repeat-binding factor 2), hPOT1 (protection of telomeres protein 1), RAP1 (repressor/activator protein 1), TPP1 (tripeptidyl peptidase 1) and TIN2 (TRF1-interacting nuclear factor 2). As shown in Figure 13.10, the shelterin complex protects chromosome ends from ATM (ataxia telangiectasia mutated protein) kinase and ATR (ataxia telangiectasia and Rad3-related protein) kinase mediated DNA damage response pathways. During replication, the hPOT1/TPP1 sub-complex recruits the telomerase subunits hTERT and hTR (human telomerase RNA) at telomeres (Nandakumar and Cech 2013; Smith et al. 2020).

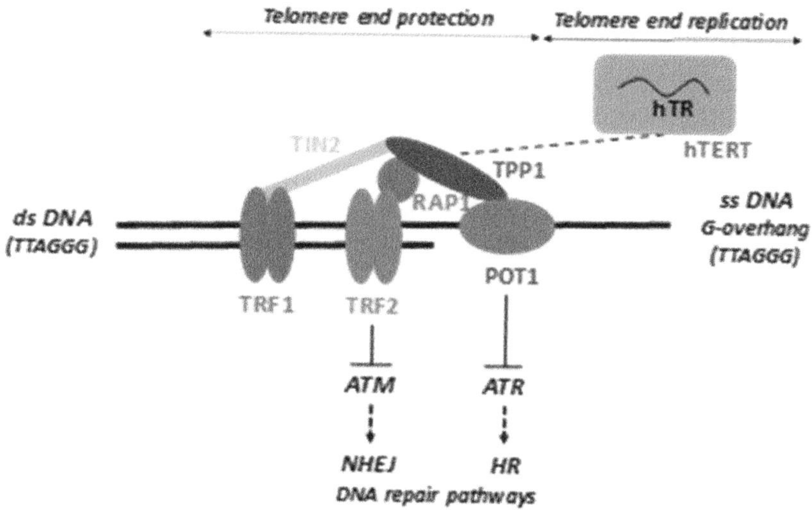

Figure 13.10. *Schematic view of the human telomere complex structure and functions. For a color version of this figure, see www.iste.co.uk/saintome/telomeres.zip*

In the last two decades, scientists designed compounds able to interact with the various components of shelterin. Such molecules could either prevent the interaction of telomeric DNA with DNA-bound proteins (TRF1, TRF2 or POT1), or impair the correct association of the shelterin complex, or even modify the nature of the shelterin proteins.

13.5.1. *Inhibitors of the shelterin proteins binding to DNA*

Crystal structures show that TRF1 and TRF2 form homodimers, which similarly recognize the same *GTTAGGGTTAGGG* minimal ds-DNA sequence with high affinity and selectivity. Recognition occurs via their respective C-terminal basic Myb domains (c-Myb) bound to opposite faces of DNA (Court et al. 2005; Hanaoka et al. 2005). Surprisingly, no c-Myb ligand, able to compete with ds-telomeric DNA, has been disclosed so far.

Crystal structures show that hPot1 specifically binds, with nanomolar range affinity, the decamer minimal *TTAGGGTTAG* sequence of telomeric 3′ G-overhang, using two close OB-fold (OB1 and OB2) elements (Lei et al. 2003, 2004).

The formation of a ternary complex DNA-POT1-TPP1, in which the central domain of TPP1 is bound to POT1, significantly improves the interaction of the telomeric G-tail with hPOT1. The association of POT1 with TPP1 is mandatory for the recruitment of telomerase at telomeres through its interaction with the OB-fold domain of TPP1 (Wang et al. 2007; Zhu et al. 2019).

Based on the available structures of POT1, virtual screenings for selective inhibitors or ligands were performed (Malik 2013; Amir et al. 2019). Today, the most potent compounds only fairly inhibit hPOT1-DNA binding *in vitro*: 18f (Figure 13.11; 56% at 100 µM) (Malik 2013) and Congo Red (Figure 13.11; 50% at 130 µM), a potent OB-fold ligand (K_D hPot1$_{OB1}$ = 0.7 µM) (Altschuler et al. 2012).

Generic OB-fold ligands (TDRL 505, TDRL 551) were initially developed against replication protein A (RPA1,2,3), which "non-sequence specifically" binds to ssDNA via its OB-fold domains, first to DBD-A and DBD-B from RPA70 subunit, then to DBD-C from RPA70 and DBD-D from RPA32 (Bochkareva et al. 2002). Usually, these generic OB-fold ligands also bind to POT1 in the same range than RPA (K_D POT1$_{OB1}$ = 10–15 µM for TDRL 505 and TDRL 551; Figure 13.11) (Anciano Granadillo et al. 2010).

Recently, the same authors reported a new series of OB-fold ligands, derived from TDRL-551, with 1 log improved affinity for RPA (Gavande et al. 2020). Since the distance between OB1 and OB2 domains from POT1 (5 aa) is different from the distance between DBD-A and DBD-B from RPA (22 aa), it should be possible to design OB-fold ligand dimers able to selectively discriminate the OB1 and OB2 domains of POT1 from the DBD-A and DBD-B domains of RPA.

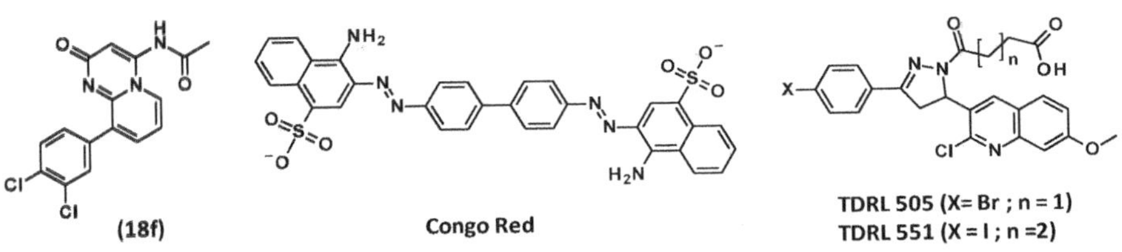

Figure 13.11. *Structures of OB-fold ligands*

13.5.2. *Inhibitors of protein–protein interactions involving the shelterin proteins*

Increasing structural data on the Shelterin complex and its assembly mechanism have been available in the last two decades.

Crystal structures have been solved of the TRFH domains of TRF1 and TRF2, which are involved in their dimerization and in the recruitment of their partners (Fairall et al. 2001). The interaction between the RBM domain of TRF2 and the RCT domain of RAP1 have also been solved (Chen et al. 2011). Molecular dynamics gave insights on the interaction of the TBM domain of TIN2 with the TRFH domains of TRF1 and TRF2 (Kalathiya et al. 2017).

The crystal structure of the fission yeast "Shelterin bridge" (TIN2-RAP1-TPP1) was solved and the cooperative mechanism, which is the driving force for the shelterin bridge assembly, was elucidated (Kim et al. 2017). Other cooperative interactions occur at shelterin: for instance, the binding of RAP1 on TRF2 induces adjustments allowing the correct positioning of TIN2 on TRF1.

Thus, based on available structural data, it was attractive to design compounds that target telomeres, either by interfering with protein–protein interactions involved in the shelterin complex, or by preventing the interaction between proteins of shelterin and their partners. Such approaches should open up new avenues in anti-telomere drug development. However, up until now, there are only a few reports on such compounds, which are so far only peptidomimetics or cyclic peptides. For instance, a stapled peptide (Figure 13.12(A)), which mimics the $\alpha 1$ helix of TRF2, efficiently binds to RAP1 *in vitro* (Kd = 7 nM), with more than 100-fold affinity than the corresponding TRF2 peptide, and thus blocks the interaction between RAP1 and TFR2 (*no cellular data reported*) (Ran et al. 2016). A series of cyclic constrained peptides (Figure 13.12(B)) was designed to bind to the TRFH domain of TRF2. *In vitro*, the best peptides (Kd = 0.12–0.2 μM) compete with the binding of Apollo on the TRFH domain TRF2. In HeLa cells, they do not modify the telomeric localization of TRF2, but they significantly increase the number of γH2AX foci at 1μM (Di Maro et al. 2014). Finally, starting from the 11-mer sequence of the TBM domain of Apollo, which binds to the TRFH domain of TRF2 (Figure 13.12(C)), a cyclic peptide (Figure 13.12(C')) was designed which exhibits the same affinity for TRF2 as the Apollo protein (Kd = 0.09 μM). Truncated analogs keep micromolar affinities for TRF2 (Chen et al. 2018; *no cellular data reported*).

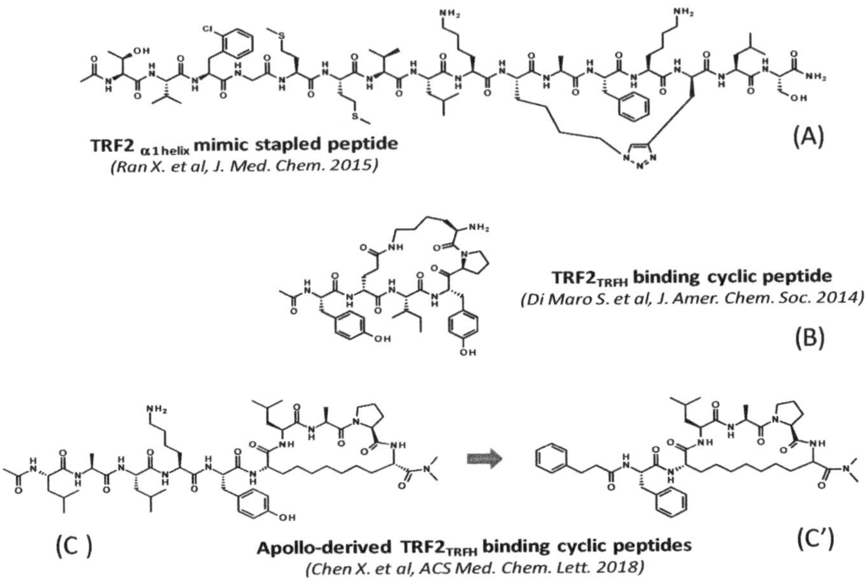

Figure 13.12. *Structures of the inhibitors of protein–protein interactions involving the shelterin complex*

13.5.3. *Chemical modulators of the shelterin proteins*

The precise chemical nature of shelterin members is important for their localization at telomeres and the recruitment of their partners or effectors during the cell cycle.

13.5.3.1. *Inhibitors of phosphorylation*

Thus, phosphorylation of TRF1 at Ser-435 by Plk1 is essential for its binding on telomeric DNA (Wu et al. 2008). Phosphorylation at Thr-122 by CK2 is mandatory for the dimerization of TRF1 and its subsequent binding on telomeres (Kim et al. 2008). Phosphorylation of TRF1 at Tyr-371 by CyclinB-Cdk1 modulates the resolution of sister chromatids in late mitosis (McKerlie and Zhu 2011). The mitotic spindle kinase BUB1, which interacts with TRF2 during S-phase, phosphorylates TRF1 in order to recruit BLM helicase needed to overcome replication stress at telomeres (Li et al. 2018). Blasco's team highlighted the role of the PI3K/AKT pathway in the regulation of TRF1 at telomeres. Decreased phosphorylation of TRF1 by AKT (at either Tyr-330, Ser-344 or Tyr248) or treatment with nanomolar concentration of specific PI3Kα inhibitors both reduces the amount of TRF1 located at telomeres and the viability of cells. *In vivo*, the PI3Kα specific inhibitor BYL719

(Figure 13.13) lowers the TRF1 level and increases DNA damage (Mendez-Pertuz et al. 2017). The same group recently developed a cell-based high-throughput screening and thus showed that multiple pathways regulate the phosphorylation status of TRF1. They also identified new kinases able to phosphorylate TRF1 on various sites (Erk1/2, mTOR and Braf^{V600E}) and are investigating TRF1-targeting combinations of kinase inhibitors (Bejarano et al. 2019). Aurora C kinase phosphorylates TRF2 at Thr-358 (Spengler 2007). Erk1/2 phosphorylates TRF2 at Ser-323 *in vitro* and in the cytoplasm of human tumor cells. The expression of non-"phosphorylable" mutant S323A of TRF2 leads to growth arrest and tumor reversion of melanoma cells (Picco et al. 2016).

As demonstrated by treatment with roscovitine (Figure 13.13), CDKs control the phosphorylation status of TRF2 at Ser-365, which modulates its telomere accessibility across the cell cycle. This TRF2-phosphoswitch regulates the transient binding and unbinding of RTEL1 helicase to telomeres, mediating the transient disassembly of T-loops during the S-phase to allow telomere replication (Sarek et al. 2019).

Akt1 inhibition prevents both the dimerization of TPP1 and the recruitment of TPP1-hPOT1 sub-complex at telomeres, thus enhancing telomere dysfunction (Han et al. 2013). Phosphorylation of TPP1 by NEK6 kinase is mandatory for the recruitment of telomerase at telomeres and telomere elongation in the G2/M phase (Hirai et al. 2016).

TIN2 is cell-cycle dependently phosphorylated by RSK2 at Ser295 and 330 (Yang and Counter 2013).

Phosphorylation of the long isoform of TIN2 (TIN2L) at Ser 396 enhances its interaction with TRF2 and is permissive for telomere elongation (Nelson et al. 2018).

The functions of the various shelterin proteins are highly dependent on their phosphorylation status across the cell cycle. Thus, it should be interesting to investigate, in cells and *in vivo*, the combination of telomeric DNA targeting drugs (G-quadruplex ligands, metal complexes, etc.) with specific (Akt1, PI3Kα, Plk1, Cdk2…) or dual Akt1/Cdk2 (A-674563, Figure 13.13) inhibitors of kinases, which modulate the phosphorylation of TRF1, TRF2, TPP1 or TIN2.

It could also be worth looking at the combination of telomeric DNA targeting drugs with checkpoint kinase inhibitors (ATM, ATR, etc.) in tumor cells with DNA repair liabilities.

13.5.3.2. *Inhibitors of polyADP-ribosylation*

Selective parylation of its N-ter domain by the tankyrase 1 (TANK1) cell cycle dependently controls the binding of TRF1 to telomeres (Cook 2002). Selective TANK1 inhibitors abolish the recruitment of repair proteins (XRCC1 and Pol β) at damaged telomeres (Yang et al. 2017). The selective TANK1 inhibitor XAV939 (Figure 13.13) synergizes with G-quadruplex ligands in tumor cells (Ivancich 2014). Recently, Birrocio's team linked a G-quadruplex ligand to a generic PARP inhibitor to design the first in class dual G4/PARP ligand. This compound (Figure 13.13) stabilizes telomeric G4 oligos *in vitro*. In BRCA2-deficient cells, it induces a DNA damage response localized at telomeres, inhibits parylation and exhibits antiproliferative effects (Salvati et al. 2017).

13.5.3.3. *Inhibitors of acetylation*

Histone acetyltransferase p300 acetylates and stabilizes TRF2, thus allowing its efficient binding at telomeres (Her and Chung 2013). SIRT6 deacetylates *in vivo* the TRFH domain of TRF2. Knockdown of SIRT6 upregulates TRF2 protein levels and counteracts its downregulation during the DNA-damage response, leading to cell survival (Rizzo et al. 2017). SIRT6 activator MDL-800 (Figure 13.13) exhibits tumor growth inhibition in a xenograft model of hepatocellular carcinoma. However, the telomere-related contribution to this antitumor activity is unknown (Huang et al. 2018).

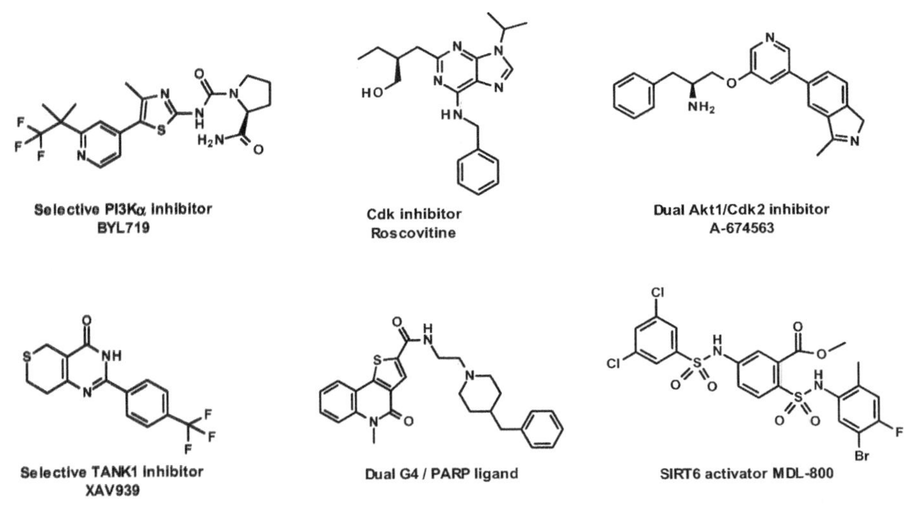

Figure 13.13. *Structures of chemical modulators of shelterin proteins*

13.6. Compounds targeting telomerase

The human telomerase, a ribonucleoprotein complex composed of an RNA-dependent DNA polymerase (hTERT) and a telomerase RNA (hTR) is mandatory for maintaining telomere length and protecting chromosomal integrity. While hTR is broadly expressed, hTERT is often silenced in somatic cells and re-activated in more than 85% cancer cells, where it confers the capacity for limitless proliferation. Thus, hTERT appeared as a compelling target for cancer inhibition. Along with its established role in telomere lengthening, telomerase also serves non-canonical extra-telomeric roles in oncogenic signaling, resistance to apoptosis, and enhanced DNA damage response (Segal-Bendirdjian and Géli 2019).

13.6.1. *Oligonucleotides as telomerase inhibitors*

Imetelsat sodium (GRN163L, Figure 13.14) is a palmitoyl-conjugated 13-mer N3' P5' thiophophoramidate deoxyribo-oligonucleotide (NPS), which exhibits a complementary sequence to hTR. Thus, GRN163L competitively inhibits telomerase and shows significant antitumor activity in various cell lines and mouse xenografts (Dikmen et al. 2005).

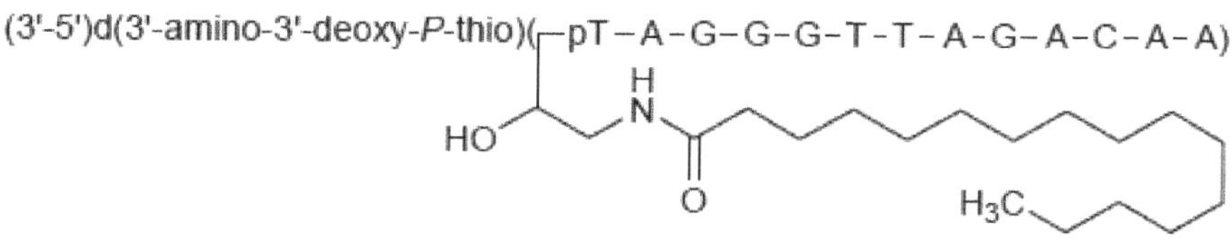

Figure 13.14. *Structure of Imetelsat or GRN163L*

Imetelsat entered clinical trials in 2014 for the treatment of myelofibrosis, myelodysplastic syndrome (MDS) and leukemiae (AML), and was further evaluated in other clinical settings. Haematological toxicities that require frequent "drug holidays" and telomerase-related deleterious side-effects on stem cells strongly limit its use in clinic as a stand-alone. **Imetelstat** moved to phase 3 trials, where it was found inactive against pediatric brain tumors. However, it is still in phase 2 trials against myelofibrosis and MDS and it received a FDA fast-track designation in 2017 for the treatment of adult patients with transfusion-dependent anemia due to MDS. Combinations of imetelsat, with either PARP inhibitors in pancreatic cancer cells (Burchett et al. 2017) or Trastuzumab (specific anti-HER2 mAb) in HER2 positive breast tumors (Koziel and Herbert 2015), which are promising in preclinical models, are starting clinical evaluation.

T-oligos: T-oligos are single strand oligonucleotides, sharing sequence homology with terminal telomeric DNA. T-oligos are often included in reviews on telomerase inhibitors. However, their mode of action does not correspond to true telomerase inhibiton. In nuclei, they are more likely to self-associate into G-quadruplex structures and thus induce disruption of the shelterin complex, as G-quadruplex ligands do. Despite the evidence that they exhibit potent telomere-related anticancer effects, they have not yet progressed to clinical trials, probably due both to the incomplete elucidation of their mechanism of action and to issues in their delivery.

13.6.2. *Small molecule as inhibitors of telomerase*

Historically, BIBR1532 (Figure 13.15) was the first non-nucleotidic small molecule that potently and selectively inhibits telomerase activity by noncompetitive binding to the active site of hTERT (Pascolo et al. 2002; Bryan et al. 2015). BIBR1532 induces telomere shortening and a senescent phenotype in telomerase positive cells. It promotes growth arrest in tumor cells and has no effect on normal haematopoietic cells. *In vivo*, mice carrying subcutaneous implants of HT1080 cells developed no or only small tumors over a 60 days treatment (Damm et al. 2001; El-Daly et al. 2005). However, the long duration of treatment, needed to observe its antitumor effects, limits its use and BIBR1532 has never started any clinical evaluation.

Nucleoside analog zidovudine (AZT, Figure 13.15), originally developed against retroviral infections, inhibits telomerase *in vitro* and leads to telomere shortening in cells (Gomez et al. 2012).

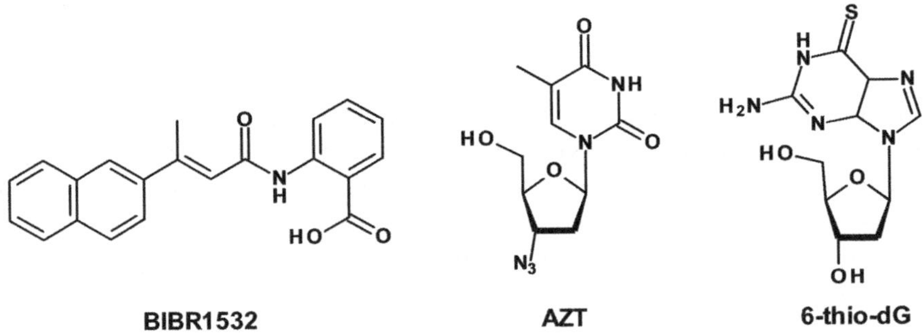

Figure 13.15. *Structures of BIBR1532, AZT and 6-thio-dG*

Another nucleoside analog, 6-thio-2′-deoxyguanosine (6-Thio-dG, Figure 13.15), is a telomerase substrate. 6-thio-dG rapidly leads to telomere uncapping and growth arrest of tumor cells (Mender et al. 2015). *In vivo*, 6-thio-dG is able to overcome resistance in non-small cell lung cancers and pediatric brain tumors (Mender et al. 2018; Sengupta et al. 2018). Despite such promising data, 6-thio-dG has not moved to clinical trials so far.

Many others small molecule inhibitors of telomerase have been disclosed. A current trend is to investigate irreversible inhibitors (Kellermann et al. 2017) as tools to study the telomeric and extra-telomeric functions of telomerase (Betori et al. 2020).

13.6.3. *Biologics as inhibitors of telomerase*

h-TERT-derived immunogenic peptides: hTERT is a tumor-associated antigen (TAA). It is overexpressed in over 85% of tumor cells, including cancer stem cells. Short and long peptides derived from hTERT, which form complexes with MHC class I and class II molecules, were identified as targets for cancer immunotherapy.

Development of h-TERT-related vaccines: vaccines based on hTERT-derived epitopes, such as GV1001 (MHC class II-restricted peptide vaccine composed of 16 amino acids from the active site of hTERT) or GRNVAC1 (dendritic cell-based vaccine from patients with mRNA encoding hTERT) are available in clinics (Mizukoshi and Kaneko 2019).

In a phase 1/2 study, GV1001 induces specific T-cell responses in advanced pancreatic and lung cancer patients. However, in a phase 3 study for patients with metastatic pancreatic cancer, the addition of GV1001 to chemotherapy does not improve the overall survival.

In phase 2, GRNVAC1 is effective on acute myeloid leukemia (AML), since it significantly improves the disease free survival of patients at high risk of relapse and induces a positive immune response to telomerase.

13.7. General conclusion

Altogether the data collected over two decades has permitted the elucidation of specific molecular and cellular mechanisms triggered by telomere-targeted G4 ligands, and reinforced the notion that this class of compounds can act as broad inhibitors of telomere-related processes and have potential as selective antitumor drugs. Yet, significant differences have been observed in the biological response of G4 ligands, which have been rationalized in terms of differences in the DNA damage pathways triggered by each ligand (De Cian et al. 2008b), and in terms of telomere status that could strongly influence the activity of G4 ligand (Rizzo et al. 2017). Finally, since G4 forming regions can be found in promoters of several cancer-related genes (oncogenes and others), G4 targeting molecules could have additional extra telomeric features, which could explain their potential as anticancer agents.

Although the first generation of G4 ligand targeting telomeres have been invaluable tools, most did not satisfy the pharmacological criteria required for drug development and none of them have reached the clinical trial phase so far. Hence, it can be expected that the next generation of G4-drugs will be discovered in the future on the grounds established by the pioneering works described in this chapter.

Finally, since G4-interactive compounds act as indirect inhibitors of telomerase and more largely of shelterin proteins by preventing binding to their DNA substrate, strategies based on combination of these agents with direct inhibitors of telomerase and of shelterin proteins, such as those described herein, represent highly promising therapeutic approaches to be developed in the future with the aim of providing synergistic antitumor activity.

13.8. Acknowledgments

The authors wish to thank all the collaborators involved in this work for their invaluable contribution and apologize to all colleagues whose work was not cited due to space limitations. A special thanks to Marc Farag and Dr. Daniela Verga for their help with the figures.

13.9. References

Abraham Punnoose, J., Ma, Y., Li, Y., Sakuma, M., Mandal, S., Nagasawa, K., Mao, H. (2017). Adaptive and specific recognition of telomeric G-quadruplexes via polyvalency induced unstacking of binding units. *J. Am. Chem. Soc.*, 139(22), 7476–7484.

Adrian, M., Ang, D.J., Lech, C.J., Heddi, B., Nicolas, A., Phan, A.T. (2014). Structure and conformational dynamics of a stacked dimeric G-quadruplex formed by the human CEB1 minisatellite. *J. Am. Chem. Soc.*, 136(17), 6297–6305.

Agrawal, S., Ojha, R.P., Maiti, S. (2008). Energetics of the human Tel-22 quadruplex-telomestatin interaction: A molecular dynamics study. *J. Phys. Chem. B,* 112(22), 6828–6836.

Altschuler, S.E., Croy, J.E., Wuttke, D.S. (2012). A small molecule inhibitor of Pot1 binding to telomeric DNA. *Biochemistry,* 51(40), 7833–7845.

Amir, M., Mohammad, T., Prasad, K., Hasan, G.M., Kumar, V., Dohare, R., Islam, A., Ahmad, F., Imtaiyaz Hassan, M. (2020). Virtual high-throughput screening of natural compounds in-search of potential inhibitors for protection of telomeres 1 (POT1). *J. Biomol. Struct. Dyn.*, 38(15), 4625–4634.

Amrane, S., Adrian, M., Heddi, B., Serero, A., Nicolas, A., Mergny, J. L., Phan, A.T. (2012). Formation of pearl-necklace monomorphic G-quadruplexes in the human CEB25 minisatellite. *J. Am. Chem. Soc.,* 134(13), 5807–5816.

Anciano Granadillo, V.J., Earley, J.N., Shuck, S.C., Georgiadis, M.M., Fitch, R.W., Turchi, J.J. (2010). Targeting the OB-folds of replication protein A with small molecules. *J. Nucleic Acids*, 304035.

Barbieri, C.M., Srinivasan, A.R., Rzuczek, S.G., Rice, J.E., LaVoie, E.J., Pilch, D.S. (2007). Defining the mode, energetics and specificity with which a macrocyclic hexaoxazole binds to human telomeric G-quadruplex DNA. *Nucleic Acids Res.,* 35(10), 3272–3286.

Baudoin, O., Gonnet, F., Teulade-Fichou, M.P., Vigneron, J.P., Tabet, J.C., Lehn, J.M. (1999). Molecular recognition of nucleotide pairs by a cyclo-bis-intercaland-type receptor molecule: A spectrophotometric and electrospray mass spectrometry study. *Chemistry, an European Journal*, 5(9), 2762–2771.

Bejarano, L., Bosso, G., Louzame, J., Serrano, R., Gomez-Casero, E., Martinez-Torrecuadrada, J., Martinez, S., Blanco-Aparicio, C., Pastor, J., Blasco, M.A. (2019). Multiple cancer pathways regulate telomere protection. *EMBO Mol. Med.,* 11(7), e10292.

Berardinelli, F., Siteni, S., Tanzarella, C., Stevens, M.F., Sgura, A., Antoccia, A. (2015). The G-quadruplex-stabilising agent RHPS4 induces telomeric dysfunction and enhances radiosensitivity in glioblastoma cells. *DNA Repair (Amst)* 25, 104–115.

Bertrand, H., Bombard, S., Monchaud, D., Teulade-Fichou, M.P. (2007a). A platinum-quinacridine hybrid as G-quadruplex ligand. *J. Biol. Inorg. Chem.,* 12, 1003–1014.

Bertrand, H., Monchaud, D., De Cian, A., Guillot, R., Mergny, J.L., Teulade-Fichou, M.P. (2007b). The importance of metal geometry in the recognition of G-quadruplex-DNA by metal-terpyridine complexes. *Org. Biomol. Chem.,* 5(16), 2555–2559.

Bertrand, H., Bombard, S., Monchaud, D., Talbot, E., Guedin, A., Mergny, J.L., Grunert, R., Bednarski, P.J., Teulade-Fichou, M.P. (2009). Exclusive platination of loop adenines in the human telomeric G-quadruplex. *Org. Biomol. Chem.,* 7(14), 2864–2871.

Betori, R.C., Liu, Y., Mishra, R.K., Cohen, S.B., Kron, S.J., Scheidt, K.A. (2020). Targeted covalent inhibition of telomerase. *ACS Chem. Biol.,* 15(3), 706–717.

Betzer, J.F., Nuter, F., Chtchigrovsky, M., Hamon, F., Kellermann, G., Ali, S., Calmejane, M.A., Roque, S., Poupon, J., Cresteil, T. et al. (2016). Linking of antitumor trans NHC-Pt(II) complexes to G-quadruplex DNA ligand for telomeric targeting. *Bioconjug. Chem.,* 27(6), 1456–1470.

Bochkareva, E., Korolev, S., Lees-Miller, S.P., Bochkarev, A. (2002). Structure of the RPA trimerization core and its role in the multistep DNA-binding mechanism of RPA. *EMBO J.,* 21(7), 1855–1863.

Bonnat, L., Bar, L., Gennaro, B., Bonnet, H., Jarjayes, O., Thomas, F., Dejeu, J., Defrancq, E., Lavergne, T. (2017). Template-mediated stabilization of a DNA G-quadruplex formed in the HIV-1 promoter and comparative binding studies. *Chemistry,* 23(23), 5602–5613.

Bryan, C., Rice, C., Hoffman, H., Harkisheimer, M., Sweeney, M., Skordalakes, E. (2015). Structural basis of telomerase inhibition by the highly specific BIBR1532. *Structure,* 23(10), 1934–1942.

Bugaut, A. and Alberti, P. (2015). Understanding the stability of DNA G-quadruplex units in long human telomeric strands. *Biochimie,* 113, 125–133.

Burchett, K.M., Etekpo, A., Batra, S.K., Yan, Y., Ouellette, M.M. (2017). Inhibitors of telomerase and poly(ADP-ribose) polymerases synergize to limit the lifespan of pancreatic cancer cells. *Oncotarget,* 8(48), 83754–83767.

Burger, A.M., Dai, F., Schultes, C.M., Reszka, A.P., Moore, M.J., Double, J.A., Neidle, S. (2005). The G-quadruplex-interactive molecule BRACO-19 inhibits tumor growth, consistent with telomere targeting and interference with telomerase function. *Cancer Res.,* 65(4), 1489–1496.

Campbell, N.H., Parkinson, G.N., Reszka, A.P., Neidle, S. (2008). Structural basis of DNA quadruplex recognition by an acridine drug. *J. Am. Chem. Soc.,* 130(21), 6722–6724.

Campbell, N.H., Karim, N.H., Parkinson, G.N., Gunaratnam, M., Petrucci, V., Todd, A.K., Vilar, R., Neidle, S. (2012). Molecular basis of structure-activity relationships between salphen metal complexes and human telomeric DNA quadruplexes. *J. Med. Chem.,* 55(1), 209–222.

Cao, Q., Li, Y., Freisinger, E., Qin, P.Z., Sigel, R.K.O., Mao, Z.-W. (2017). G-quadruplex DNA targeted metal complexes acting as potential anticancer drugs. *Inorganic Chemistry Frontiers,* 4(1), 10–32.

Charif, R., Granotier-Beckers, C., Bertrand, H.C., Poupon, J., Ségal-Bendirdjian, E., Teulade-Fichou, M.P., Boussin, F.D., Bombard, S. (2017). Association of a platinum complex to a G-quadruplex ligand enhances the telomere disruption. *Chem. Res. Tox,* 38(8), 1629–1640.

Chen, Y., Rai, R., Zhou, Z.R., Kanoh, J., Ribeyre, C., Yang, Y., Zheng, H., Damay, P., Wang, F., Tsujii, H. et al. (2011). A conserved motif within RAP1 has diversified roles in telomere protection and regulation in different organisms. *Nat. Struct. Mol. Biol.,* 18(2), 213–221.

Chen, X., Liu, L., Chen, Y., Yang, Y., Yang, C.Y., Guo, T., Lei, M., Sun, H., Wang, S. (2018). Cyclic peptidic mimetics of apollo peptides targeting telomeric repeat binding factor 2 (TRF2) and apollo interaction. *ACS Med. Chem. Lett.,* 9(5), 507–511.

Cheng, M.K., Modi, C., Cookson, J.C., Hutchinson, I., Heald, R.A., McCarroll, A.J., Missailidis, S., Tanious, F., Wilson, W.D., Mergny, J.L. et al. (2008). Antitumor polycyclic acridines. 20. Search for DNA quadruplex binding selectivity in a series of 8,13-dimethylquino[4,3,2-kl]acridinium salts: Telomere-targeted agents. *J. Med. Chem.,* 51(4), 963–975.

Chung, W.J., Heddi, B., Hamon, F., Teulade-Fichou, M.P., Phan, A.T. (2014). Solution structure of a G-quadruplex bound to the bisquinolinium compound Phen-DC(3). *Angew. Chem. Int. Ed. Engl.,* 53(4), 999–1002.

Collie, G.W., Promontorio, R., Hampel, S.M., Micco, M., Neidle, S., Parkinson, G.N. (2012). Structural basis for telomeric G-quadruplex targeting by naphthalene diimide ligands. *J. Am. Chem. Soc.,* 134(5), 2723–2731.

Cook, B.D., Dynek, J.N., Chang, W., Shostak, G., Smith, S. (2002). Role for the related poly(ADP-Ribose) polymerases tankyrase 1 and 2 at human telomeres. *Mol. Cell. Biol.,* 22(1), 332–342.

Court, R., Chapman, L., Fairall, L., Rhodes, D. (2005). How the human telomeric proteins TRF1 and TRF2 recognize telomeric DNA: A view from high-resolution crystal structures. *EMBO Rep.,* 6(1), 39–45.

Cuenca, F., Greciano, O., Gunaratnam, M., Haider, S., Munnur, D., Nanjunda, R., Wilson, W.D., Neidle, S. (2008). Tri- and tetra-substituted naphthalene diimides as potent G-quadruplex ligands. *Bioorg. Med. Chem. Lett.,* 18(5), 1668–1673.

Damm, K., Hemmann, U., Garin-Chesa, P., Hauel, N., Kauffmann, I., Priepke, H., Niestroj, C., Daiber, C., Enenkel, B., Guilliard, B. et al. (2001). A highly selective telomerase inhibitor limiting human cancer cell proliferation. *EMBO J.,* 20(24), 6958–6968.

De Cian, A., Delemos, E., Mergny, J.L., Teulade-Fichou, M.P., Monchaud, D. (2007a). Highly efficient G-quadruplex recognition by bisquinolinium compounds. *J. Am. Chem. Soc.,* 129(7), 1856–1857.

De Cian, A., Guittat, L., Kaiser, M., Sacca, B., Amrane, S., Bourdoncle, A., Alberti, P., Teulade-Fichou, M.P., Lacroix, L., Mergny, J.L. (2007b). Fluorescence-based melting assays for studying quadruplex ligands. *Methods,* 42(2), 183–195.

De Cian, A., Grellier, P., Mouray, E., Depoix, D., Bertrand, H., Monchaud, D., Teulade-Fichou, M.P., Mergny, J.L., Alberti, P. (2008a). Plasmodium telomeric sequences: Structure, stability and quadruplex targeting by small compounds. *ChemBioChem,* 9(16), 2730–2739.

De Cian, A., Lacroix, L., Douarre, C., Temime-Smaali, N., Trentesaux, C., Riou, J.F., Mergny, J.L. (2008b). Targeting telomeres and telomerase. *Biochimie,* 90(1), 131–155.

Di Maro, S., Zizza, P., Salvati, E., De Luca, V., Capasso, C., Fotticchia, I., Pagano, B., Marinelli, L., Gilson, E., Novellino, E. et al. (2014). Shading the TRF2 recruiting function: A new horizon in drug development. *J. Am. Chem. Soc.,* 136(48), 16708–16711.

Dikmen, Z.G., Gellert, G.C., Jackson, S., Gryaznov, S., Tressler, R., Dogan, P., Wright, W.E., Shay, J.W. (2005). *In vivo* inhibition of lung cancer by GRN163L: A novel human telomerase inhibitor. *Cancer Res,* 65(17), 7866–7873.

Dixon, I.M., Lopez, F., Tejera, A.M., Esteve, J.P., Blasco, M.A., Pratviel, G., Meunier, B. (2007). A G-quadruplex ligand with 10000-fold selectivity over duplex DNA. *J. Am. Chem. Soc.,* 129(6), 1502–1503.

Doria, F., Salvati, E., Pompili, L., Pirota, V., D'Angelo, C., Manoli, F., Nadai, M., Richter, S.N., Biroccio, A., Manet, I. et al. (2019). Dyads of G-quadruplex ligands triggering DNA damage response and tumour cell growth inhibition at subnanomolar concentration. *Chemistry – A European Journal,* 25(47), 11085–11097.

Drosopoulos, W.C., Kosiyatrakul, S.T., Schildkraut, C.L. (2015). BLM helicase facilitates telomere replication during leading strand synthesis of telomeres. *J. Cell Biol.,* 210(2), 191–208.

El-Daly, H., Kull, M., Zimmermann, S., Pantic, M., Waller, C.F., Martens, U.M. (2005). Selective cytotoxicity and telomere damage in leukemia cells using the telomerase inhibitor BIBR1532. *Blood,* 105(4), 1742–1749.

Fairall, L., Chapman, L., Moss, H., de Lange, T., Rhodes, D. (2001). Structure of the TRFH dimerization domain of the human telomeric proteins TRF1 and TRF2. *Mol. Cell,* 8(2), 351–361.

Gavande, N.S., VanderVere-Carozza, P.S., Pawelczak, K.S., Vernon, T.L., Jordan, M.R., Turchi, J.J. (2020). Structure-guided optimization of replication protein A (RPA)-DNA interaction inhibitors. *ACS Medicinal Chemistry Letters*, 11(6), 1118–1124.

Gavathiotis, E., Heald, R.A., Stevens, M.F., Searle, M.S. (2003). Drug recognition and stabilisation of the parallel-stranded DNA quadruplex d(TTAGGGT)4 containing the human telomeric repeat. *J. Mol. Biol.,* 334(1), 25–36.

Georgiades, S.N., Karim, N.H.A., Kogularamanan S., Vilar, R. (2010). Interaction of metal complexes with G-quadruplex DNA. *Angew. Chem. Int. Ed.,* 49, 4020–4034.

Gomez, D., Paterski, R., Lemarteleur, T., Shin-Ya, K., Mergny, J.L., Riou, J.F. (2004). Interaction of telomestatin with the telomeric single-strand overhang. *J. Biol. Chem.,* 279(40), 41487–41494.

Gomez, D.E., Armando, R.G., Alonso, D.F. (2012). AZT as a telomerase inhibitor. *Front. Oncol.,* 2(113), 1–5.

Gowan, S.M., Heald, R., Stevens, M.F., Kelland, L.R. (2001). Potent inhibition of telomerase by small-molecule pentacyclic acridines capable of interacting with G-quadruplexes. *Mol. Pharmacol.,* 60(5), 981–988.

Granotier, C., Pennarun, G., Riou, L., Hoffschir, F., Gauthier, L.R., De Cian, A., Gomez, D., Mandine, E., Riou, J.F., Mergny, J.L. et al. (2005). Preferential binding of a G-quadruplex ligand to human chromosome ends. *Nucleic Acids Res.,* 33(13), 4182–4190.

Hampel, S.M., Sidibe, A., Gunaratnam, M., Riou, J.F., Neidle, S. (2010). Tetrasubstituted naphthalene diimide ligands with selectivity for telomeric G-quadruplexes and cancer cells. *Bioorg. Med. Chem. Lett.,* 20(22), 6459–6463.

Han, X., Liu, D., Zhang, Y., Li, Y., Lu, W., Chen, J., Songyang, Z. (2013). Akt regulates TPP1 homodimerization and telomere protection. *Aging Cell,* 12(6), 1091–1099.

Hanaoka, S., Nagadoi, A., Nishimura, Y. (2005). Comparison between TRF2 and TRF1 of their telomeric DNA-bound structures and DNA-binding activities. *Protein Science,* 14(1), 119–130.

Harrison, R.J., Cuesta, J., Chessari, G., Read, M.A., Basra, S.K., Reszka, A.P., Morrell, J., Gowan, S.M., Incles, C.M., Tanious, F.A. et al. (2003). Trisubstituted acridine derivatives as potent and selective telomerase inhibitors. *J. Med. Chem.,* 46(21), 4463–4476.

Her, Y.R. and Chung, I.K. (2013). p300-mediated acetylation of TRF2 is required for maintaining functional telomeres. *Nucleic Acids Research,* 41(4), 2267–2283.

Hirai, Y., Tamura, M., Otani, J., Ishikawa, F. (2016). NEK6-mediated phosphorylation of human TPP1 regulates telomere length through telomerase recruitment. *Genes Cells,* 21(8), 874–889.

Huang, Z., Zhao, J., Deng, W., Chen, Y., Shang, J., Song, K., Zhang, L., Wang, C., Lu, S., Yang, X. et al. (2018). Identification of a cellularly active SIRT6 allosteric activator. *Nat. Chem. Biol.,* 14(12), 1118–1126.

Hwang, I.P., Mailliet, P., Hossard, V., Riou, J.F., Bugaut, A., Roger, L. (2019). Investigating the effect of mono- and dimeric 360A G-quadruplex ligands on telomere stability by single telomere length analysis (STELA). *Molecules,* 24(3), 577–586.

Ivancich, M. (2014). Targeting the telomere with T-oligo, G quadruplex stabilizers, and tankyrase inhibitors. *Journal of Cancer Science & Therapy,* 06(10), 429–432.

Kalathiya, U., Padariya, M., Baginski, M. (2017). Molecular basis and quantitative assessment of TRF1 and TRF2 protein interactions with TIN2 and Apollo peptides. *Eur. Biophys. J.*, 46(2), 171–187.

Kellermann, G., Dingli, F., Masson, V., Dauzonne, D., Segal-Bendirdjian, E., Teulade-Fichou, M.P., Loew, D., Bombard, S. (2017). Exploring the mechanism of inhibition of human telomerase by cysteine-reactive compounds. *FEBS Lett.*, 591(6), 863–874.

Kim, M.Y., Vankayalapati, H., Shin-Ya, K., Wierzba, K., Hurley, L.H. (2002). Telomestatin, a potent telomerase inhibitor that interacts quite specifically with the human telomeric intramolecular g-quadruplex. *J. Am. Chem. Soc.*, 124(10), 2098–2099.

Kim, M.K., Kang, M.R., Nam, H.W., Bae, Y.S., Kim, Y.S., Chung, I.K. (2008). Regulation of telomeric repeat binding factor 1 binding to telomeres by casein kinase 2-mediated phosphorylation. *J. Biol. Chem.*, 283(20), 14144–14152.

Kim, J.K., Liu, J., Hu, X., Yu, C., Roskamp, K., Sankaran, B., Huang, L., Komives, E.A., Qiao, F. (2017). Structural basis for shelterin bridge assembly. *Mol. Cell,* 68(4), 698–714.e695.

Kotar, A., Kocman, V., Plavec, J. (2020). Intercalation of a heterocyclic ligand between quartets in a G-rich tetrahelical structure. *Chemistry,* 26(4), 814–817.

Koziel, J.E. and Herbert, B.S. (2015). The telomerase inhibitor imetelstat alone, and in combination with trastuzumab, decreases the cancer stem cell population and self-renewal of HER2+ breast cancer cells. *Breast Cancer Res. Treat.*, 149(3), 607–618.

Lee, H.S., Carmena, M., Liskovykh, M., Peat, E., Kim, J.H., Oshimura, M., Masumoto, H., Teulade-Fichou, M.P., Pommier, Y., Earnshaw, W.C. et al. (2018). Systematic analysis of compounds specifically targeting telomeres and telomerase for clinical implications in cancer therapy. *Cancer Res.*, 78(21), 6282–6296.

Lei, M., Podell, E.R., Baumann, P., Cech, T.R. (2003). DNA self-recognition in the structure of Pot1 bound to telomeric single-stranded DNA. *Nature,* 426(6963), 198–203.

Lei, M., Podell, E.R., Cech, T.R. (2004). Structure of human POT1 bound to telomeric single-stranded DNA provides a model for chromosome end-protection. *Nat. Struct. Mol. Biol.*, 11(12), 1223–1229.

Lejault, P. (2017). Ligands de la jonction double-brin/simple-brin de l'ADN télomérique comme sondes moléculaires pour étudier la biologie des télomères. MNHN thesis, University of Paris Sorbonne.

Li, F., Kim, H., Ji, Z., Zhang, T., Chen, B., Ge, Y., Hu, Y., Feng, X., Han, X., Xu, H. et al. (2018). The BUB3-BUB1 complex promotes telomere DNA replication. *Mol. Cell,* 70(3), 395–407.e394.

Lista, M.J., Martins, R.P., Billant, O., Contesse, M.A., Findakly, S., Pochard, P., Daskalogianni, C., Beauvineau, C., Guetta, C., Jamin, C. et al. (2017). Nucleolin directly mediates Epstein-Barr virus immune evasion through binding to G-quadruplexes of EBNA1 mRNA. *Nat. Commun.,* 8, 16043–16056.

Ma, T.Z., Zhang, M.J., Liao, T.C., Li, J.H., Zou, M., Wang, Z.M., Zhou, C.Q. (2020). Dimers formed with the mixed-type G-quadruplex binder pyridostatin specifically recognize human telomere G-quadruplex dimers. *Org. Biomol. Chem.*, 18(5), 920–930.

Maleki, P., Ma, Y., Iida, K., Nagasawa, K., Balci, H. (2017). A single molecule study of a fluorescently labeled telomestatin derivative and G-quadruplex interactions. *Nucleic Acids Res.,* 45(1), 288–295.

Malik, A.M. (2013). Design, synthesis and evaluation of inhibitors of POT1-DNA interactions. Thesis for Doctor of Philosophy, University of Nottingham.

Mancini, J., Rousseau, P., Castor, K.J., Sleiman, H.F., Autexier, C. (2016). Platinum(II) phenanthroimidazole G-quadruplex ligand induces selective telomere shortening in A549 cancer cells. *Biochimie,* 121, 287–297.

Mandal, S., Kawamoto, Y., Yue, Z., Hashiya, K., Cui, Y., Bando, T., Pandey, S., Hoque, M.E., Hossain, M.A., Sugiyama, H. et al. (2019). Submolecular dissection reveals strong and specific binding of polyamide-pyridostatin conjugates to human telomere interface. *Nucleic Acids Res.,* 47(7), 3295–3305.

Marchand, A., Granzhan, A., Iida, K., Tsushima, Y., Ma, Y., Nagasawa, K., Teulade-Fichou, M.P., Gabelica, V. (2015). Ligand-induced conformational changes with cation ejection upon binding to human telomeric DNA G-quadruplexes. *J. Am. Chem. Soc.,* 137(2), 750–756.

McKerlie, M. and Zhu, X.D. (2011). Cyclin B-dependent kinase 1 regulates human TRF1 to modulate the resolution of sister telomeres. *Nat. Commun.,* 2(371), 1372–1384.

Mender, I., Gryaznov, S., Dikmen, Z.G., Wright, W.E., Shay, J.W. (2015). Induction of telomere dysfunction mediated by the telomerase substrate precursor 6-thio-2'-deoxyguanosine. *Cancer Discov.,* 5(1), 82–95.

Mender, I., LaRanger, R., Luitel, K., Peyton, M., Girard, L., Lai, T.P., Batten, K., Cornelius, C., Dalvi, M.P., Ramirez, M. et al. (2018). Telomerase-mediated strategy for overcoming non-small cell lung cancer targeted therapy and chemotherapy resistance. *Neoplasia,* 20(8), 826–837.

Mendez-Pertuz, M., Martinez, P., Blanco-Aparicio, C., Gomez-Casero, E., Belen Garcia, A., Martinez-Torrecuadrada, J., Palafox, M., Cortes, J., Serra, V., Pastor, J. et al. (2017). Modulation of telomere protection by the PI3K/AKT pathway. *Nat. Commun.,* 8(1), 1278–1295.

Mergny, J.L., Lacroix, L., Teulade-Fichou, M.P., Hounsou, C., Guittat, L., Hoarau, M., Arimondo, P.B., Vigneron, J.P., Lehn, J.M., Riou, J.F. et al. (2001). Telomerase inhibitors based on quadruplex ligands selected by a fluorescence assay. *Proc. Natl. Acad. Sci. USA,* 98(6), 3062–3067.

Merle, P., Gueugneau, M., Teulade-Fichou, M.P., Müller-Barthélémy, M., Amiard, S., Chautard, E., Guetta, C., Dedieu, V., Communal, Y., Mergny, J.L. et al. (2015). Highly efficient radiosensitization of human glioblastoma and lung cancer cells by a G-quadruplex DNA binding compound. *Sci. Rep.,* 5, 16255–16266.

Micco, M., Collie, G.W., Dale, A.G., Ohnmacht, S.A., Pazitna, I., Gunaratnam, M., Reszka, A.P., Neidle, S. (2013). Structure-based design and evaluation of naphthalene diimide G-quadruplex ligands as telomere targeting agents in pancreatic cancer cells. *J. Med. Chem.,* 56(7), 2959–2974.

Mizukoshi, E. and Kaneko, S. (2019). Telomerase-targeted cancer immunotherapy. *Int. J. Mol. Sci.,* 20(8), 1823–1839.

Monchaud, D. and Teulade-Fichou, M.P. (2008). A hitchhiker's guide to G-quadruplex ligands. *Org. Biomol. Chem.,* 6(4), 627–636.

Moore, M.J., Schultes, C.M., Cuesta, J., Cuenca, F., Gunaratnam, M., Tanious, F.A., Wilson, W.D., Neidle, S. (2006). Trisubstituted acridines as G-quadruplex telomere targeting agents. Effects of extensions of the 3,6- and 9-side chains on quadruplex binding, telomerase activity, and cell proliferation. *J. Med. Chem.,* 49(2), 582–599.

Morel, E., Poyer, F., Vaslin, L., Bombard, S., Teulade-Fichou, M.P. (2016). Photoactivatable platinum(II) terpyridine derivatives for G-quadruplex DNA double anchoring. *Inorganica Chimica Acta,* 452, 152–158.

Müller, S., Sanders, D.A., Di Antonio, M., Matsis, S., Riou, J.F., Rodriguez, R., Balasubramanian, S. (2012). Pyridostatin analogues promote telomere dysfunction and long-term growth inhibition in human cancer cells. *Org. Biomol. Chem.,* 10(32), 6537–6546.

Nandakumar, J. and Cech, T.R. (2013). Finding the end: Recruitment of telomerase to telomeres. *Nat. Rev. Mol. Cell Biol.,* 14(2), 69–82.

Neidle, S. (2010). Human telomeric G-quadruplex: The current status of telomeric G-quadruplexes as therapeutic targets in human cancer. *FEBS J.,* 277(5), 1118–1125.

Neidle, S. (ed.) (2012). The biology and pharmacology of telomeric quadruplex ligands. In *Therapeutic Applications of Quadruplex Nucleic Acids*, Academic Press, Boston.

Neidle, S. (2016). Quadruplex nucleic acids as novel therapeutic targets. *J. Med. Chem.,* 59(13), 5987–6011.

Neidle, S. and Parkinson, G. (2002). Telomere maintenance as a target for anticancer drug discovery. *Nat. Rev. Drug Discov.*, 1(5), 383–393.

Nelson, N.D., Dodson, L.M., Escudero, L., Sukumar, A.T., Williams, C.L., Mihalek, I., Baldan, A., Baird, D.M., Bertuch, A.A. (2018). The C-terminal extension unique to the long isoform of the shelterin component TIN2 enhances its interaction with TRF2 in a phosphorylation- and dyskeratosis congenita cluster-dependent fashion. *Mol. Cell. Biol.*, 38(12), e-00025-00018.

Ohnmacht, S.A. and Neidle, S. (2014). Small-molecule quadruplex-targeted drug discovery. *Bioorg. Med. Chem. Lett.*, 24(12), 2602–2612.

Pascolo, E., Wenz, C., Lingner, J., Hauel, N., Priepke, H., Kauffmann, I., Garin-Chesa, P., Rettig, W. J., Damm, K., Schnapp, A. (2002). Mechanism of human telomerase inhibition by BIBR1532, a synthetic, non-nucleosidic drug candidate. *J. Biol. Chem.*, 277(18), 15566–15572.

Pennarun, G., Granotier, C., Gauthier, L.R., Gomez, D., Hoffschir, F., Mandine, E., Riou, J.F., Mergny, J.L., Mailliet, P., Boussin, F.D. (2005). Apoptosis related to telomere instability and cell cycle alterations in human glioma cells treated by new highly selective G-quadruplex ligands. *Oncogene,* 24(18), 2917–2928.

Pennarun, G., Granotier, C., Hoffschir, F., Mandine, E., Biard, D., Gauthier, L.R., Boussin, F.D. (2008). Role of ATM in the telomere response to the G-quadruplex ligand 360A. *Nucleic Acids Res.,* 36(5), 1741–1754.

Pennarun, G., Hoffschir, F., Revaud, D., Granotier, C., Gauthier, L.R., Mailliet, P., Biard, D. S., Boussin, F.D. (2010). ATR contributes to telomere maintenance in human cells. *Nucleic Acids Res.,* 38(9), 2955–2963.

Picco, V., Coste, I., Giraud-Panis, M.J., Renno, T., Gilson, E., Pagès, G. (2016). ERK1/2/MAPK pathway-dependent regulation of the telomeric factor TRF2. *Oncotarget*, 7(29), 46615–46627.

Platella, C., Pirota, V., Musumeci, D., Rizzi, F., Iachettini, S., Zizza, P., Biroccio, A., Freccero, M., Montesarchio, D., Doria, F. (2020). Trifunctionalized naphthalene diimides and dimeric analogues as G-quadruplex-targeting anticancer agents selected by affinity chromatography. *Int. J. Mol. Sci.,* 21(6), 1964–1986.

Rakers, V., Cadinu, P., Edel, J.B., Vilar, R. (2018). Development of microfluidic platforms for the synthesis of metal complexes and evaluation of their DNA affinity using online FRET melting assays. *Chemical Science,* 9(14), 3459–3469.

Ran, X., Liu, L., Yang, C.Y., Lu, J., Chen, Y., Lei, M., Wang, S. (2016). Design of high-affinity stapled peptides to target the repressor activator protein 1 (RAP1)/telomeric repeat-binding factor 2 (TRF2) protein-protein interaction in the shelterin complex. *J. Med. Chem.*, 59(1), 328–334.

Rao, L. and Bierbach, U. (2007). Kinetically favored platination of adenine in the g-rich human telomeric repeat. *J. Am. Chem. Soc.*, 129(51), 15764–15765.

Rao, L., Dworkin, J.D., Nell, W.E., Bierbach, U. (2011). Interactions of a platinum-modified perylene derivative with the human telomeric G-quadruplex. *J. Phys. Chem. B,* 115(46), 13701–13712.

Read, M., Harrison, R.J., Romagnoli, B., Tanious, F.A., Gowan, S.H., Reszka, A.P., Wilson, W.D., Kelland, L.R., Neidle, S. (2001). Structure-based design of selective and potent G quadruplex-mediated telomerase inhibitors. *Proc. Natl. Acad. Sci. USA,* 98(9), 4844–4849.

Reed, J.E., Neidle, S., Vilar, R. (2007). Stabilisation of human telomeric quadruplex DNA and inhibition of telomerase by a platinum-phenanthroline complex. *Chem Commun (Camb),* 42, 4366–4368.

Reznichenko, O., Quillevere, A., Martins, R.P., Loaec, N., Kang, H., Lista, M.J., Beauvineau, C., Gonzalez-Garcia, J., Guillot, R., Voisset, C. et al. (2019). Novel cationic bis(acylhydrazones) as modulators of Epstein-Barr virus immune evasion acting through disruption of interaction between nucleolin and G-quadruplexes of EBNA1 mRNA. *Eur. J. Med. Chem.,* 178, 13–29.

Riou, J.F., Guittat, L., Mailliet, P., Laoui, A., Renou, E., Petitgenet, O., Megnin-Chanet, F., Helene, C., Mergny, J.L. (2002). Cell senescence and telomere shortening induced by a new series of specific G-quadruplex DNA ligands. *Proc. Natl. Acad. Sci. USA,* 99(5), 2672–2677.

Rizzo, A., Salvati, E., Porru, M., D'Angelo, C., Stevens, M.F., D'Incalci, M., Leonetti, C., Gilson, E., Zupi, G., Biroccio, A. (2009). Stabilization of quadruplex DNA perturbs telomere replication leading to the activation of an ATR-dependent ATM signaling pathway. *Nucleic Acids Res.,* 37(16), 5353–5364.

Rizzo, A., Iachettini, S., Zizza, P., Cingolani, C., Porru, M., Artuso, S., Stevens, M., Hummersone, M., Biroccio, A., Salvati, E., et al. (2014). Identification of novel RHPS4-derivative ligands with improved toxicological profiles and telomere-targeting activities. *J. Exp. Clin. Cancer Res.,* 33, 81–89.

Rizzo, A., Iachettini, S., Salvati, E., Zizza, P., Maresca, C., D'Angelo, C., Benarroch-Popivker, D., Capolupo, A., Del Gaudio, F., Cosconati, S. et al. (2017). SIRT6 interacts with TRF2 and promotes its degradation in response to DNA damage. *Nucleic Acids Res.*, 45(4), 1820–1834.

Rocca, R., Talarico, C., Moraca, F., Costa, G., Romeo, I., Ortuso, F., Alcaro, S., Artese, A. (2017). Molecular recognition of a carboxy pyridostatin toward G-quadruplex structures: Why does it prefer RNA? *Chem. Biol. Drug Des.,* 90(5), 919–925.

Rodriguez, R., Müller, S., Yeoman, J.A., Trentesaux, C., Riou, J.F., Balasubramanian, S. (2008). A novel small molecule that alters shelterin integrity and triggers a DNA-damage response at telomeres. *J. Am. Chem. Soc.,* 130(47), 15758–15759.

Rosu, F., Gabelica, V., Smargiasso, N., Mazzucchelli, G., Shin-Ya, K., De Pauw, E. (2010). Cation involvement in telomestatin binding to g-quadruplex DNA. *J. Nucleic Acids*, 1–7.

Sabater, L., Nicolau-Travers, M.L., De Rache, A., Prado, E., Dejeu, J., Bombarde, O., Lacroix, J., Calsou, P., Defrancq, E., Mergny, J.L. et al. (2015). The nickel(II) complex of guanidinium phenyl porphyrin, a specific G-quadruplex ligand, targets telomeres and leads to POT1 mislocalization in culture cells. *J. Biol. Inorg. Chem.,* 20(4), 729–738.

Saha, A., Duchambon, P., Masson, V., Loew, D., Bombard, S., Teulade-Fichou, M.P. (2020). Nucleolin discriminates drastically between long-loop and short-loop quadruplexes. *Biochemistry,* 59(12), 1261–1272.

Saintome, C., Alberti, P., Guinot, N., Lejault, P., Chatain, J., Mailliet, P., Riou, J.F., Bugaut, A. (2018). Binding properties of mono- and dimeric pyridine dicarboxamide ligands to human telomeric higher-order G-quadruplex structures. *Chem Commun (Camb),* 54(15), 1897–1900.

Saker, L., Ali, S., Masserot, C., Kellermann, G., Poupon, J., Teulade-Fichou, M.P., Segal-Bendirdjian, E., Bombard, S. (2018). Platinum complexes can bind to telomeres by coordination. *Int. J. Mol. Sci.,* 19(7), 1951–1966.

Salvati, E., Leonetti, C., Rizzo, A., Scarsella, M., Mottolese, M., Galati, R., Sperduti, I., Stevens, M.F., D'Incalci, M., Blasco, M. et al. (2007). Telomere damage induced by the G-quadruplex ligand RHPS4 has an antitumor effect. *J. Clin. Invest.*, 117(11), 3236–3247.

Salvati, E., Botta, L., Amato, J., Di Leva, F.S., Zizza, P., Gioiello, A., Pagano, B., Graziani, G., Tarsounas, M., Randazzo, A. et al. (2017). Lead discovery of dual G-quadruplex stabilizers and poly(ADP-ribose) polymerases (PARPs) inhibitors: A new avenue in anticancer treatment. *J. Med. Chem.,* 60(9), 3626–3635.

Sarek, G., Kotsantis, P., Ruis, P., Van Ly, D., Margalef, P., Borel, V., Zheng, X.F., Flynn, H.R., Snijders, A.P., Chowdhury, D. et al. (2019). CDK phosphorylation of TRF2 controls t-loop dynamics during the cell cycle. *Nature,* 575(7783), 523–527.

Segal-Bendirdjian, E. and Géli, V. (2019). Non-canonical roles of telomerase: Unraveling the imbroglio. *Front. Cell Dev. Biol.,* 7, 332–344.

Sengupta, S., Sobo, M., Lee, K., Senthil Kumar, S., White, A.R., Mender, I., Fuller, C., Chow, L.M.L., Fouladi, M., Shay, J.W. et al. (2018). Induced telomere damage to treat telomerase expressing therapy-resistant pediatric brain tumors. *Mol. Cancer Ther.*, 17(7), 1504–1514.

Shi, D.F., Wheelhouse, R.T., Sun, D., Hurley, L.H. (2001). Quadruplex-interactive agents as telomerase inhibitors: Synthesis of porphyrins and structure-activity relationship for the inhibition of telomerase. *J. Med. Chem.*, 44(26), 4509–4523.

Shin-Ya, K., Wierzba, K., Matsuo, K., Ohtani, T., Yamada, Y., Furihata, K., Hayakawa, Y., Seto, H. (2001). Telomestatin, a novel telomerase inhibitor from Streptomyces anulatus. *J. Am. Chem. Soc.,* 123(6), 1262–1263.

Smith, E.M., Pendlebury, D.F., Nandakumar, J. (2020). Structural biology of telomeres and telomerase. *Cell. Mol. Life Sci.,* 77(1), 61–79.

Spengler, D. (2007). The protein kinase Aurora C phosphorylates TRF2. *Cell Cycle,* 6(20), 2579–2580.

Stafford, V.S., Suntharalingam, K., Shivalingam, A., White, A.J., Mann, D.J., Vilar, R. (2015). Syntheses of polypyridyl metal complexes and studies of their interaction with quadruplex DNA. *Dalton Trans.,* 44(8), 3686–3700.

Sun, D., Thompson, B., Cathers, B.E., Salazar, M., Kerwin, S.M., Trent, J.O., Jenkins, T.C., Neidle, S., Hurley, L.H. (1997). Inhibition of human telomerase by a G-quadruplex-interactive compound. *J. Med. Chem.,* 40(14), 2113–2116.

Suntharalingam, K., White, A.J.P., Vilar, R. (2009). Synthesis, structural characterization, and quadruplex DNA binding studies of platinum(II)-terpyridine complexes. *Inorganic Chemistry,* 48(19), 9427–9435.

Tahara, H., Shin-Ya, K., Seimiya, H., Yamada, H., Tsuruo, T., Ide, T. (2006). G-Quadruplex stabilization by telomestatin induces TRF2 protein dissociation from telomeres and anaphase bridge formation accompanied by loss of the 3' telomeric overhang in cancer cells. *Oncogene,* 25(13), 1955–1966.

Tera, M., Ishizuka, H., Takagi, M., Suganuma, M., Shin-Ya, K., Nagasawa, K. (2008). Macrocyclic hexaoxazoles as sequence- and mode-selective G-quadruplex binders. *Angew Chem. Int. Ed. Engl.,* 47(30), 5557–5560.

Teulade-Fichou, M.P., Perrin, D., Boutorine, A., Polverari, D., Vigneron, J.P., Lehn, J.M., Sun, J.S., Garestier, T., Helene, C. (2001). Direct photocleavage of HIV-DNA by quinacridine derivatives triggered by triplex formation. *J. Am. Chem. Soc.,* 123(38), 9283–9292.

Teulade-Fichou, M.P., Carrasco, C., Guittat, L., Bailly, C., Alberti, P., Mergny, J.L., David, A., Lehn, J.M., Wilson, W.D. (2003). Selective recognition of G-Quadruplex telomeric DNA by a bis(quinacridine) macrocycle. *J. Am. Chem. Soc.,* 125(16), 4732–4740.

Trajkovski, M., Morel, E., Hamon, F., Bombard, S., Teulade-Fichou, M.P., Plavec, J. (2015). Interactions of Pt-ttpy with G-quadruplexes originating from promoter region of the c-myc gene deciphered by NMR and gel electrophoresis analysis. *Chemistry,* 21(21), 7798–7807.

Verga, D., Hamon , F., Nicoleau, C., Guetta, C., Wu, T.D., Guerquin-Kern, J.L., Marco, S., Teulade-Fichou, M.P. (2017). Chemical imaging by NanoSIMS provides high-resolution localization of the G-quadruplex interactive drug (Br)-PhenDC3 on human chromosomes. *J. Mol. Biol. Mol. Imaging,* 4(1), 1029–1035.

Vilar, R. (2018). Nucleic acid quadruplexes and metallo-drugs. *Met. Ions Life Sci.,* 18, 325–349.

Wang, Y. and Patel, D.J. (1993). Solution structure of the human telomeric repeat d[AG3(T2AG3)3] G-tetraplex. *Structure,* 1(4), 263–282.

Wang, F., Podell, E.R., Zaug, A.J., Yang, Y., Baciu, P., Cech, T.R., Lei, M. (2007). The POT1-TPP1 telomere complex is a telomerase processivity factor. *Nature,* 445(7127), 506–510.

White, E.W., Tanious, F., Ismail, M.A., Reszka, A.P., Neidle, S., Boykin, D.W., Wilson, W.D. (2007). Structure-specific recognition of quadruplex DNA by organic cations: Influence of shape, substituents and charge. *Biophys. Chem.,* 126(1–3), 140–153.

Wu, Z.Q., Yang, X., Weber, G., Liu, X. (2008). Plk1 phosphorylation of TRF1 is essential for its binding to telomeres. *J. Biol. Chem.,* 283(37), 25503–25513.

Xie, X., Reznichenko, O., Chaput, L., Martin, P., Teulade-Fichou, M.P., Granzhan, A. (2018). Topology-selective, fluorescent "light-up" probes for G-quadruplex DNA based on photoinduced electron transfer. *Chemistry – A European Journal,* 24(48), 12638–12651.

Xu, Y. (2011). Chemistry in human telomere biology: Structure, function and targeting of telomere DNA/RNA. *Chemical Society Reviews,* 40(5), 2719–2740.

Yang, S. and Counter, C.M. (2013). Cell cycle regulated phosphorylation of the telomere-associated protein TIN2. *PLoS One,* 8(8), e71697.

Yang, L., Sun, L., Teng, Y., Chen, H., Gao, Y., Levine, A.S., Nakajima, S., Lan, L. (2017). Tankyrase1-mediated poly(ADP-ribosyl)ation of TRF1 maintains cell survival after telomeric DNA damage. *Nucleic Acids Res.,* 45(7), 3906–3921.

Zheng, X.H., Mu, G., Zhong, Y.F., Zhang, T.P., Cao, Q., Ji, L.N., Zhao, Y., Mao, Z.W. (2016). Trigeminal star-like platinum complexes induce cancer cell senescence through quadruplex-mediated telomere dysfunction. *Chem. Commun. (Camb),* 52(98), 14101–14104.

Zhou, C.Q., Liao, T.C., Li, Z.Q., Gonzalez-Garcia, J., Reynolds, M., Zou, M., Vilar, R. (2017). Dinickel-Salphen complexes as binders of human telomeric dimeric G-quadruplexes. *Chemistry – A European Journal.,* 23(19), 4713–4722.

Zhu, J., Liu, W., Chen, C., Zhang, H., Yue, D., Li, C., Zhang, L., Gao, L., Huo, Y., Liu, C. et al. (2019). TPP1 OB-fold domain protein suppresses cell proliferation and induces cell apoptosis by inhibiting telomerase recruitment to telomeres in human lung cancer cells. *J. Cancer Res. Clin. Oncol.*, 145(6), 1509–1519.

List of Authors

Patrizia ALBERTI
Structure and Instability of Genomes
National Museum of Natural History
CNRS, INSERM
Paris
France

Aurélia BARASCU
Institute of Physico-Chemical Biology
PSL, CNRS
Sorbonne University
Paris
France

Sophie BOMBARD
Institute Curie
Paris-Saclay University
Orsay
France

Julia COOPER
Department of Biochemistry and Molecular Genetics
University of Colorado Anschutz Medical Campus
Aurora, CO
USA

Stéphane COULON
CNRS UMR7258, INSERM UMR1068
Institut Paoli-Calmettes
CRCM
University of Aix-Marseille
France

Laure CRABBE
Molecular, Cellular and Developmental Biology Department
Integrative Biology Center
University of Toulouse, CNRS, UPS
France

Anabelle DECOTTIGNIES
Genetic and Epigenetic Alterations of Genomes
Duve Institute
UCLouvain
Brussels
Belgium

Mathilde GAUCHIER
Institute of Human Genetics
CNRS
University of Montpellier
France

Vincent GÉLI
Institute of Paoli-Calmettes
CNRS, INSERM
University of Aix-Marseille
France

Claire GHILAIN
Medicine Faculty
CNRS, INSERM, IRCAN
Côte d'Azur University
Nice
France

Eric GILSON
Medicine Faculty
CNRS, INSERM, IRCAN
Côte d'Azur University
Nice
France

Marie-Josèphe GIRAUD-PANIS
Medicine Faculty
CNRS, INSERM, IRCAN
Côte d'Azur University
Nice
France

Caroline KANNENGIESSER
APHP Department of Genetics
Hôpital Bichat
Paris Cité University
France

Arturo LONDOÑO-VALLEJO
Telomere and Cancer Lab
Institute Curie, CNRS
PSL Research University
Paris
France

Patrick MAILLIET
Structure and Instability of Genomes
National Museum of Natural History
CNRS, INSERM
Paris
France

Patrick REVY
Imagine Institute, INSERM
Paris Cité University
France

Lauréline ROGER
Structure and Instability of Genomes
National Museum of Natural History
CNRS, INSERM
Paris
France

Carole SAINTOMÉ
Structure and Instability of Genomes
National Museum of Natural History
CNRS, INSERM
and
UFR 927,
Sorbonne University
Paris
France

Sonia STINUS
Molecular, Cellular and Developmental
Biology Department
Integrative Biology Center
University of Toulouse, CNRS, UPS
France

Maria Teresa TEIXEIRA
Institute of Physico-Chemical Biology
PSL, CNRS
Sorbonne University
Paris
France

Marie-Paule TEULADE-FICHOU
Institute Curie
Paris-Saclay University
Orsay
France

Index

I, K, L

M, N, O

P, Q

R

W, Y

Printed and bound by CPI Group (UK) Ltd, Croydon, CR0 4YY

13/09/2026

14960997-0008